Studies in Computational Intelligence

Volume 774

Series editor

Janusz Kacprzyk, Polish Academy of Sciences, Warsaw, Poland
e-mail: kacprzyk@ibspan.waw.pl

The series "Studies in Computational Intelligence" (SCI) publishes new developments and advances in the various areas of computational intelligence—quickly and with a high quality. The intent is to cover the theory, applications, and design methods of computational intelligence, as embedded in the fields of engineering, computer science, physics and life sciences, as well as the methodologies behind them. The series contains monographs, lecture notes and edited volumes in computational intelligence spanning the areas of neural networks, connectionist systems, genetic algorithms, evolutionary computation, artificial intelligence, cellular automata, self-organizing systems, soft computing, fuzzy systems, and hybrid intelligent systems. Of particular value to both the contributors and the readership are the short publication timeframe and the world-wide distribution, which enable both wide and rapid dissemination of research output.

More information about this series at http://www.springer.com/series/7092

El-Ghazali Talbi · Amir Nakib
Editors

Bioinspired Heuristics for Optimization

 Springer

Editors
El-Ghazali Talbi
Parc Scientifique de la Haute Borne
INRIA Lille Nord Europe
Villeneuve-d'Ascq, France

Amir Nakib
Université Paris Est Laboratoire Images
Signaux et Systèmes Intelligents (LISSI)
Vitry-sur-Seine, France

ISSN 1860-949X   ISSN 1860-9503   (electronic)
Studies in Computational Intelligence
ISBN 978-3-030-06978-0   ISBN 978-3-319-95104-1   (eBook)
https://doi.org/10.1007/978-3-319-95104-1

This Springer imprint is published by the registered company Springer Nature Switzerland AG
The registered company address is: Gewerbestrasse 11, 6330 Cham, Switzerland

Preface

In the last decade, metaheuristics have become standard tools for solving industrial and healthcare optimization problems. In the literature, different metaheuristics were developed, and this domain is under continuous progress. For instance, the most famous examples are Tabu search (TS), evolutionary algorithms (EAs), simulated annealing (SA), ant colony optimization (ACO), particle swarm optimization (PSO), and memetic algorithms.

The editors, both leading experts in this field, have selected 19 chapters dedicated to the design of new metaheuristics and their different applications to complex optimization problems.

This book *Bioinspired Heuristics for Optimization* aims at providing a review of recent work on the design of bioinspired and heuristics algorithms. Moreover, the book is oriented towards both theoretical and applications aspects of these algorithms on real-world and complex problems. Chapters are best papers selected from the 6th international conference on metaheuristics and nature inspired computing that held at Marrakech (Morocco) from 27th to 31st.

Paris, France
January 2018

Amir Nakib
El-Ghazali Talbi

Contents

Chapter 1
Possibilistic Framework
for Multi-objective Optimization
Under Uncertainty

Oumayma Bahri, Nahla Benamor and El-Ghazali Talbi

Abstract Optimization under uncertainty is an important line of research having today many successful real applications in different areas. Despite its importance, few works on multi-objective optimization under uncertainty exist today. In our study, we address combinatorial multi-objective problem under uncertainty using the possibilistic framework. To this end, we firstly propose new Pareto relations for ranking the generated uncertain solutions in both mono-objective and multi-objective cases. Secondly, we suggest an extension of two well-known Pareto-base evolutionary algorithms namely, SPEA2 and NSGAII. Finally, the extended algorithms are applied to solve a multi-objective Vehicle Routing Problem (VRP) with uncertain demands.

1.1 Introduction

Most real-world decision problems are multi-objective in nature as they require the simultaneous optimization of multiple and usually conflicting objectives. These multi-objective problems are a very important and widely discussed research topic. Yet, despite the massive number of existing resolution methods and techniques for multi-objective optimization, there still many open questions in this area. In fact, there is no consideration of uncertainty in the classical multi-objective concepts

O. Bahri · N. Benamor
LARODEC Lab. University Tunis, Le Bardo 2000 Tunis, Tunisia
e-mail: oumayma.b@gmail.com

N. Benamor
e-mail: nahla.benamor@gmx.fr

O. Bahri · E.-G. Talbi (✉)
INRIA CNRS Lab., 59650 Villeneuve d'Ascq, Lille, France
e-mail: el-ghazali.talbi@lifl.fr

© Springer International Publishing AG, part of Springer Nature 2019
E.-G. Talbi and A. Nakib (eds.), *Bioinspired Heuristics for Optimization*,
Studies in Computational Intelligence 774,
https://doi.org/10.1007/978-3-319-95104-1_1

and techniques, which makes their application to real-life optimization problems impossible.

Moreover, uncertainty characterizes almost all practical applications, in which the big amount of data provides certainly some unavoidable imperfections. This imperfection might result from using unreliable information sources caused by inputting data incorrectly, faulty reading instruments or bad analysis of some training data. It may also be the result of poor decision-maker opinions due to any lack of its background knowledge or even due to the difficulty of giving a perfect qualification for some costly situations. The classical way to deal with uncertainty is the probabilistic reasoning, originated from the middle of the 17th century [18]. However, probability theory was considered for a long time as a very good quantitative tool for uncertainty treatment, but as good as it is, this theory is only appropriate when all numerical data are available, which is not always the case. Indeed, there are some situations such as the case of total ignorance, which are not well handled and which can make the probabilistic reasoning unsound [24]. Therefore, a panoply of non-classical theories of uncertainty have recently emerged such as fuzzy sets theory [30], possibility theory [31] and evidence theory [23]. Among the aforementioned theories of uncertainty, our interest will focus on possibility theory which offers a natural and simple model to handle uncertain data and presents an appropriate framework for experts to express their partial beliefs numerically or qualitatively. Nevertheless, while the field of optimization under uncertainty has gained considerable attention during several years in the mono-objective context, only few studies have been focused on treating uncertain optimization problems within a multi-objective setting. This chapter addresses the multi-objective optimization problems under uncertainty in the possibilistic setting.

The remainder of the chapter is organized as follows. Section 1.2 recalls the main concepts of deterministic multi-objective optimization. Section 1.3 gives an overview of existing approaches for multi-objective optimization under uncertainty. Section 1.4 presents in detail our proposed possibilistic framework after briefly recalling the basics of possibility theory. Finally, Sect. 1.5 describes an illustrative example on a multi-objective vehicle routing problem with uncertain demands and summarizes the obtained results.

1.2 Background on Deterministic Multi-objective Optimization

Deterministic multi-objective optimization is the process of optimizing systematically and simultaneously two or more conflicting objectives subject to certain constraints. In contrast to mono-objective optimization, a multi-objective optimization problem does not restrict to find a unique global solution but it aims to find the most preferred exact solutions among the best ones.

Formally, a basic multi-objective optimization problem (MOP), defined in the sense of minimization of all the objectives, consists of solving a mathematical program of the form:

$$MOP = \begin{cases} Min\ F(x) = (f_1(x), f_2(x), \ldots, f_n(x)) \\ s.t.\ x \in S \end{cases} \tag{1.1}$$

where n ($n \geq 2$) is the number of objectives and $x = \{x_1, \ldots, x_k\}$ is the set of decision variables from the decision space S, which represents the set of feasible solutions associated with equality and inequality constraints. $F(x)$ is the vector of independent objectives to be minimized. This vector F can be defined as a cost function in the objective space by assigning an objective vector $\vec{y}$ which represents the quality of the solution (or fitness).

$$F : X \rightarrow Y \subseteq R^n, \quad F(x) = \vec{y} = \begin{pmatrix} y_1 \\ \ldots \\ y_n \end{pmatrix} \tag{1.2}$$

In order to identify better solutions of a given MOP, other concepts of optimality should be applied such as *Pareto dominance*, *Pareto optimality*, *Pareto optimal set* and *Pareto front*. Without loss of generality, we assume that the sense of minimization of all the objectives is considered in the following concepts definition:

An objective vector $\mathbf{x} = (x_1, \ldots, x_n)$ is said to *Pareto dominate* another objective vector $\mathbf{y} = (y_1, \ldots, y_n)$ (denoted by $\mathbf{x} \prec_p \mathbf{y}$) if and only if no component of $\mathbf{y}$ is smaller than the corresponding component of $\mathbf{x}$ and at least one component of $\mathbf{x}$ is strictly smaller:

$$\forall i \in 1, \ldots, n : x_i \leq y_i \wedge \exists i \in 1, \ldots, n : x_i < y_i. \tag{1.3}$$

For a minimization $MOP(F, S)$, a solution $x^* \in X$ is *Pareto optimal* (also known as efficient, non-dominated or non-inferior) if for every $x \in X$, $F(x)$ does not dominate $F(x^*)$, that is, $F(x) \not\prec_p F(x^*)$.

A *Pareto optimal set* P^* is defined as:

$$P^* = \{x \in X / \exists x' \in X, F(x') \not\prec_p F(x)\}. \tag{1.4}$$

The image of this *Pareto optimal set* P^* in the objective space is called *Pareto front* PF^* defined as:

$$PF^* = \{F(x), x \in P^*\}. \tag{1.5}$$

Yet, finding the true *Pareto front* of a general MOP is NP-hard. Thus, the main goal of multi-objective optimization is to identify a good approximation of the *Pareto front*, from which the decision maker can select an optimal solution based on the current situation. The approximated front should satisfy two properties: (1) convergence or closeness to the exact Pareto front and (2) uniform diversity of the obtained solutions around the Pareto front. Figure 1.1 illustrates an example of approximated front

Fig. 1.1 Example of Pareto
front with uniform diversity
and bad convergence

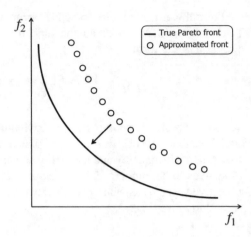

having a very good spread of solutions (uniform diversity) but a bad convergence,
since the solutions are far from the true Pareto front.

There are several deterministic optimization methods to deal with multi-objective
combinatorial problems, such as the metaheuristics, which mark a great revolution
in the field of optimization. A review of various metaheuristics can be found in
[10]. Among the well-know metaheuristics, evolutionary algorithms seem particularly suitable for both theoretical and practical MOPs, since they have the ability to
search partially ordered spaces for several alternative trade-offs [4–6]. Some of the
most popular multi-objective evolutionary algorithms (MOEAs) are: Multi-Objective
Genetic Algorithm (MOGA) [11], Niched-Pareto Genetic Algorithm (NPGA) [14],
Pareto-Archived Evolutionary Strategy (PAES) [17], Strength Pareto Evolutionary
Algorithms (SPEA, SPEA2) [32, 33] and Non-dominated Sorting Genetic Algorithms (NSGA, NSGAII) [7, 8]. Such algorithms are based on three main components
namely, Fitness assignment, Diversity preserving and Elitism.

Fitness assignment: allows to guide the search algorithm toward Pareto optimal
solutions for a better convergence. The fitness assignment procedure assigns to each
objective vector, a scalar-valued fitness that measures the quality of solution. According to the fitness assignment strategy, four different categories can be identified:

- Pareto-based assignment: based on the concept of dominance and Pareto optimality to guide the search process. The objective vectors are scalarized using the
  dominance relation.
- Scalar-based assignment: based on the MOP transformation into a mono-objective
  problem by using for example aggregation methods and weighted metrics.
- Criterion-based assignment: based on the separate handling of various non com-
  mensurable objectives by performing a sequential search according to a given
  preference order of objectives or by handling the objectives in parallel.
- Indicator-based assignment: based on the use of performance quality indicators to
  drive the search toward the Pareto front.

Diversity preserving: used to generate a diverse set of Pareto solutions. According to the strategy of density estimation, three categories can be distinguished:

- Distance-based density assessment: based on the distance between individuals in the feature space. Examples of techniques are, Niche sharing, Clustering, K-th nearest neighbor and Crowding.
- Grid-based density assessment: based on the way in which a number of individuals residing within predetermined cells are located. Histogram method is an example.
- Distribution-based density assessment: based on the probability density of individuals using for example probability density estimation functions.

Elitism: consists in archiving the best solutions found (e.g, Pareto optimal solutions) in order to prevent the loss of good solutions during the search process. Archiving process can be done using an archive (elite population) or an external population and its strategy of update usually relies on size, convergence and diversity criteria. Depending on the manner in which the archiving process is performed, MOEAs can be classified into two categories, namely non-elitist and elitist MOEAs. Moreover, almost all MOEAs follow the same basic steps in the search process [12], as outlined in the following pseudo code:

Generic MOEA Framework
Initialize random population P **While** (Stopping condition is not satisfied) Fitness evaluation of solutions in P; Environmental selection of "good" solutions; Diversity preserving of candidate solutions; Update and store elite solutions into an external population or archive; Mating selection to create the mating pool for variation; Variation by applying crossover and mutation operators; **End While**

An MOEA begins its search with a population of solutions usually generated at random. Thereafter, an iterative optimization process takes place by the use of six search operators: evaluation of the population individuals, environmental selection to choose better solutions based on their fitness, diversity preservation of candidate solutions, updating and archiving the solutions into an external population or archive, mating selection operator in which solutions are picked from the updated population to fill an intermediate mating pool and finally variation operator to generate new solutions. The process stops when one or more pre-specified stopping conditions are met.

All the above concepts and techniques of deterministic multi-objective optimization are widely used and applied successfully to several combinatorial decision problems in many interesting areas, but their application to real-life decision making situations often faces some difficulties. Yet, most of real-world optimization problems are naturally subject to various types of uncertainties caused by many sources such as missing information, forecasting, data approximation or noise in measurements. These uncertainties are very difficult to avoid in practical applications and so should be taken into account within the optimization process. Therefore, a variety of method-ologies and approaches for handling optimization problems under uncertainty have been proposed in the last years. Unfortunately, almost all of them have been devoted to solve such problems in the mono-objective context, while only few studies have been performed in the multi-objective setting. A review of some existing approaches for uncertain multi-objective optimization will be summarized in the next section.

1.3 Existing Approaches for Uncertain Multi-objective Optimization

Uncertain multi-objective optimization has gained more and more attention in recent years, since it closely reflects the reality of many real-world problems. Such problems, known as multi-objective problems under uncertainty, are naturally characterized by the necessity of optimizing simultaneously several objectives subject to a set of constraints and while considering that some input data are ill-known and without knowing what their full effects will be. In these problems, the set of objectives and/or constraints to be satisfied can be affected by the uncertainty of input data or uncontrollable problem parameters. Hence, the aim of optimization in this case will be to find solutions of a multi-objective problem that are not only feasible and optimal but also their objectives and/or constraints are allowed to have some acceptable (or minimal) uncertainties. These uncertainties can take different forms in terms of distribution, bounds, and central tendency.

Yet, considering the uncertainty in the objective functions seems to be very appli-cable but highly critical, since the propagation of input uncertainties to the objectives may have a major impact on the whole optimization process and consequently on the problem solutions. In most of the existing approaches for dealing with multi-objective problems under uncertainty, the objective functions to be optimized are transformed into different forms in order to simplify their resolution by eliminating one of the two basic characteristics of such problems: multi-objectivity and uncertainty propa-gation. In fact, some of these approaches have been often limited to simply reduce the problem to mono-objective context by considering the set of objectives as if there's only one, using for example an aggregation function (a weighted sum) of all the objectives [13] or preferring only one objective to be optimized (based on a preference indicator) and fixing the remaining objectives as constraints [22]. The con-sidered single objective is then optimized using appropriate mono-objective methods for uncertainty treatment.

Some other approaches have been focused on treating the problem as multi-objective but with ignoration of uncertainty propagation to the objective functions by converting them into deterministic functions using statistical properties. For exam-ple, in [27], expectation values are used to approximate the observed interval-valued objectives and so the goal became to optimize the expected values of these objec-tives. In [1], the average value per objective is firstly computed and then a ranking method based on the average values of objectives is proposed. Similarly, [7] suggested to consider the mean value for each objective vectors and then to apply classical deterministic multi-objective optimizers. Nevertheless, the uncertainty of objective values must not be ignored during the optimization process, because if the input data or parameters are highly uncertain, how can the optimizer simply state that the uncertainty of outputs is completely certain? It may be feasible only for simplicity or other practical reasons as long as the algorithm performance will not be affected.

To this end, some distinct approaches have been suggested to handle the problem as-is without erasing any of its multi-objective or uncertain characteristics by introducing a particular multi-objective optimizer for this purpose. Indeed, [20, 21] proposed to display uncertainty in objective functions through intervals of belief functions and then introduced an extensions of Pareto dominance for ranking the generated interval-valued objectives. [15, 16] suggested to express uncertainty in the objectives via special types of probability distributions and then independently proposed a stochastic extension of Pareto dominance. Our interest in this chapter will focus on handling multi-objective problems under uncertainty in the possibilistic setting while considering the uncertainty propagation to the set of objectives to be optimized.

1.4 Proposed Possibilistic Framework for Multi-objective Problems Under Uncertainty

This section provides firstly a brief background on possibility theory and then presents in detail the proposed possibilistic framework for solving multi-objective problems with uncertain data. The framework is composed of three main stages: an adaptation of possibilistic setting, a new Pareto optimality and an extension of some optimization algorithms to our uncertain context.

1.4.1 Basics on Possibility Theory

Possibility theory, issued from Fuzzy Sets theory, was introduced by Zadeh [31] and further developed by Dubois and Prade [9]. This theory offers a flexible tool for representing uncertain information such as expressed by humans. Its basic building block is the notion of possibility distribution, denoted by π and defined as the following:

Let $V = \{X_1, \ldots, X_n\}$ be a set of state variables whose values are ill-known. We denote by x_i any instance of X_i and by D_{X_i} the domain associated with X_i. $\Omega = D_{X_1} \times \cdots \times D_{X_n}$ denotes the universe of discourse, which is the cartesian product of all variable domains V. Vectors $\omega \in \Omega$ are often called realizations or simply "states" (of the world). The agent's knowledge about the value of the x_i's can be encoded by a possibility distribution π that corresponding to a mapping from the universe of discourse Ω to the scale $[0, 1]$, i.e. $\pi : \Omega \rightarrow [0, 1]$; $\pi(\omega) = 1$ means that the realization ω is totally possible and $\pi(\omega) = 0$ means that ω is an impossible state. It is generally assumed that there exist at least one state ω which is totally possible -π is said then to be *normalized*. Extreme cases of knowledge are presented by:

- *complete knowledge* i.e. $\exists \omega_0 \in \Omega$, $\pi(\omega_0) = 1$ and $\forall \omega \neq \omega_0$, $\pi(\omega) = 0$.
- *total ignorance* i.e. $\forall \omega \in \Omega$, $\pi(\omega) = 1$ (all values in Ω are possible).

From π, one can describe the uncertainty about the occurrence of an event $A \subseteq \Omega$ via two dual measures: the possibility $\Pi(A)$ and the necessity $N(A)$ expressed by:

$$\Pi(A) = sup_{\omega \in A} \pi(\omega). \tag{1.6}$$

$$N(A) = 1 - \Pi(\neg A) = 1 - sup_{\omega \notin A} \pi(\omega) \tag{1.7}$$

Measure $\Pi(A)$ corresponds to the possibility degree (i.e. the plausibility) of A and it evaluates to what extent A is consistent (i.e. not contradictory) with the knowledge represented by π. Yet, the expression "*it is possible that A is true*" does not entail anything about the possibility nor the impossibility of A. Thus, the description of uncertainty about the occurrence of A needs its dual measure $N(A)$ which corresponds to the extent to which A is impossible and it evaluates at which level A is certainly implied by the π (the certainty degree of A). Main properties of these two dual measures are summarized in Table 1.1. The particularity of the possibilistic scale is that it can be interpreted in two manners: in an ordinal manner, i.e. when the possibility degrees reflect only an ordering between the possible values and in a numerical manner, i.e. when the handled values make sense in the ranking scale.

Technically, a possibility distribution is a normal fuzzy set (at least one membership grade equals 1). Indeed, all fuzzy numbers can be interpreted as specific possibility distributions. More precisely, given a variable X whose values are restricted by a fuzzy set F characterized by its membership function μ_F, so that π_X is taken as equal to the membership function $\mu_F(x)$. Thus, the possibility and necessity measures will be expressed in terms of supremum degrees of the μ_F, i.e. $\Pi(X) = sup_{x \in X} \mu_F(x)$ and $N(X) = 1 - sup_{x \notin X} \mu_F(x)$. In this work, we are interested in a particular form of possibility distributions, namely those represented by triangular fuzzy numbers and commonly known as triangular possibility distributions. A triangular possibility distribution π_X is defined by a triplet $[\underline{x}, \widehat{x}, \overline{x}]$, as shown in Fig. 1.2, where $[\underline{x}, \overline{x}]$ is the interval of possible values called its bounded support and $\widehat{x}$ denotes its kernel value (the most plausible value).

Table 1.1 Possibility measure Π and Necessity measure N

$\Pi(A) = 1$ and $\Pi(\overline{A}) = 0$	$N(A) = 1$ and $N(\overline{A}) = 0$	A is certainly true
$\Pi(A) = 1$ and $\Pi(\overline{A}) \in]0, 1[$	$N(A) \in]0, 1[$ and $N(\overline{A}) = 0$	A is somewhat certain
$\Pi(A) = 1$ and $\Pi(\overline{A}) = 1$	$N(A) = 0$ and $N(\overline{A}) = 0$	Total ignorance

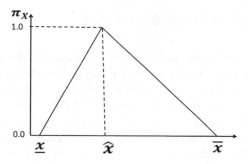

Fig. 1.2 Triangular possibility distribution

In the remaining, we use $X = [\underline{x}, \widehat{x}, \overline{x}] \subseteq \mathbb{R}$ to denote the triangular fuzzy number X, meaning that X is represented by a triangular possibility distribution π_X. This representation is characterized by a membership function μ_X which assigns a value within $[0, 1]$ to each element in $x \in X$. Its mathematical definition is given by:

$$\mu_X(x) = \begin{cases} \frac{x - \underline{x}}{\widehat{x} - \underline{x}}, & \underline{x} \leq x \leq \widehat{x} \\ 1, & x = \widehat{a} \\ \frac{\overline{x} - x}{\overline{x} - \widehat{x}}, & \widehat{x} \leq x \leq \overline{x} \\ 0, & otherwise. \end{cases} \tag{1.8}$$

However, in practical use of triangular fuzzy numbers, a ranking procedure needs to be applied for decision-making. In other words, one triangular fuzzy number needs to be evaluated and compared with the others in order to make a choice among them. Indeed, all possible topological relations between two triangular fuzzy numbers $A = [\underline{a}, \widehat{a}, \overline{a}]$ and $B = [\underline{b}, \widehat{b}, \overline{b}]$ may be covered by only four different situations, which are: Fuzzy disjoint, Fuzzy weak overlapping, Fuzzy overlapping and Fuzzy inclusion [19]. These situations, illustrated in Fig. 1.3 should be taken into account for ranking triangular fuzzy numbers.

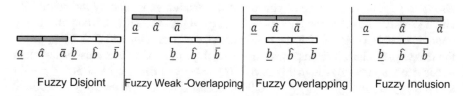

Fig. 1.3 Possible topological situations for two TFNs

1.4.2 Adaptation of Possibilistic Setting

In the following, we choose to express the uncertain data of multi-objective problems under uncertainty using triangular possibility distributions (i.e. triangular fuzzy numbers) as defined in the previous subsection. Then, as a multi-objective optimization problem under uncertainty involves the simultaneous satisfaction of several objectives respecting a set of constraints and while considering some input data uncertainties, we assume that the observed objectives and some constraints (especially those depends on uncertain variables) are affected by the used form of these uncertainties.

Thus, as in our case, uncertainty is represented by a triangular fuzzy form, the uncertain constraints in such a problem may be disrupted by this fuzzy form and so will be fuzzy constraints. Yet, the satisfaction of such constraints cannot be directly predicted since it is difficult to estimate directly that a fuzzy constraint is fully satisfied or fully violated. At this level, we propose firstly to use the two measures of possibility theory Π and N in order to express the satisfaction of a given fuzzy constraint, as follows:

Let $X = [\underline{x}, \widehat{x}, \overline{x}] \subseteq \mathbb{R}$ be a triangular fuzzy variable, let x be any instance of X, let v be a given fixed value and let $C = (X \leq v)$ be a fuzzy constraint that depends only on the value of X and whose membership function is $\mu(x)$, then we have the measures $\Pi(X \leq v)$ and $N(X \leq v) = 1 - \Pi(X > v)$ are equal to:

$$\Pi(\widetilde{X} \leq v) = \sup \mu_{x \leq v}(x) = \begin{cases} 1 & \text{if } v > \widehat{x} \\ \frac{v - \underline{x}}{\widehat{x} - \underline{x}} & \text{if } \underline{x} \leq v \leq \widehat{x} \\ 0 & \text{if } v < \underline{x}. \end{cases} \qquad (1.9)$$

$$N(\widetilde{X} \leq v) = 1 - \sup \mu_{x > v}(x) = \begin{cases} 1 & \text{if } v > \overline{x} \\ \frac{v - \widehat{x}}{\overline{x} - \widehat{x}} & \text{if } \widehat{x} \leq v \leq \overline{x} \\ 0 & \text{if } v < \widehat{x}. \end{cases} \qquad (1.10)$$

These formulas will be used to express the degrees that a solution satisfies the fuzzy constraint.

Example 1.1 As an example of constraint satisfaction expressed by the possibility and necessity measures, we have $Q = [\underline{q}, \widehat{q}, \overline{q}] = [20, 45, 97]$ is a triangular fuzzy quantity of objects, $M = 50$ is the maximum size of a package and $C = (Q \leq M)$ is the fuzzy constraint which imposes that the total quantity of objects must be less than or equal to the package size. In this case, $\Pi(Q \leq M) = 1$ because $M = 50 > \widehat{q} = 45$ and $N(Q \leq M) = \frac{M - \widehat{q}}{\overline{q} - \widehat{q}} = \frac{50 - 45}{97 - 45} = 0.096$ because $\widehat{q} = 45 \leq M = 50 \leq \overline{q} = 97$.

Note that, a constraint may fail even though its possibility achieves 1 and holds even though its necessity is 0. In addition, an often used definition says that the possibility measure Π gives always the best case and shows the most optimist attitude, while the necessity N gives the worst case and shows the most pessimist attitude. Then, as presented above, Π and N are related to each others by a dual relationship. Therefore, a combination of these two measures allows the expression of both optimistic and pessimistic attitude of the decision maker. From these remarks, we can conclude that it is more efficient at this step to use the linear combination of possibility and necessity measures proposed by [3], rather than treating each measure separately. This linear combination is defined as the following:

Given a constraint A, its weight denoted by $W(A)$ which corresponds to the combination of the weighted possibility and necessity, is expressed by:

$$W(A) = \lambda\, \Pi(A) + (1 - \lambda)\, N(A) \geq \alpha. \qquad (1.11)$$

where the parameter $\lambda \in [0, 1]$, measures the degree of optimism or confidence of the decision maker such that:

$$\lambda = \begin{cases} 1 & \text{Total optimistic case} \\ 0 & \text{Total pessimistic case} \\ 0.5 & \text{Neither optimistic nor pessimistic.} \end{cases} \qquad (1.12)$$

and $\alpha \in [0, 1]$ is a given threshold of satisfaction fixed by the decision maker. This formula indicates that the weight measure $W(A)$ must be higher than a given threshold α. The higher it is, the greater the constraint will be satisfied.

Secondly, knowing that propagating the uncertainty of multi-objective problem's data through the resolution model leads often to uncertain formulation of objective functions and as in our case the uncertain data are represented by triangular fuzzy numbers, the objective functions will be consequently disrupted by this fuzzy form. Let us assume that, a multi-objective triangular-valued function can be mathematically defined as:

$$F : X \rightarrow Y \subseteq (R \times R \times R)^n,$$

$$F(x) = \overrightarrow{y} = \begin{pmatrix} y_1 = [\underline{y_1}, \widehat{y_1}, \overline{y_1}] \\ \dots \\ y_n = [\underline{y_n}, \widehat{y_n}, \overline{y_n}] \end{pmatrix} \qquad (1.13)$$

Clearly, in this case, the classical multi-objective techniques cannot be applied since they are only meant for deterministic case. Therefore, a need for special optimization methods techniques to handle the generated triangular-valued functions is evident. To this end, we first introduce a new Pareto dominance over triangular fuzzy numbers, in both mono-objective and multi-objective cases.

1.4.3 New Pareto Optimality over Triangular Fuzzy Numbers

In this section, we first present new mono-objective dominance relations between two TFNs. Then based on these mono-objective dominance, we define a new Pareto

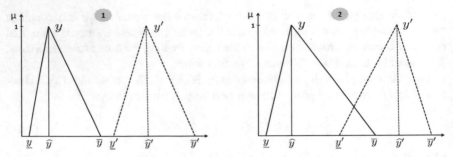

Fig. 1.4 Total dominance and Partial strong-dominance

dominance between vectors of TFNs, for multi-objective case. Note that, the minimization sense is considered in all our definitions.

Mono-objective dominance relations: In the mono-objective case, three dominance relations over triangular fuzzy numbers are defined: Total dominance ($\prec_t$), Partial strong-dominance ($\prec_s$) and Partial weak-dominance ($\prec_w$).

Definition 1.1 (*Total dominance*) Let $y = [\underline{y}, \widehat{y}, \overline{y}] \subseteq \mathbb{R}$ and $y' = [\underline{y}', \widehat{y}', \overline{y}'] \subseteq \mathbb{R}$ be two triangular fuzzy numbers. y *dominates* y' *totally or certainly* (denoted by $y \prec_t y'$) if: $\overline{y} < \underline{y}'$.

This dominance relation represents the fuzzy disjoint situation between two triangular fuzzy numbers and it imposes that the upper bound of y is strictly inferior than the lower bound of y' as shown by case (1) in Fig. 1.4.

Definition 1.2 (*Partial strong-dominance*) Let $y = [\underline{y}, \widehat{y}, \overline{y}] \subseteq \mathbb{R}$ and $y' = [\underline{y}', \widehat{y}', \overline{y}']$ $\subseteq \mathbb{R}$ be two triangular fuzzy numbers. y *strong dominates* y' *partially or uncertainly* (denoted by $y \prec_s y'$) if:

$$(\overline{y} \geq \underline{y}') \wedge (\widehat{y} \leq \underline{y}') \wedge (\overline{y} \leq \widehat{y}').$$

This dominance relation appears when there is a fuzzy weak-overlapping between both triangles and it imposes that firstly there is at most one intersection between them and secondly this intersection should not exceed the interval of their kernel values $[\widehat{y}, \widehat{y}']$, as shown by case (2) in Fig. 1.4.

Definition 1.3 (*Partial weak-dominance*) Let $y = [\underline{y}, \widehat{y}, \overline{y}] \subseteq \mathbb{R}$ and $y' = [\underline{y}', \widehat{y}', \overline{y}']$ $\subseteq \mathbb{R}$ be two triangular fuzzy numbers. y *weak dominates* y' *partially or uncertainly* (denoted by $y \prec_w y'$) if:

1. Fuzzy overlapping

$$[(\underline{y} < \underline{y}') \wedge (\overline{y} < \overline{y}')] \wedge$$
$$[((\widehat{y} \leq \underline{y}') \wedge (\overline{y} > \widehat{y}')) \vee ((\widehat{y} > \underline{y}') \wedge (\overline{y} \leq \widehat{y}')) \vee ((\widehat{y} > \underline{y}') \wedge (\overline{y} > \widehat{y}'))].$$

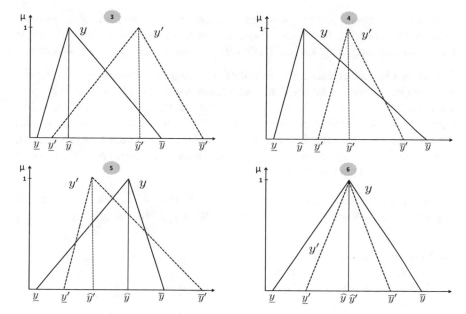

Fig. 1.5 Partial weak-dominance

2. Fuzzy Inclusion

$$(\underline{y} < \underline{y}') \wedge (\overline{y} \geq \overline{y}').$$

In this dominance relation, the two situations of fuzzy overlapping and inclusion may occur. Figure 1.5 presents four examples of possible cases, where in (3) and (5) y and y' are overlapped, while, in (4) and (6) y' is included in y.

Yet, the partial weak-dominance relation cannot discriminate all possible cases and leads often to some incomparable situations as for cases (5) and (6) in Fig. 1.5. These incomparable situations can be distinguished according to the kernel value positions in fuzzy triangles. Thus, we propose to consider the kernel values configuration as condition to identify the cases of incomparability, as follows:

$$\widehat{y} - \widehat{y}' = \begin{cases} < 0, \ y \prec_w y' \\ \geq 0, \ y \text{ and } y' \text{ can be incomparable.} \end{cases}$$

Subsequently, to handle the identified incomparable situations (with kernel condition $\widehat{y} - \widehat{y}' \geq 0$), we introduce another comparison criterion, which consists in comparing the discard between both fuzzy triangles as follows:

$$y \prec_w y' \Leftrightarrow (\underline{y}' - \underline{y}) \leq (\overline{y}' - \overline{y})$$

Similarly, it is obvious that: $y' \prec_w y \Leftrightarrow (\underline{y}' - \underline{y}) > (\overline{y}' - \overline{y})$.

It is easy to check that in the mono-objective case, we obtain a total pre-order between two triangular fuzzy numbers, contrarily to the multi-objective case, where the situation is more complex and it is common to have some cases of indifference.

Pareto dominance relations: In the multi-objective case, we propose to use the mono-objective dominance relations, defined previously, in order to rank separately the triangular fuzzy solutions of each objective function. Then, depending on the types of mono-objective dominance founded for all the objectives, we define the Pareto dominance between the vectors of triangular fuzzy solutions. In this context, two Pareto dominance relations: *Strong Pareto dominance* ($\prec_{SP}$) and *Weak Pareto dominance* ($\prec_{WP}$) are introduced.

Definition 1.4 (*Strong Pareto dominance*) Let $\overrightarrow{y}$ and $\overrightarrow{y}'$ be two vectors of triangular fuzzy numbers. $\overrightarrow{y}$ strong Pareto dominates $\overrightarrow{y}'$ (denoted by $\overrightarrow{y} \prec_{SP} \overrightarrow{y}'$) if:

(a) $\forall i \in 1, \ldots, n : y_i \prec_t y'_i \lor y_i \prec_s y'_i$.
(b) $\exists i \in 1, \ldots, n : y_i \prec_t y'_i \land \forall j \neq i : y_j \prec_s y'_j$
(c) $\exists i \in 1, \ldots, n : (y_i \prec_t y'_i \lor y_j \prec_s y'_j) \land \forall j \neq i : y_j \prec_w y'_j$.

The strong Pareto dominance holds if either y_i total dominates or partial strong dominates y'_i in all the objectives (Fig. 1.6a: $y_1 \prec_t y'_1$ *and* $y_2 \prec_t y'_2$), either y_i total dominates y'_i in one objective and partial strong dominates it in another

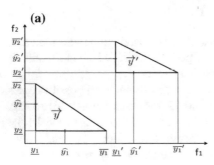
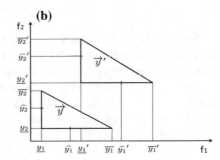
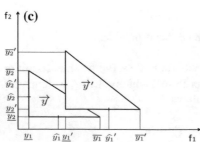
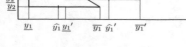
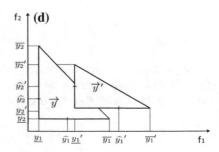

Fig. 1.6 Strong Pareto dominance

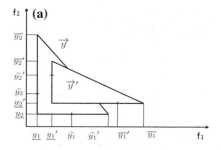 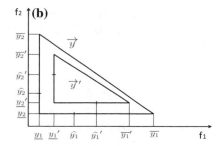

Fig. 1.7   Weak Pareto dominance and Case of indifference

(Fig. 1.6b:$y_1 \prec_s y'_1$ *and* $y_2 \prec_t y'_2$), or at least y_i total or partial strong dominates y'_i in one objective and weak dominates it in another (Fig. 1.6c, d: $y_1 \prec_s y'_1$ *and* $y_2 \prec_w y'_2$).

Definition 1.5 (*Weak Pareto dominance*) Let $\overrightarrow{y}$ and $\overrightarrow{y}'$ be two vectors of triangular fuzzy numbers. $\overrightarrow{y}$ weak Pareto dominates $\overrightarrow{y}'$ (denoted by $\overrightarrow{y} \prec_{WP} \overrightarrow{y}'$) if: $\forall i \in 1, \ldots, n : y_i \prec_w y'_i$.

The weak Pareto dominance holds if y_i weak dominates y'_i in all the objectives (Fig. 1.7e). Yet, a case of indifference (defined below) can occur if there is a weak dominance with inclusion type in all the objectives (Fig. 1.7f).

Definition 1.6 (*Case of indifference*) Two vectors of triangular fuzzy numbers are indifferent or incomparable (denoted by $\overrightarrow{y} \parallel \overrightarrow{y}'$) if: $\forall i \in 1, \ldots, n : y_i \subseteq y'_i$.

The proposed Pareto dominance in bi-dimensional objective space can easily be generalized for ranking more than two objectives. Note that, if the considered triangular objectives are non-independent, the estimation in a bi-dimensional space can have different distributions (non-triangular) like linear shapes. Finally, the issue now is how integrate this dominance in the research process of multi-objective optimization algorithms.

1.4.4   Extended Optimization Algorithm

In the following, we present an extension of two well-known Pareto-based multi-objective evolutionary algorithms: SPEA2(Strength Pareto Evolutionary Algorithm2) [32] and NSGAII (Non-dominated Sorting Genetic Algorithm II) [7], in order to enable them handling a multi-objective problem with triangular-valued objectives. Both algorithms have proved to be very powerful tools for multi-objective optimization. Due to their population-based nature, they are able to generate multiple optimal solutions in a single run with respect to the good convergence and diversification of obtained solutions. We call our two extended algorithms respectively, ESPEA2 and ENSGAII.

ESPEA2:

SPEA2 is an improved version of the Strength Pareto Evolutionary Algorithm SPEA initially proposed by [33]. This evolutionary algorithm uses mainly 3 techniques: a dominance based approach as fitness assignment strategy, a nearest neighbor technique that allows a good diversity preservation and an archive with fixed size that guarantees the elitist storage of optimal solutions. To extend such techniques to triangular fuzzy context, we propose firstly to replace the classical dominance approach by the new Pareto dominance approach proposed for ranking triangular-valued objectives. Secondly, an adaptation of the nearest neighbor technique is introduced. Indeed, in SPEA2, this technique is based on *Euclidean distance* to estimate the density in its neighborhood and it consists in calculating for each solution (objective vector) the distance to its k-nearest neighbor and then adding the reciprocal value to the fitness vector. Yet, as in our case the solutions are triangular objective vectors and knowing that the *Euclidean distance* should be applied only between two exact vectors, we propose to use the expected value as a defuzzification method [29] in order to approximate the considered triangular vectors, such that for each triangular fuzzy number $y_i = [\underline{y_i}, \widehat{y_i}, \overline{y_i}]$, the expected value is defined by:

$$E(y_i) = (\underline{y_i} + 2 \times \widehat{y_i} + \overline{y_i})/4 \qquad (1.14)$$

Then, the Euclidean distance between two triangular vectors $\vec{y} = (y_1, \ldots, y_n)$ and $\vec{y}' = (y'_1, \ldots, y'_n)$ can be applied as follows:

$$D(\vec{y}, \vec{y}') = D(E(\vec{y}), E(\vec{y}')) = \sqrt{\sum_{i=1..n} (E(y_i) - E(y'_i))^2} \qquad (1.15)$$

Finally, we adapt the SPEA2 archive to triangular space in order to enable it keeping the obtained triangular solutions. These extensions are integrated into the research process of SPEA2 by modifying the following steps:

- Evaluation: Rank individuals using the new Pareto dominance $\prec_{TP}$.
- Environmental selection:
  1. Copy all non-dominated individuals having fitness values lower than one in the triangular archive A with fixed size N.
  2. if A is too large (size(A) > N) then, reduce A by means of truncation operator based on Nearest neighbor method to keep only the non-dominated individuals with good spread.
  3. else if A is too small (size(A) < N) then, fill A with the best dominated individuals.
  4. otherwise (size(A) = N), the environmental selection is completed.

- Mating selection: Perform binary tournament selection with replacement on the archive A in order to fill the mating pool.

ENSGAII:

NSGAII is an extension of an elitism PMOEA called Non-dominated Sorting Genetic Algorithm NSGA, originally proposed by [8]. Unlike the SPEA2 algorithm, NSGAII uses a crowded-comparison operator as diversity preservation technique in order to maintain a uniformly spread front by front. In addition, it does not use an explicit archive for the elitism operation, it only consider the population as a repository to store both elitist and non-elitist solutions. To extend NSGAII to triangular context, we propose at the first step to use the new Pareto dominance between triangular-valued objectives in order to ensure the fitness assignment procedure, in which a dominance depth strategy is applied. At the second stage, we provide an adaptation of the crowded-comparison operator. Indeed, this operator uses the *Crowding Distance* that serves to get a density estimation of individuals surrounding a particular individual in the population. More precisely, the total *Crowding Distance CD* of an individual is the sum of its individual objectives' distances, that in turn are the differences between the individual and its closest neighbors. For the ith objective function y_i, this distance is expressed by:

$$CD(i) = \sum_{i=1..n} (f_{y_i}(i+1) - f_{y_i}(i-1))/(f_{y_i}^{max} - f_{y_i}^{min}) \qquad (1.16)$$

Where f_{y_i} is the fitness value of its neighbors $(i-1)$ and $(i+1)$, $f_{y_i}^{max}$ and $f_{y_i}^{min}$ are respectively the maximum and minimum value of y_i.

However, as in our case, the objective functions are represented by triangular fuzzy values, we propose also to approximate these triangular numbers by calculating their expected values (Equation 1.14) before applying the Crowding distance. Finally, it is necessary to adapt both Evaluation and Selection steps in NSGAII, like in SPEA2 algorithm. The distinctive features of NSGAII lie in using the crowding comparison procedure as truncation operator to reduce the population in the environmental selection step and also in considering it as a second selection criteria when two solutions have the same rank in the tournament selection step.

1.5   Application on a Multi-objective Vehicle Routing Problem

The Vehicle Routing Problem (VRP) is an important combinatorial optimization problem, widely used in a large number of real-life applications [28]. The classical VRP consists in finding optimal routes used by a set of identical vehicles, stationed at a central depot, to serve a given set of customers geographically distributed and with known demands. Through the years, many variants and models derived from the basic VRP have been discussed and examined in the literature. In this work, we are

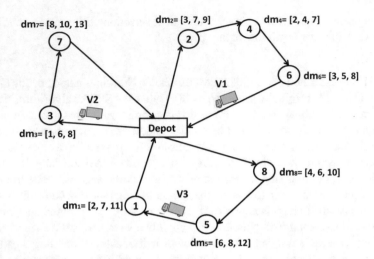

Fig. 1.8 Example of Mo-VRPTW-UD

interested in a well-known variant of VRP, the so-called Multi-objective VRP with
Time Windows and Uncertain Demands (MO-VRPTW-UD). This variant is based
firstly on the principle of classical VRP, where all the data are deterministic, excepting the customer demands which are uncertain, meaning that the actual demand is
only known when the vehicle arrives at the customer location. Several researchers
have tried to solve this problem and proposed to deal with the uncertainty of demands
using different ways such as probability distributions, dempster belief functions and
possibility distributions [?][13, 26]. In our case, the uncertainty of demands is represented via triangular fuzzy numbers (defined previously) and the objectives to be
optimized are respectively, the minimization of the total traveled distance and the
total tardiness time. Formally, a MO-VRPTW-UD may be defined as follows:

Let $G(N, A)$ be a weighted directed graph with an arc set A and a node set
$N_i = \{N_0, \ldots, N_n\}$ where the node N_0 is the central depot and the other nodes $N_i \neq N_0$
represent the customers. For each customer is associated an uncertain demand dm_i.
Only one vehicle k with a limited capacity Q, is allowed to visit each customer.
A feasible vehicle route R is represented by the set of served customers, starting
and ending at the central depot: $R_k = (N_0, N_1, \ldots, N_n, N_0)$. X_{ij}^k denotes the decision variable which is equal to 1 if the vehicle k travels directly from node N_i to
node N_j and to 0 otherwise. d_{ij} denotes the symmetric traveled distance between
two nodes (N_i, N_j). This distance is proportional to the corresponding travel time
t_{ij}. Figure 1.8 illustrates an example of MO-VRPTW-UD, with a central depot, 3
vehicles ($V1, V2, V3$) having a maximum capacity $Q = 10$ and a set of 8 customers
represented by nodes. Each customer $i = 1 \ldots 8$ has an uncertain demand expressed
in our case by a triangular fuzzy number $dm = [\underline{dm_i}, \widehat{dm}, \overline{dm}]$ (Ex: the fuzzy demand
of the customer 1 is $dm_1 = [2, 7, 11]$).

The main constraints of this problem are: Vehicle capacity constraint, Distance constraint and Time windows constraint.

(1) Vehicle capacity constraint: This constraint imposes that the sum of customer demands in each route must not exceed the limited capacity of associated vehicle. It may be defined as: $\sum_{i=1}^{n} dm_{N_i} \leq Q$. Yet, as in our case the customers demands are fuzzy values $\widetilde{dm} = [\underline{dm}, \widehat{dm}, \overline{dm}]$, we cannot directly verify if the capacity constraint is satisfied or not and so clearly the constraint satisfaction changes to fuzzy. For example, consider the customer 7 with fuzzy demand $dm_7 = [8, 10, 13]$ shown in Fig. 1.8, we cannot check if dm_7 is lower, equal or higher than $Q = 10$ in order to estimate the transportation costs in terms of time spent and traveled distance. Thus, to handle the satisfaction of this constraint, we propose to use firstly the two measures Π and N of fuzzy constraint satisfaction defined previously. For this example, we obtain $\Pi(dm \leq Q) = 1$ and $N(dm \leq Q) = 0$. Then, by applying the linear combination given by equation (8) (with for example $\lambda = 0.5$ and $\alpha = 0.2$), we can conclude that the satisfaction of the fuzzy capacity constraint is possible ($W(dm \leq Q) = 0.5 > 0.2$).

(2) Distance constraint: imposes that each vehicle with a limited capacity Q must deliver goods to the customers according to their uncertain demands dm, with the minimum transportation costs in term of traveled distance. In other words, if the capacity constraint of a vehicle is not satisfied, the delivery fails and causes wasted costs. Therefore, to calculate the traveled distance on a route R defined a priori, three different situations may be found:

- *The demand of a customer is lower than the vehicle capacity ($\sum_{i=1}^{f} dm_{N_i} < Q$):* In this case, the vehicle will serve the current customer f and then move to the next one ($f + 1$).
- *The demand of a customer is equal to the vehicle capacity ($\sum_{i=1}^{f} dm_{N_i} = Q$):* In this case, the priori optimization strategy is used. In fact, the vehicle leaves the depot to serve the first customer f with its total capacity. As it becomes empty, this vehicle will return to the depot to load and serve the next customer ($f + 1$). Thus, the traveled distance will be: $D(R) = d_{N_0 N_1} + \sum_{i=1}^{f-1} d_{N_i N_{i+1}} + d_{N_f N_0} + d_{N_0 N_{f+1}} + \sum_{i=f+1}^{n-1} d_{N_i N_{i+1}} + d_{N_n N_0}$.
- *The demand of a customer is higher than the vehicle capacity ($\sum_{i=1}^{f} dm_{N_i} > Q$):* In this case, the vehicle will serve the customer f with its total capacity ($Q - \sum_{i=1}^{f-1} d_{N_i}$), go to the depot to load, return back to the same customer f to deliver the remaining quantity and then move to the next customer ($f + 1$). Thus, the traveled distance will be: $D(R) = d_{N_0 N_1} + \sum_{i=1}^{n-1} d_{N_i N_{i+1}} + d_{N_f N_0} + d_{N_0 N_f} + d_{N_n N_0}$.

Yet as in our case the demands are represented by a triplet of fuzzy values, we propose to calculate separately the distance for each value of the triangular fuzzy demand based on the three situations presented above. Consequently, the traveled distance will be calculated three times and so obtained as triangular number $D = [\underline{D}, \widehat{D}, \overline{D}]$.

(3) Time windows constraint: imposes that each customer will be served within its time window that represents the interval of time planned for receiving the vehicle service. This means that, if the vehicle arrives too soon, it should wait until the arrival time of its time window to serve the customer, while if it arrives too late (after the fixed departure time), wasted cost in term of tardiness time appears. The time windows constraint uses the following notations:

- The central depot has a time window $[0, l_0]$, meaning that each vehicle that leaves the depot at a time 0 goes back to the depot before the time l_0.
- Each customer i will be served within his time window $[e_i, l_i]$ by exactly one vehicle, where the lower bound e_i represents the earliest arrival time and the upper bound l_i represents the latest arrival time for the visit of vehicles.
- A waiting time W_i means that the vehicles must arrive before the lower bound of the window e_i.
- A_i, B_i refers respectively to the arrival and the departure times to the customer i.
- Each customer imposes a service time S_i^k that corresponds to the goods loading/unloading time used by the vehicle.
- t_{ij} refers to the travel time from customer i to j.

Firstly, the time needed to serve two consecutive customers i and j is defined as follows:

$$x_{ij}^k(S_i^k + t_{ij} + S_j^k) \leq 0 \quad with \quad A_i + S_i^k \leq B_i.$$

Besides, a vehicle must arrive at a customer i between the time window $[e_i, l_i]$, but if it arrives before the lower bound e_i, it must wait a while W_i. This waiting time is calculated as follows:

$$W_i = \begin{cases} 0 & \text{if } A_i \geq e_i \\ e_i - A_i & \text{otherwise.} \end{cases}$$

where, the arrival time at customer i is equal to: $A_i = B_{i-1} + t_{i,i-1}$ and the departure time is equal to: $B_i = A_i + W_i + S_i$. While, if the vehicle arrives at a customer i after the upper bound of its time window l_i, a tardiness time must be calculated as follows:

$$T_i = \begin{cases} 0 & \text{if } A_i \leq l_i \\ A_i - l_i & \text{otherwise.} \end{cases}$$

In the case of routes failure, wasted costs in term of tardiness time will appear. Yet, knowing that the travel time depends mainly on the traveled distance and as in our case the obtained distance is a triangular value, the time spent to serve customers will be disrupted by this triangular form and consequently the tardiness time will be also obtained as triangular fuzzy number $T = [\underline{T}, \widehat{T}, \overline{T}]$. Finally, all these constraints combined with the constraints of classical VRP model the MO-VRPTW-UD problem.

To solve the MO-VRPTW-UD problem, the two extended SPEA2 and NSGAII algorithms based on our new Pareto optimality are applied. These algorithms are implemented with the version 1.3-beta of ParadisEO under Linux, especially with the ParadisEO-MOEO module dedicated to multi-objective optimization [2]. Subsequently, to validate the proposed algorithms, we choose to test our VRP application using the Solomon's benchmark, which is considered as a basic reference for the evaluation of several VRP resolution methods [25]. More precisely, six different Solomon's instances are used in our experimentation, namely, C101, C201, R101, R201, RC101 and RC201. Yet, in these instances, all the input values are exact and so the uncertainty of customer demands is not taken into account. At this level, we propose to generate for each instance the triangular fuzzy version of crisp demands in the following manner. Firstly, the kernel value $(\widehat{dm})$ for each triangular fuzzy demand dm is kept the same as the current crisp demand dm_i of the instance. Then, the lower $(\underline{dm})$ and upper $(\overline{dm})$ bounds of this triangular fuzzy demand are uniformly sampled at random in the intervals $[50\%dm, 95\%dm]$ and $[105\%dm, 150\%dm]$, respectively. This fuzzy generation manner ensures the quality and reliability of generated fuzzy numbers. Finally, each of the six sampled fuzzy instances is tested on the both algorithms executed 30 times. Since 30 runs have been performed on each algorithm SPEA2 and NSGAII, we obtained for each instance, 30 sets of optimal solutions that represent the Pareto fronts of our problem. Each solution shows the lowest traveled distance and tardiness time, which are represented by triangular numbers. Examples of two Pareto fronts obtained for one execution of the instance C101 using each algorithm are shown in Figs. 1.10 and 1.11, where the illustrated fronts are composed by a set of triangles, such that each triangle represents one Pareto optimal solution. For instance, the bold triangular (in Fig. 1.10) represents an optimal solution with minimal distance (the green side) equal to [2413, 2515, 2623] and tardiness time (the red side) equal to [284312, 295280, 315322]. Note that, both algorithms converge to optimal fronts approximation in a very short run-time (Approx. 0.91 minutes for SPEA2 and 2.30 minutes for NSGAII). However, we cannot compare results with the obtained results of other proposed approaches for solving MO-VRPTW-UD because of incompatibilities between the objectives to be optimized.

To assess the performance of our both algorithms, we propose to use two well-known unary quality indicators:

(i) *Hypervolume Indicator* (I_H) [34], considered one of the few indicators that measures the approximation quality in terms of convergence and diversity simultaneously. This intuitive quality indicator needs the specification of a reference point Z^{Max} that denotes an upper bound over all the objectives and a reference set Z_N^* of non-dominated solutions. In our case, the quality of a given output set A in compar-

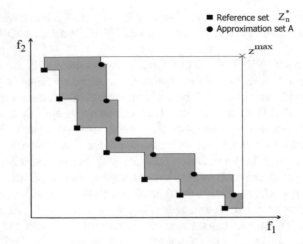

Fig. 1.9 Hypervolume difference indicator

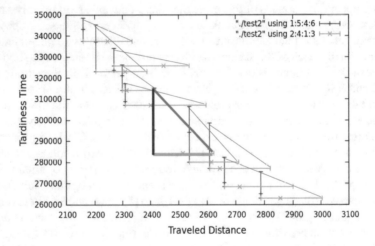

Fig. 1.10 Pareto front (C101-ESPEA2)

ison to Z_N^* is measured using the Hypervolume difference metric I_H^-. As shown in Fig. 1.9, this indicator computes the difference between these two sets by measuring the portion of the objective space weakly dominated by Z_N^* and not by A.

   (ii) *Epsilon Indicator* (I_ϵ) [35], dedicated to the measure of approximations quality in term of convergence. More explicitly, this indicator is used to compare non-dominated approximations and not the solutions. In our case, we use the additive ϵ-indicator $(I_{\epsilon+})$ which is a distance based indicator that gives the minimum factor by which an approximation A has to be translated in the criterion space to weakly dominate the reference set Z_N^*. This indicator can be defined as follows:

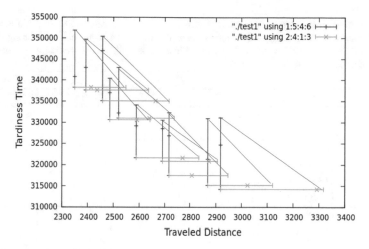

Fig. 1.11 Pareto front (C101-ENSGAII)

$$I_{\epsilon+}^1(A) = I_{\epsilon+}(A, Z_N^*) \tag{1.17}$$

where

$$I_{\epsilon+}(A, B) = min\{\forall z \in B, \exists z' \in A : z_i' - \epsilon \le z_i, \forall\, 1 \le i \le n\} \tag{1.18}$$

However, these two indicators are only meant to evaluate the quality of deterministic Pareto front approximations. Thus, to enable them evaluating our uncertain approximations (i.e, Triangular fuzzy solutions), we propose to consider the expected values of the triangular solutions (Function) as the sample of values to be used for the qualification of our both algorithms. In other words, the both indicators are simply applied on the samples of expected values computed for each instance. Therefore, as in our case 30 runs per algorithm have been performed, we obtain 30 Hypervolume differences and 30 epsilon measures for each tested sample. Once all these values are computed, we need to use statistical analysis to be able to compare our two algorithms. To this end, we choose to use Wilcoxon statistical test described in [35].

Table 1.2 gives a comparison of SPEA2 and NSGAII algorithms for the six tested instances. This comparison based on the results of I_H^- and $I_{\epsilon+}$ indicators, shows that the SPEA2 algorithm is significantly better than the NSGAII algorithm on all the instances, excepting the instances R201 and RC201, where for I_H^- there is no significant difference between the approximations of both algorithms.

Table 1.2 Algorithms comparison using Wilcoxon test with a P-value = 0.5%. According to the metric under consideration (I_H^- or $I_{\epsilon+}$), either the algorithm located at a specific row is significantly better ($\prec$) than the algorithm located at a specific column, either it is worse ($\succ$) or there is no significant difference between both ($\equiv$)

Instances	Algorithms	I_H^-		$I_{\epsilon+}$	
		ESPEA2	ENSGAII	ESPEA2	ENSGAII
C101	ESPEA2	-	$\prec$	-	$\prec$
	ENSGAII	$\succ$	-	$\succ$	-
C201	ESPEA2	-	$\prec$	-	$\prec$
	ENSGAII	$\succ$	-	$\succ$	-
R101	ESPEA2	-	$\prec$	-	$\prec$
	ENSGAII	$\succ$	-	$\succ$	-
R201	ESPEA2	-	$\equiv$	-	$\prec$
	ENSGAII	$\equiv$	-	$\succ$	-
RC101	ESPEA2	-	$\prec$	-	$\prec$
	ENSGAII	$\succ$	-	$\succ$	-
RC201	ESPEA2	-	$\equiv$	-	$\prec$
	ENSGAII	$\equiv$	-	$\succ$	-
C102	ESPEA2	-	$\prec$	-	$\prec$
	ENSGAII	$\succ$	-	$\succ$	-
C202	ESPEA2	-	$\prec$	-	$\prec$
	ENSGAII	$\succ$	-	$\succ$	-
R102	ESPEA2	-	$\prec$	-	$\prec$
	ENSGAII	$\succ$	-	$\succ$	-
R202	ESPEA2	-	$\equiv$	-	$\prec$
	ENSGAII	$\equiv$	-	$\succ$	-
RC102	ESPEA2	-	$\prec$	-	$\prec$
	ENSGAII	$\succ$	-	$\succ$	-
RC202	ESPEA2	-	$\equiv$	-	$\prec$
	ENSGAII	$\equiv$	-	$\succ$	-

1.6   Conclusion

This chapter addresses the multi-objective problems with fuzzy data, in particular, with triangular-valued objective functions. To solve such problems, we have proposed an extension of two multi-objective evolutionary algorithms: SPEA2 and NSGAII by integrating a new triangular Pareto dominance. The implemented algorithms have been applied on a multi-objective vehicle routing problem with uncertain demands and then experimentally validated on the Solomon's benchmark. Subsequently, we have obtained an encouraging results. As a future work, we intend to refine the algorithmic features by introducing for example a new fuzzy distance for the density estimation techniques and to extend the proposed Pareto dominance for ranking other fuzzy shapes like trapezoidal fuzzy numbers. Another perspective will be the extension of multi-objective performance metrics to uncertain context (i.e, fuzzy context).

References

1. Babbar, M., Lakshmikantha, A., & Goldberg, D. E. (2003). A modified NSGA-II to solve noisy multiobjective problems. *Genetic and Evolutionary Computation, 2723*, 2127.
2. Basseur, M., Liefooghe, A. L. J., & El-Ghazali, T. (2007). ParadisEO-MOEO: A framework for evolutionary multi-objective optimization. In *Evolutionary multi-criterion optimization* (pp. 386–400).
3. Brito, J., Morino, J. A., & Verdegay, J. L. (2009). Fuzzy optimization in vehicle routing problems. In *IFSA-EUSFLAT*.
4. Coello, C. A. C., & Lamont, G. B. (2004). *Applications of multi-objective evolutionary algorithms*. World Scientific.
5. Coello, C. A. C., Lamont, G. B., & Van Veldhuisen, D. A. (2007). Evolutionary algorithms for solving multi-objective problems. *Springer, 5*, 79–104.
6. Deb, K. (2001). *Multi-objective optimization using evolutionary algorithms*. Wiley.
7. Deb, K., Agrawal, S., et al. (2000). A fast elitist non-dominated sorting genetic algorithm for multi-objective optimization: NSGA-II. *Evolutionary Computation, 6*.
8. Deb, K., & Srinivas, N. (1994). Multiobjective optimization using nondominated sorting in genetic algorithms. *Evolutionary Computation, 3*, 221248.
9. Dubois, D., & Prade, H. (1998). An introductory survey of possibility theory and its recent developments. *Fuzzy Theory and Systems, 10*, 21–42.
10. El-Ghazali, T. (2009). *Metaheuristics: From design to implementation*. Wiley.
11. Fonseca, C. M., Fleming, P. J. (1993). Genetic algorithms for multiobjective optimization: formulation, discussion and generalization. In: *Proceedings of the Fifth International Conference on Genetic Algorithms* (pp. 416–423).
12. Goh, C. K., & Tan, K. C. (2009). *Evolutionary multi-objective optimization in uncertain environments: Issues and algorithms*. Springer.
13. Goncalves, G., Hsu, T., & Xu, J. (2009). Vehicle routing problem with time windows and fuzzy demands: An approach based on the possibility theory. *Inderscience, 4*, 312–330.
14. Horn, J., Nafpliotis, N., & Goldberg, D. (1994). A Niched Pareto genetic algorithm for multi-objective optimization. *Evolutionary Computation, 1*, 82–87.
15. Hughes, E. (2001). Evolutionary multi-objective ranking with uncertainty and noise. In *Evolutionary Multi-criterion optimization* (p. 329343).
16. Hughes, E. J. (2001). *Constraint handling with uncertain and noisy multi-objective evolution*. Seoul, Korea.
17. Knowles, J. D., & Corne, D. W. (2000). Approximating the nondominated front using the pareto archived evolution strategy. *Evolutionary Computation, 2*, 149–172.
18. Kolmogorov, A. N. (1960). *Foundations of the theory of probability* (2nd ed.). Chelsea Pub Co.
19. Laarabi, M., Sacile, R., et al. (2013). Ranking triangular fuzzy numbers using fuzzy set. In *Fuzzy logic and applications* (pp. 100–108).
20. Limbourg, P. (2005). Multi-objective optimization of problems with epistemic uncertainty. In *Evolutionary multi-criterion optimization* (Vol. 35).
21. Limbourg, P., & Daniel, E. (2005). An optimization algorithm for imprecise multi-objective problem functions. *Evolutionary Computation, 1*, 459–466.
22. Paquete, L., & Sttzle, T. (2007). Stochastic local search algorithms for multiobjective combinatorial optimization: Methods and analysis. *Handbook of approximation algorithms and metaheuristics, 13*, 79–104.

23. Shafer, G. (1976). *A mathematical theory of evidence*. University Press: Princeton.
24. Smets, P. (1989). Constructing the Pignistic Probability Function in a Context of Uncertainty. In *Proceeding 5th Conference on Uncertainty in Artificial intelligence*.
25. Solomon, M. (1987). Algorithms for the vehicle routing and scheduling problem with time window constraints. *Operations Research*, **35**.
26. Sulieman, D., Jourdan, L., & El-Ghazali, T. (2010). Using multiobjective metaheuristics to solve VRP with uncertain demands. *Evolutionary Computation*, 1–8.
27. Teich, J. (2001). Pareto-front exploration with uncertain objectives. In *Evolutionary multi-criterion optimization* (p. 314328).
28. Toth, P., & Vigo, D. (2002). *The vehicle routing problem*. Siam.
29. Wang, Z., & Tian, F. (2010). A Note of the expected value and variance of Fuzzy variables. *International Journal of Nonlinear Science*, **9**.
30. Zadeh, L. (1965). Fuzzy sets. *Information and Control*.
31. Zadeh, L. (1999). Fuzzy sets as a basis for a theory of possibility. *Fuzzy Sets and Systems*, *100*, 9–34.
32. Zitzler, E., Laumans, M., & Thiele, L. (2001). SPEA2: Improving the strength Pareto evolutionary algorithm. Technical Report 103, Computer Engineering and Networks Laboratory (TIK).
33. Zitzler, E., & Thiele, L. (1999). Multiobjective evolutionary algorithms: A comparative case study and the strength Pareto approach. *Evolutionary Computation*, *4*, 257271.
34. Zitzler, E., Thiele, L., & Knowles, J. (2005). A tutorial on the performance assessment of stochastic multiobjective optimizers. In *Evolutionary multi-criterion optimization*.
35. Zitzler, E., Thiele, L., et al.: Performance assessment of multiobjective optimizers: An analysis and review. *Evolutionary Computation*, 117–132.

Chapter 2
Solving the Uncapacitated Single Allocation p-Hub Median Problem on GPU

A. Benaini, A. Berrajaa, J. Boukachour and M. Oudani

Abstract This chapter presents a parallel GPU implementation for solving the Uncapacitated Single Allocation p-Hub Median Problem with genetic algorithm. Starting from an initial solution that locates hubs at center nodes, our GPU implementation quickly reaches optimal or near-optimal solutions. It efficiently exploits the computing power of the GPU and outperforms, in cost and in computing time, the other approaches proposed in the literature for this problem. We compare it to the best recent solutions on the benchmarks up to 1000 nodes and on large instances up to 6000 nodes generated by us. Moreover, we solved instance problems so far unsolved.

2.1 Introduction

Hubs are sort of facilities that serve to transfer transshipment and sort in many-to-many complex distribution networks. They find applications in airline passengers, telecommunications, distributed computing and postal delivery networks. In air traffic, hubs are the central airports for long haul by cargo planes. In the telecommunication and distributed computing networks, hubs may be concentrators, servers, routers or multiplexers, etc. In postal distribution networks, hubs are sorting and the cross dock centers. The hub location problem can be stated as follows. Given a set of N nodes and the corresponding cost flows w_{ij} between nodes i and j, $1 \leq i, j \leq N$,

A. Benaini (✉) · J. Boukachour
Normandie University, LMAH, Le Havre, France
e-mail: abdelhamid.benaini@univ-lehavre.fr

J. Boukachour
e-mail: jaouad.boukachour@univ-lehavre.fr

A. Berrajaa
Normandie University and University Mohammed 1er, LaRI, Oujda, Morocco
e-mail: berrajaa.achraf@gmail.com

M. Oudani
Private University of Marrakech, Marrakech, Morocco
e-mail: oudani.mustapha@gmail.com

© Springer International Publishing AG, part of Springer Nature 2019
E.-G. Talbi and A. Nakib (eds.), *Bioinspired Heuristics for Optimization*,
Studies in Computational Intelligence 774,
https://doi.org/10.1007/978-3-319-95104-1_2

the objective is to locate p appropriate nodes in a network to be hubs and to allocate the non hub nodes to hubs such that the total flow-weighted costs across the network is minimized. The success of this type of network is due to the economy of scale achieved by consolidating the traffic through hub-hub arcs. Indeed, in this network any flow between pair of nodes can only take place through the hubs. The cost from an origin non-hub i to a destination non-hub j through two hubs k and l is expressed as $C_{ijkl} = \chi w_{ik} + \alpha w_{kl} + \gamma w_{lj}$ where $\alpha < 1$ represents the scale economy generated by consolidating flows between hubs, while χ and γ represent the distribution and the collection costs respectively and are often greater than 1.

Several studies concerning hub location problems have been developed since 1980. Different variants of hub location problems have been defined and classified according to the method of allocation into two main categories. The single allocation, where each non hub is assigned to exactly one hub, while the multiple variant enables the non-hubs to be allocated to several hubs. The problem is said to be uncapacitated if the hubs have infinite capacities; otherwise it is capacitated. Moreover, if the number of hubs p is given then the problem is called Uncapacitated Single Allocation p-Hub Median Problem (USApHMP); otherwise the problem is said to be Uncapacitated Single Allocation Hub Location Problem (USAHLP). In this last case, the number of hubs and their locations are decision variables that must be defined. This is formulated by adding an additional term in the objective function, aiming to minimize the fixed cost of installing hubs.

Other versions of hub problems have been introduced in the literature such as (1) p-hub center problem [1] where the objective is to minimize the maximum travel time between two demand centers (2) The hub arc problem [2] which aims to overtake the shortcomings of the p-hub median problem by introducing the bridges arcs between hubs without any discount factor (3) The dynamic hub location problem [3] where either cost, demands or resources may vary in the planning horizon (4) With congestion, where economic penalties are imposed if hubs with congestion levels are used; (5) With non-linear costs with stochastic elements; (6) With vehicle routing constraints [4]. Reviews, synthesis and methods for solving these problems are presented earlier [5].

As the USApHMP is NP-hard, it cannot be solved to optimality (by exact methods) for realistically sized instances in reasonable time. Exact methods provide optimal solutions for smaller size problem instances with $N \leq 200$ nodes. So, many of the solution techniques suggested for this problem are heuristics. Among the most promising and successful heuristics are Tabu Search, Simulated Annealing, Genetic Algorithm and Greedy methods. This chapter will focus on USApHMP for which we propose a parallel genetic algorithm and it GPU implementation. To our knowledge, this is the first GPU implementation that solves this problem. We show under tests on known benchmarks the efficiency of our approach. Note that the GPUs and the underlying parallel programming model Cuda are actually available in most personal computers. So, massively parallel computing has become a commodity technology.

The remainder of this chapter is organized as follows. Related works are provided in Sect. 2.2. In Sect. 2.3, we present the mathematical formulation of the problem. The GA description is presented in Sect. 2.4, followed by the GPU implementation in Sect. 2.5. Computational results are reported in Sect. 2.6 and Concluding remarks are given in Sect. 2.7.

2.2   Related Works

The first mathematical formulation for the USApHMP as a quadratic integer program was presented in [6]. Two heuristics were developed to solve it with numerical results for CAB (Civilian Aeronautics Board) data. Reference [1] presented integer formulations for the p-hub median problem (pHMP), p-hub center problem and hub covering problems. Possible extensions with flow thresholds were also studied. Also [2] introduced the p-HMP with two heuristics for solving it tested on a dataset with 10-40 nodes and up to 8 hubs. Reference [7] studied the special case of the single allocation two-hub location problem. In this particular case, the quadratic program is transformed to a linear program and to a minimum cut problem. Reference [8] proposed an hybrid genetic algorithm and tabu search heuristic and reported the results on the CAB dataset. Reference [9] presented a solving approach for the multiple allocation p-HMP version and described how this could be adapted to the single allocation case. Their method is based on the shortest-path algorithm to find lower bounds for a branch-and-bound scheme that find the exact solution. The results are reported on AP data for a multiple allocation case up to 200 nodes. Reference [10] studied four models: the first was concerned with a capacitated network, the second focus on the minimum threshold model, the third determined the numbers of open hubs and the last one introduced a flow-dependent cost function.

Reference [11] proposed a model implemented in a GIS environment to prove that hub networks may emerge naturally on traffic networks to take advantage of economies of scale. Reference [12] studied the polyhedral properties of the single allocation hub location problem and proposed a Branch-and-Cut algorithm for solving this problem. Reference [13] proposed a hybrid heuristic to solve the USAHLP. Their method is based on a combination of an upper bound method search, simulated annealing and tabu list heuristic and was tested on CAB data and AP data up to 200 nodes. Reference [14] proposed three variants of tabu search heuristics and a two-stage integrated tabu search to solve the USAHLP. These authors used the multi-start principle to generate different initial solutions that are improved by the tabu search. They reported the results for new introduced larger instances with 300 and 400 nodes. Reference [15] proposed a general variable neighborhood search for the USApHMP. They reported the results on large AP data, PlanetLab instances and Urand instances up to 1000 nodes. Reference [4] introduced the single allocation hub location problem under congestion with a generalized Benders decomposition algorithm to solve AP instances.

Reference [16] proposed a memetic algorithm to solve the USAHLP. Their evolutionary approach is based on two local search heuristics. They tested their algorithm on the well-known benchmark and generated larger scale instances up to 900 nodes. They gave the optimal solutions of AP data up to 200 nodes. Reference

[17] proposed Discrete Particle Swarm Optimization to solve the USAHLP. They obtained the optimal solutions on all CAB datasets and on AP data up to 200 nodes. Reference [18] introduced a planar version of the USAHLP with the particularity that a hub could be located anywhere in the plan. Reference [19] proposed a threshold accepting algorithm to solve the USAHLP and reported the results on the AP and CAB benchmarks.

Reference [20] made use of dataset structures to propose a new linearization of the quadratic formulation of the problem. They obtained optimal solutions on the AP data up to 200 nodes. Reference [21] introduced a new version of USApHMP, where the discount factor between hubs is replaced by a decision variable. They proposed a branch-and-bound algorithm with Lagrangian relaxation to solve the problem. Recently [22] proposed a Tabu Search heuristic for solving the USAHLP and reported the results both on CAB data and an AP dataset up to 400 nodes. Reference [23] presented a greedy deterministic and randomized algorithms for constructing an initial feasible solution and improves it by GRASP heuristic.

References [24, 25] proposed an efficient GA for solving the uncapacited single and multiple allocation hub problems. Binary encoding and adapted genetic operators to these problems are used (only allocation hubs are given as the solution). They showed, under experimental results on ORLIB instances with up to 200 nodes, that the GA approach quickly reaches all known optimal solutions. In this study, we use similar integer encoding and genetic operators to conceive a parallel GA that solves efficiently the USApHMP.

Several studies suggest different strategies to implement GAs on different parallel machines [26–29]. It appears that are three major types of parallel GAs: (1) the master-slave model, (2) the island model and (3) the fine-grained model. In the master-slave model, the master node holds the population and performs most of the GA operations. The fitness evaluation, crossover, correction and mutation operations on groups of individuals are made by each slave. In a coarse-grained model, the population is divided into several nodes. Each node then has a subpopulation on which it executes GA operations. In fine-grained models, each node only has a single individual and can only communicate with several neighboring nodes. In this case, the population is the collection of all the individuals in each node. There are conflicting reports over whether multiple independent runs of GAs with small populations can reach solutions of higher quality, or can find acceptable solutions faster, than a single run with a large population. We adopt the fine-grained model, which is more suitable for the GPU, to conceive the GA for the USApHMP and we solve the USAHLP by solving in parallel all the USApHMP for $p = 1, \ldots, N$. The solution of the USAHLP is the one of the USApHMP with the minimum flow cost among these N solutions.

2.3 The Problem Formulation

The USAHLP can be stated as follows. Given N nodes $1, \ldots, N$. Locate p hubs and allocate each non-hub to a single hub in order to minimize the total flow cost. There are three levels of decisions in the USAHLP : (i) the number p of hubs to be located; (ii) which nodes will be hubs (iii) how the non-hub nodes will be allocated to hubs. In the USApHMP, the decision concerns only (ii) and (iii), since the hub number p is given.

The USAHLP is formulated as a MIP by Ernst et al. [30] as follows. Let w_{ij} the flow originated from i to j, d_{ij} the distance between i and j and F_k the cost for opening a hub at the node k. Let z_{ik} the binary variable that is equal to 1 if node i is assigned to the hub k and 0 otherwise (with $z_{kk} = 1$ for each hub k). y_{kl}^i the total flow emanating from i that is routed between hubs k and l. $O_i = \sum_j w_{ij}$ the flow departing from origin i and $D_i = \sum_j w_{ji}$ the total cost destined to i.

$$minimize \ \sum_i \sum_k d_{ik} z_{ik} (\chi O_i + \gamma D_i) + \alpha \sum_i \sum_k \sum_l d_{kl} y_{kl}^i + \sum_k F_k z_{kk} \quad (2.1)$$

Subject to:

$$\sum_k z_{ik} = 1, \ 1 \leq i \leq N \quad (2.2)$$

$$z_{ik} \leq z_{kk}, \ 1 \leq i, k \leq N \quad (2.3)$$

$$\sum_l y_i^{kl} - \sum_l y_i^{lk} = O_i z_{ik} - \sum_j w_{ij} z_{jk}, \ 1 \leq i, k \leq N \quad (2.4)$$

$$z_{ik} \in \{0, 1\}, \ 1 \leq i, k \leq N \quad (2.5)$$

$$y_i^{kl} \geq 0, \ 1 \leq i, k, l \leq N \quad (2.6)$$

The objective function (1) minimizes the total cost of flow transportation between all origin- destination nodes plus the cost of establishing the hubs. Constraint (2) imposes each non-hub to be allocated to exactly one hub. Constraint (3) enforces that flow is only sent via open hubs, preventing direct transmission between non-hub nodes. Constraint (4) is the flow conservation constraint and finally constraint (5) and (6) reflect non-negative and binary nature of variables y_{kl}^i and z_{ik} respectively.

The USApHMP (the number of hubs p is given) is formulated as follows:

$$minimize \ \sum_i \sum_k d_{ik} z_{ik} (\chi O_i + \gamma D_i) + \alpha \sum_i \sum_k \sum_l d_{kl} y_{kl}^i \quad (2.7)$$

subject to

$$\sum_k z_{kk} = p \quad (2.8)$$

and to constraints (2)–(6). Unfortunately these problems are NP-Hard and hence difficult to solve for optimally. So, heuristic and/or parallel approaches are needed to solve large size intances. Figure 2.1a shows a simple example of the USApHMP with $N = 5$ nodes presented by their distances (numbers on the arcs). The number

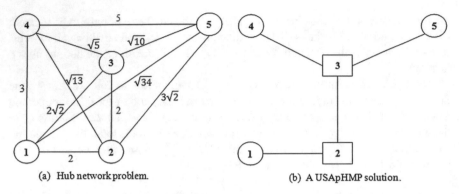

(a) Hub network problem.   (b) A USApHMP solution.

Fig. 2.1 Example of USApHMP solution given in [26]

of hubs that will be located is $p = 2$ and the parameters $\chi = \gamma = 1$ and $\alpha = 0.25$. The amount of flow w_{ij} from a node i to a node j is equal to 1, even for $i = j$. So $O_i = D_i = 5$ for all i. Figure 2.1b presents a solution for this network. The hubs are nodes 2 and 3. The nodes 1 and 2 are allocated to hub 2, while nodes 3, 4 and 5 are allocated to hub 3 (each hub is allocated to itself).The overall transportation cost of this solution is 79.983.

2.4   The Genetic Algorithm

In the following, we outline the GA for solving USApHMP and we show how to use it to solve USAHLP.

Encoding:
Since an encoding of a solution of the USApHMP must express the locations/ allocations, we use a simple encoding by an N-array also used among others in [24, 25, 31]. That is, the encoding S is defined by $S[i] = k$ if node i is allocated to the hub k and $S[k] = k$ if k is hub (each hub is allocated to itself). For instance, the corresponding encoding of the solution of Fig.2.1b is $S = (2, 2, 3, 3, 3)$.

Initial solutions for the parallel GA:
In order to quickly reach an optimal or near-optimal solution, we generate a feasible initial solution with p hubs, say P_p, for the GA as follow.

- Compute the p center nodes i.e. the p hubs k with the smallest distances $\sum_{j=1}^{N} d_{kj}$ to other nodes and locate the p hubs at these p nodes.
- Allocate each non hub node to its nearest hub.

Note that other heuristics for generating initial feasible solutions are proposed in the literature depending on the nature of the problem studied [23].

We use *randon_generate(s)* that randomly permutes in the solution s one hub with one non hub and allocates each non-hub to its nearest hub to generate a new individual (solution) from s. We apply R times *randon_generate*() to the intial solution P_p, generated as mentionned above, to get R initial feasible solutions $P_p^i, 0 \leq i < R$. Then we run in parallel $GA(P_p^0), \ldots GA(P_p^{R-1})$ to get the solution of the USApHMP; here $GA(P_p^i)$ is the parallel GA with initial solution P_p^i.

2.4.1   Genetic Operators

Given a current solution s, the genetic operators used here consist of:

- *Generate*(s) which generates p_1 and p_2 from the current solution s by computing $p_1 = random\_generate(s)$ and $p_2 = random\_generate(s)$.
- *Crossover*(p_1, p_2) which creates two children ch_1 and ch_2 from (p_1, p_2) by performing a random cut in p_1 and in p_2, then copying the first part of p_1 to ch_1 and the first part of p_2 to ch_2, interchanging for both p_1 and p_2 the offsprings informations.
- *Correction*(s) which performes a correction to ensure the validity of the solution s, in terms of the number of hubs if different of p. If the number of hubs is greater than p then keep the p hubs with the greatest $O_i + D_i$. If the number of hubs is less than p then the missing hubs are randomly chosen among the non-hubs with the greatest amount of flow $O_i + D_i$. We apply the *correction*() operator to p_1 and p_2 to get c_1 and c_2 each with exactly p hubs (located at nodes where the traffic density is).
- *Mutation*() whose objective is to enlarge the search space and to avoid the convergence to the same solution. The mutation operator used here consists of permuting one hub with its neighbor non-hub.

2.4.2   Solution Evaluation

The following definitions of fitness are used in the standard benchmarks to evaluate the solutions. Let $A = \sum_i \sum_k d_{ik} z_{ik} \chi O_i + \gamma D_i + \sum_i \sum_k \sum_l \alpha d_{kl} y_i^{kl}, B = \sum_i \sum_j w_{ij}$, and $C = \sum_k F_k z_{kk}$.

- For the USAHLP: The fitness for CAB data is given by $\frac{A}{B} + C$ and the fitness for all other data instances except PlanetLab is given by $10^{-3}(A + C)$.
- For the USApHMP: The fitness for CAB data is given by $\frac{A}{B}$ and the fitness for all other data instances except PlanetLab is given by $10^{-3}A$.

Note that the reason of multiplying by 10^{-3} is to obtain the unit cost for flow transportation. This was discovered when we tried to reproduce the optimal solutions and this was confirmed by contacting M.R. Silva [14].

2.5   The GPU Implementation

Graphics Processing Units (GPUs) are available in most personal computers. They are used to accelerate the execution of a variety of problems. The smallest unit in the GPU that can be executed is called a thread. Threads (all executing the same code and synchronized) are grouped into blocks of equal size that are grouped in grids (blocks are independent and cannot be synchronized).

The memory hierarchy of the GPU consists of three levels: 1) the global memory that is accessible by all threads 2) the shared memory accessible by all threads of a block and 3) the local memory (register) accessible by a thread. Shared memory has a low latency (2 cycles) and is of limited size. Global memory has a high latency (400 cycles) and is of large size (4 GB for the Quadro).

An entire block is assigned to a single SM (Stream Multiprocessor). Each SM is composed of 32 streaming processors that share a same shared memory. Several blocks can run on the same SM. Each block is divided into Warps (32 threads by Warp) that are executed in parallel. The programmer must control the block sizes, the number of Warps and the different memory access.

A typical CUDA program is a C program where the functions are distinguished based on whether they are meant for execution on the CPU or on the GPU. Those executed on the GPU are called kernels and are executed by several threads. We implemented the previously presented algorithm on GPU (Nvidia Quadro with 4 GB and 384 cores running under CUDA 7.5 environment) and compare it to the best-known results in terms of computational time and quality of solutions.

2.5.1   The GPU Implementation of the GA

The GA creates R subpopulations each with n individuals ($R < 100$ for our implementation). So, the GPU is partitioned into R blocks; each with n threads (we define the master thread of each block as the thread 0 and the global master thread as the thread 0 of $block_0$). The matrices W and D are stored in the global memory. The $block_i$, $0 \leq i < R$ stores in its shared memory the genetic code of the initial solution P_p^i and executes the GA with P_p^i as initial solution as follow. At the beginning P_p (generated as indicated in Sect. 2.4) is duplicated in all $block_i$ and assigned to P_p^i, $0 \leq i < R$. Let $T_0^i, \ldots, T_{n-1}^i$ be the threads of the $block_i$. Each thread T_j^i generates a new solution (individual) $p_j^i = random\_generate(P_p^i)$. Then $(p_0^i, \ldots, p_{n-1}^i)$ becomes the initial subpopulation of the GA executed by the $block_i$. Note that the subpopulation size n is the same for all blocks. Each thread T_{2j}^i generates two children, namely ch_1 and ch_2 by crossing the parents (p_{2j}^i, p_{2j+1}^i) then T_{2j}^i applies the correction to ch_1 to obtain a new feasible individual (say c_{2j}) and T_{2j+1}^i applies the correction to ch_2 to obtain a new feasible individual (say c_{2j+1}). To enlarge the search space (to avoide local optimum), we use the mutation operator with a probability equal to 10^{-5} to the subpopulation $(c_0^i, \ldots, c_{n-1}^i)$ which consists of exchanging one hub with its

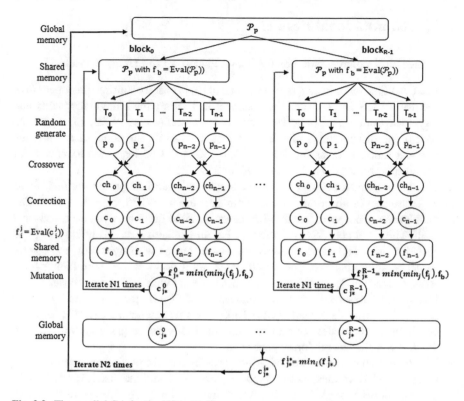

Fig. 2.2 The parallel GA for the USApHMP

neighbor non-hub. Next, each T_j^i allocates the non hubs of c_j^i to their nearest hubs and computes $f_j^i = fitness(c_j^i)$.

All c_j^i and f_j^i are stored in the shared memory of the $block_i$. Therefore, the master thread of the $block_i$ selects the individual c_{j*}^i with $min(min_j(f_j^i), f_b^i)$ where f_b^i is the cost of P_p^i (the initial solution) and updates the ancestor P_p^i as c_{j*}^i for the next iteration if any. This inner-loop of GA terminates after $N1$ iterations (the same for all the blocks).

The c_j, $0 \le i < R$, are asynchronously computed and copied in the global memory and the individual c_{j*}^{i*} with $min_i(f_{j*}^i)$ is selected as the final solution or as the new value of the ancestor P_p for the next iteration, if any, of the outer-loop. The process is repeated $N2$ times. This implementation is summarized in the Fig. 2.2 and the pseudo CUDA code executed by the CPU is the following.
1. Generate(P_p) and copy it in the global memory of the GPU;
2. Define the blocks and the grid as: dim3 dimBlock(n,1); dim3 dimGrid(R,1);
3. Launch the kernel $GA(P_p)$: GA<<< $dimGrid$, $dimBlock$ >>> (P_p);

The R blocks create R subpopulations each with n individuals to find a solution for the USApHMP. Therefore, $N \times R$ blocks are needed to compute the N solutions for the USApHMP ($p = 1, \ldots, N$) to get the solution of the USAHLP.

2.6   Computational Results

We tested our implementation on CAB, AP, PlanetLab and Urand datasets.

- The CAB dataset of instances, introduced in [6] is based on airline passenger flow between 25 US cities. It contains distances that satisfy the triangle inequality and a symmetric flow matrix between cities. The instances are of 10, 15, 20 and 25 nodes. The distribution and collection factors γ and χ are equal to 1, while the discount factor α takes the values 0.2, 0.4, 0.6, 0.8 or 1 and the fixed cost for establishing hubs (F_k) is 10, 150, 200, or 250.
- The AP (Australian Post) dataset is a real-world dataset representing mail flows in Australia. For all instances, the parameters are $\alpha = 0.75$, $\gamma = 3$, $\chi = 2$. The mail flows are not symmetric and there are possible flows between each node and itself. The fixed costs F_k are different. For all problem sizes, there are two fixed costs for hubs Tight and Loose costs. The instances with tight costs are mostly harder to solve, because nodes with larger total flows are set with higher fixed costs, because nodes with larger total flows are set with higher fixed costs.
- The Urand dataset are random instances up to 400 nodes generated by Meyer et al. [32], while the instance with 1000 nodes was generated by [15]. In these benchmarks, the nodes coordinates were randomly generated from 0 to 10^5 and the flow matrix was randomly generated.
- The PlanetLab instances are node-to-node delay data for performing internet measurements [15]. In these networks, $\alpha = \gamma = \chi = 1$ and the distance matrix does not respect the triangle inequality due to missing links (i.e. firewall restrictions). The flow $w_i j$ is equal to 0 if $i = j$ and 1 otherwise.

We report the results for the four datasets presented above. We compare our results to the recent work of [22] in terms of computing time for USAHLP and to the results of [15] for the USApHMP. For our tests, the GA was executed 10 times on each instance. The same solution was obtained for the multiple executions of each instance. The time transfer between the CPU and GPU varies according to the number of nodes. The given times in all tables include the time of generating the initial solutions, the time of data transfers CPU-GPU and the computation of the solutions on the GPU.

The following notations are used in Tables 2.1, 2.2, 2.3, 2.4, 2.5 and 2.6. N is the number of nodes in the instance. *BestSol* is the best solution if known else is written. *GpuSol* is the solution obtained by the GPU with opt if the solution is the optimum. *Topt* is the best time (in seconds) in the literature. $TGpu$ is the time (in seconds) for our implementation. p the number of hubs. Tight (T) and Loose (L).The new results are shown in these tables in bold type.

All instances of CAB are solved to optimality in times $< 1s$ for the USApHMP as in [14, 22]. So, we will report these results only for the USAHLP.

Tables 2.1, 2.2, 2.3, 2.4 and 2.5 give a comparison of our results to the best results obtained in the literature to solve USApHMP in all these benchmarks. As shown in Table 2.1 for the AP instances we obtained optimal solutions for 100 and 200 nodes

Table 2.1 Solutions of the USApHMP on AP instances

N	p	BestSol	GpuSol	TGpu
100	5	136929.444	Opt	1.310
	10	106469.566	Opt	1.310
	15	90533.523	Opt	1.49
	20	80270.962	Opt	1.63
200	5	140062.647	Opt	3.602
	10	110147.657	Opt	3.722
	15	94459.201	Opt	3.783
	20	84955.328	Opt	3.841
300	5	–	**174914.73**	5.631
	10	–	**134773.55**	5.711
	15	–	**114969.85**	5.896
	20	–	**103746.44**	5.876
400	5	–	**176357.92**	6.741
	10	–	**136378.19**	6.846
	15	–	**117347.10**	7.102
	20	–	**104668.27**	7.423

Table 2.2 Solutions of the USApHMP on PlanetLab instances

Instance	N	p	BestSol	GpuSol	TBest	TGpu
01-2005	127	12	2927946	**2904434**	148.9	0.5
02-2005	321	19	18579238	**18329984**	462.7	6.9
03-2005	324	18	20569390	**20284132**	543.8	7.5
04-2005	70	9	739954	**730810**	0.6	0.3
05-2005	374	20	25696352	**25583240**	622.6	8.3
06-2005	365	20	22214156	**22191592**	581.7	7.9
07-2005	380	20	30984986	**30782956**	546.6	8.5
08-2005	402	21	30878576	**30636170**	637.6	8.7
09-2005	419	21	32959078	**32649752**	684.9	9.3
10-2005	414	21	32836162	**32687796**	731.9	9.1
11-2005	407	21	27787880	**27644374**	588.3	9.2
12-2005	414	21	28462348	**28213748**	680.3	9.1

in a very short time (for nodes up to 50, we obtained the optimal solutions in time < 0.3 s). To our knowledge, the results on AP data instances for the pHMP with 300 and 400 nodes have not hitherto been reported in the literature. Hence, we claim that our results are the best solutions for these instances. We report our results for PlanetLab instances in Table 2.2. We can see that our approach outperforms those of [15] both in terms of cost and computing time. The state-of-the-art solutions given

Table 2.3 Solutions of the USApHMP on PlanetLab instances for other values of p

N	p	GpuSol	TGpu	N	p	GpuSol	TGpu
127	2	**3523236**	0.4	370	2	**812646**	0.2
	3	**3126396**	0.4		3	**784390**	0.2
	4	**3042004**	0.4		4	**758676**	0.2
	5	**2994190**	0.4		5	**752170**	0.2
	10	**2911194**	0.5		10	**728174**	0.3
	15	**3220752**	0.5		15	**719762**	0.3
	20	**2886548**	0.7		20	**713926**	0.3
321	2	**22060080**	5.9	374	2	**30871424**	8.0
	3	**21198104**	6.1		3	**29430588**	8.0
	4	**20364616**	6.1		4	**28198958**	8.1
	5	**19807670**	6.2		5	**27352248**	8.2
	10	**18981110**	6.7		10	**26214866**	8.2
	15	**18614864**	6.8		15	**25760530**	8.3
	20	**18423908**	7.0		20	**26802004**	8.3
324	2	**24139848**	7.1	365	2	**26192564**	7.5
	3	**23193758**	7.2		3	**25416392**	7.5
	4	**22536298**	7.2		4	**24528130**	7.6
	5	**22159678**	7.3		5	**23977334**	7.6
	10	**21339900**	7.4		10	**22708070**	7.7
	15	**20607754**	7.5		15	**22398924**	7.8
	20	**20672764**	7.6		20	**22190348**	7.9

Table 2.4 Solutions of the USApHMP on Urand large instances

N	p	BestSol	GpuSol	Topt	TGpu
1000	2	198071412.53	**8184986.50**	1.7245	9.321
	3	169450816.35	**7024184.00**	8.1550	9.785
	4	150733606.87	**6184749.01**	2.2240	10.431
	5	142450250.26	**5860994.06**	58.6070	10.89
	10	114220373.07	**4752317.00**	187.8385	13.7
	15	–	**4228256.88**	–	15.23
	20	198071412.53	**3928617.48**	403.4280	17.923

in [15] report results for couples of (N, p) with $p \approx \sqrt{N}$. We report in Table 2.3 the results for other possible values of p.

For the Urand instances, our GPU implementation gives the best solutions for instances with up to 400 nodes and outperforms those of [15] for instances with 1000 nodes. Concerning the computing times, our approach is very faster and gives the solutions in times $< 18s$ for all instances, while the best-known time is more than 7 minutes. Moreover, it increases rapidly with p in [15] whereas it does not change

Table 2.5 Solutions of the USApHMP on larger Urand instances generated by us

N	p	GpuSol	TGpu	N	p	GpuSol	TGpu
1500	20	454787506	196	4000	20	3234999192	3076
	30	407155164	286		30	2983891783	3276
	40	380114045	423		40	2769550514	3365
	50	363586538	574		50	2644606684	3648
2000	20	805749722	477	5000	20	5085803132	4662
	30	733375448	580		30	4656787498	4720
	40	686515363	714		40	4353561395	4996
	50	655938000	965		50	4143849388	5112
3000	20	1804950952	1157	6000	20	7398401957	5614
	30	1642145354	1544		30	6675723961	5748
	40	1538548764	1869		40	6293053841	5964
	50	1468780124	2086		50	5999780197	6212

Table 2.6 Solutions of the USAHLP on AP instances

N	Fi-type	BestSol	Hubs of BestSol	GpuSol	Hubs of GpuSol	TGpu
10	L	224250.05	3, 4, 7	224250.05	3, 4, 7	0.024
20	L	234690.95	7, 14	234690.95	7, 14	0.061
25	L	236650.62	8, 18	236650.62	8, 18	0.069
40	L	240986.23	14, 28	240986.23	14, 28	0.243
50	L	237421.98	15, 36	237421.98	15, 36	0.490
100	L	238016.28	29, 73	238016.28	29, 73	1.624
200	L	233803.02	–	**233801.35**	**43, 148**	3.316
300	L	263913.15	26, 79, 170, 251, 287	**258823.42**	**79, 126 , 170, 192, 287**	8.183
400	L	267873.65	99, 180, 303, 336	**259416.31**	**146, 303, 336**	13.32
10	T	263399.94	4, 5, 10	263399.94	4, 5, 10	0.026
20	T	271128.18	7, 19	271128.18	7, 19	0.072
25	T	295667.84	13	295667.84	13	0.085
40	T	293164.83	19	293164.83	19	0.297
50	T	300420.98	24	300420.96	24	0.548
100	T	305097.96	52	305097.93	52	1.864
200	T	272237.78	–	**272188.10**	**54, 122**	4.961
300	T	276023.35	30, 154, 190	**266030.76**	**30, 154, 190**	9.742
400	T	284037.25	101, 179, 372	**275769.09**	**101, 179, 372**	15.98

much for the GPU implementation. Clearly, our solutions are far better in term of cost than those of [15] as can be seen in Table 2.4, $GpuSol \approx \frac{BestSol}{24}$.

We report in Table 2.5 results for larger instances that we have generated using the same generation procedure as for the Urand instances [32]. These new challenging instances consist of large instances of up to 6000 nodes that have not hitherto been solved.

We solved the USAHLP on all 80 CAB data instances introduced in the literature, in execution time <0.04 s. In particular, we confirm the value of 1081.05 found in [22] for $N = 10, \alpha = 1, F_k = 150$, whose value is incorrect in certain works [14]. On the other hand, the results for the standard AP data are reported in Table 2.6. We obtained optimal solutions up to 200 nodes for the two types of fixed cost for hubs (T) and (L). Our approach widely surpasses the best-known solution [14] for the 200, 300 and 400 node instances and gives solutions for unsolved instances.

2.7 Conclusion

We implemented a parallel genetic algorithm on GPU for solving the Uncapacitated Single Allocation p-Hub Median Problem. We used this implementation as basic kernel to solve the Uncapacitated Single Allocation Hub Location Problem. We showed the effectiveness of our implementation on well-known benchmarks. Indeed, our approach improves the best-known solutions in cost and in computing times for well-known benchmark instances up to 1000 nodes. Moreover, we solved problems instances up to 6000 nodes so far unsolved.

Our futur works concern the design and implementation of an exact parallel tree-based algorithm to solve the studied hub problems as these algorithm structures seem to be suitable for parallel computers. Other developments concern the extension of this work to other versions of the hub location problem.

Acknowledgements We thank Dr. M. OKelly, M.R. Silva, A. Ilic' and R. Abyazi-Sani for the instance datasets provided. This work was supported by the FEDER CLASSE2 Project.

References

1. Campbell, J. F. (1994). Integer programming formulations of discrete hub location problems. *European Journal of Operational Research*, *72*(2), 387–405.
2. Campbell, J. F., Ernst, A. T., & Krishnamoorthy, M. (2005). Hub arc location problems: part i, introduction and results. *Management Science*, *51*(10), 1540–1555.
3. Contreras, I., Cordeau, J. F., & Laporte, G. (2011). The dynamic uncapacitated hub location problem. *Transportation Science*, *45*(1), 18–32.
4. de Camargo, R. S., & Miranda, G. (2012). Single allocation hub location problem under congestion: Network owner and user perspectives. *Expert Systems with Applications*, *39*(3), 3385–3391.

5. Campbell, J. F., & O'Kelly, M. E. (2012). Twenty-five years of hub location research. *Transportation Science*, *46*(2), 153–169.
6. O'kelly, M. E. (1987). A quadratic integer program for the location of interacting hub facilities. *European Journal of Operational Research*, *32*(3), 393–404.
7. Sohn, J., & Park, S. (1998). Efficient solution procedure and reduced size formulations for p-hub location problems. *European Journal of Operational Research*, *108*(1), 118–126.
8. Abdinnour-Helm, S. (1998). A hybrid heuristic for the uncapacitated hub location problem. *European Journal of Operational Research*, *106*(2), 489–499.
9. Ernst, A. T., & Krishnamoorthy, M. (1998). Exact and heuristic algorithms for the uncapacitated multiple allocation p-hub median problem. *European Journal of Operational Research*, *104*(1), 100–112.
10. Bryan, D. (1998). Extensions to the hub location problem: Formulations and numerical examples. *Geographical Analysis*, *30*(4), 315–330.
11. Horner, M. W., & O'Kelly, M. E. (2001). Embedding economies of scale concepts for hub network design. *Journal of Transport Geography*, *9*(4), 255–265.
12. Labbe, M., Yaman, H., & Gourdin, E. (2005). A branch and cut algorithm for hub location problems with single assignment. *Mathematical Programming*, *102*(2), 371–405.
13. Chen, J. F. (2007). A hybrid heuristic for the uncapacitated single allocation hub location problem. *Omega*, *35*(2), 211–220.
14. Silva, M. R., & Cunha, C. B. (2009). New simple and efficient heuristics for the uncapacitated single allocation hub location problem. *Computers and Operations Research*, *36*(12), 3152–3165.
15. Ilic, A., et al. (2010). A general variable neighborhood search for solving the uncapacitated single allocation p-hub median problem. *European Journal of Operational Research*, *206*(2), 289–300.
16. Maric, M., Stanimirovic, Z., & Stanojevic, P. (2013). An efficient memetic algorithm for the uncapacitated single allocation hub location problem. *Soft Computing*, *17*(3), 445–466.
17. Bailey, A., Ornbuki-Berrnan, B., & Asobiela, S. (2013). Discrete pso for the uncapacitated single allocation hub location problem. In *Computational Intelligence in Production and Logistics Systems (CIPLS)* (pp. 92–98).
18. Damgacioglu, H., Dinler, D., Ozdemirel, N. E., & Iyigun, C. (2015). A genetic algorithm for the uncapacitated single allocation planar hub location problem. *Computers and Operations Research*, *62*, 224–236.
19. Ting, C. J., & Wang, H. J. (2014). A threshold accepting algorithm for the uncapacitated single allocation hub location problem. *Journal of the Chinese Institute of Engineers*, *37*(3), 300–312.
20. Meier, J. F., & Clausen, U. (2015). Solving classical and new single allocation hub location problems on Euclidean data. In *Optimisation Online*, 03-4816.
21. Rostami, B., et al. (2015). Lower bounding procedures for the single allocation hub location problem. In *Electronic notes in discrete mathematics* (Vol. 320).
22. Abyazi-Sani, R., & Ghanbari, R. (2016). An efficient tabu search for solving the uncapacitated single allocation hub location problem. *Computers and Industrial Engineering*, *93*, 99–109.
23. Shobeiri, A. (2015). Grasp metaheuristic for multiple allocation p-hub location problem. Ph.D. thesis, Concordia University, Montreal, Canada
24. Kratica, J., et al. (2012). Genetic algorithm for solving uncapacitated multiple allocation hub location problem. *Computing and Informatics*, *24*(4), 415–426.
25. Topcuoglu, H., et al. (2005). Solving the uncapacitated hub location problem using genetic algorithms. *Computers and Operations Research*, *32*(4), 967–984.
26. Benaini, A., & Berrajaa, A. (2017, January). Gpu based algoithm for the capacitated single allocation hub location problem. Submitted for publication.
27. Pospichal, P., Jaros, J., & Schwarz, J. (2010). Parallel genetic algorithm on the cuda architecture. In *Applications of evolutionary computation* (pp. 442–451). Berlin: Springer.
28. Talbi, E. G. (2013). Metaheuristics on gpu. *Journal of Parallel Distributed Computing*, *73*(1), 1–3.

29. Luong, T. V., Melab, N., & Talbi, E. G. (2013). Gpu computing for parallel local search metaheuristic algorithms. *IEEE Transactions on Computers*, *62*(1), 173–185.
30. Ernst, A. T., & Krishnamoorthy, M. (1996). Efficient algorithms for the uncapacitated single allocation p-hub median problem. *Location Science*, *4*(3), 139–154.
31. Naeem, M., & Ombuki-Berman, B. (2010). An efficient genetic algorithm for the uncapacitated single allocation hub location problem. In *IEEE Congress on Evolutionary Computation*.
32. Meyer, T., Ernst, A. T., & Krishnamoorthy, M. (2009). A 2-phase algorithm for solving the single allocation p-hub center problem. *Computers and Operations Research*, *36*(12), 31433151.

Chapter 3
Phase Equilibrium Description of a Supercritical Extraction System Using Metaheuristic Optimization Algorithms

Ivan Amaya, Cristian Jiménez and Rodrigo Correa

Abstract This chapter describes the ongoing work dealing with the prediction and estimation of vapor-liquid thermodynamic properties using global optimization algorithms. For the present case, phase equilibrium parameters for the system of supercritical carbon dioxide (sCO_2) and some essential oils, were estimated using the corrected version of van der Waals and Wong-Sandler mixing rules, the Peng-Robinson state equation and the more common thermodynamic models for non-linear parameter estimation in equilibrium modeling, namely, Van Laar, NRTL and UNIQUAC. We propose using a variant of the traditional harmony search algorithm, i.e. self-regulated harmony search (SFHS), for this task. Here, we include preliminary simulation results for the system, sCO_2: α-pineno using the Wong-Sandler rule and the Van Laar model. Results show a good agreement between the experimental results reported in the literature, and the predictions using the SFHS algorithm. Furthermore, SFHS seems to be a promising algorithm for processing phase equilibrium data.

I. Amaya
Escuela de Ingeniería y Ciencias, Tecnologico de Monterrey,
Ave. Eugenio Garza Sada 2501, Monterrey, NL, Mexico
e-mail: iamaya2@itesm.mx

C. Jiménez · R. Correa (✉)
Escuela de Ingenierías Eléctrica, Electrónica y de Telecomunicaciones,
Universidad Industrial de Santander, Cra. 27 Cll. 9, Bucaramanga, Colombia
e-mail: crcorrea@uis.edu.co

C. Jiménez
e-mail: cristian.jimenez@correo.uis.edu.co

© Springer International Publishing AG, part of Springer Nature 2019
E.-G. Talbi and A. Nakib (eds.), *Bioinspired Heuristics for Optimization*,
Studies in Computational Intelligence 774,
https://doi.org/10.1007/978-3-319-95104-1_3

3.1 Introduction

Supercritical fluid extraction (SFE) is now one of the well-known green technologies. In fact, SFE is a unit operation for liquid–liquid (solute-solvent) extraction that uses relatively high pressures to obtain fractionation (via a gas above its critical properties). Many expectations have revolved around SFE since its invention, such as being an energy conservation technology, as well as being a replacement for existing extraction strategies, [14, 18, 25, 32].

Nowadays, literature has shown several unique qualities of SFE, but by no means, it is operative for all the extraction purposes. Aside from its visible advantages, it still has some drawbacks, such as a high initial capital investment (due to the use of expensive components). Thus, a high-pressure equipment for large scale applications, becomes heavy and expensive. Besides, most of the SFE is done as batch processes, and there is a very limited availability of design data for a given extraction system. Although SFE is now a positioned green technology, the motivation behind its use still comes from increasingly tightening environmental regulations on conventional solvent emissions and residues, and public health apprehensions over usage of potential hazard solvents in the production of food and pharmaceuticals [15, 17, 20, 25].

One important factor defining the application and the characteristics of a SFE process is the appropriate selection of the supercritical solvent. There are some important physical and chemical factors involved in the choice of this key element, but the main ones are: corrosiveness, density, ease of recovery, flammability, latent heat, selectivity, solubility, surface tension, viscosity and obviously, cost and toxicity. Although there are several alternatives, carbon dioxide is one of the most commonly used supercritical fluids because it is safe, inexpensive, readily available and has been used effectively replacing petroleum-based solvents, which are well known as hazardous and toxic materials. Figure 3.1 shows the schematic of its phase diagram.

As observed, its critical temperature and pressure are low. Moreover, the SFE using supercritical carbon dioxide (sCO_2) does not leave any toxic residue, and incorporates the properties and advantages of both, gas and liquid fluids. sCO_2 behaves likes a gas, in the sense that it expands to fill the vessel it occupies. Besides, it offers solubility without the presence of surface tension, making sensitive processes feasible. Moreover, it creates no potential environmental hazards, and have been demonstrated that it has very low energy consumption and low operating costs [20, 34]. Additionally, sCO_2 is very responsive to variations in pressure and temperature, making possible to manipulate the density, selectivity and solubility of the solute-solvent system, just by fine-tuning the process operating conditions. sCO_2 has liquid-like density, and its transport properties, such as diffusivity and viscosity, are intermediate between those of a liquid and a gas. SFE, in general, has several global variables which must be controlled to optimize the process: pressure, temperature, flow rate, and extraction (residence) time [8].

Ahmadian et al. [2] used SFE with sCO_2 for the extraction of antioxidant compounds from Crocus sativus petals. In their work, they employed a Box-Behnken

Fig. 3.1 P-T diagram for
carbon dioxide

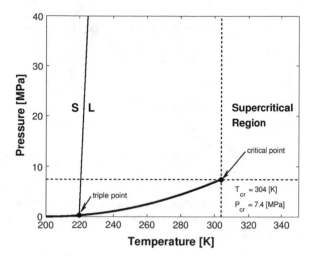

design (BBD) to optimize the process parameters such as the extraction temperature, extraction time, and the extraction pressure. Furthermore, they found optimal
operational conditions using the traditional response surface methodology, and investigated the effect of extraction conditions on total phenolic content, total flavonoid
content, and total anthocyanin content. The authors determined a relationship among
extracted antioxidants parameters by using principal component analysis. As a result,
they found an optimized experimental condition for simultaneously maximizing
extraction of the above materials. Similarly, Subroto et al. [29] reported the optimal experimental conditions for SFE using sCO₂ for extraction of candlenut oil.
Their study was oriented to evaluate the pressure, temperature, flow rate and particle size effects on oil solubility, using a Taguchi experimental design. The solubility
parameter, according to their results, showed a boost when pressure, temperature and
particle size were increased. Also, they compared the quality of oil obtained by SFE
with that obtained by traditional Soxhlet extraction using n-hexane as a solvent. The
former had better quality. Sodeifian et al. [28], used SFE with sCO₂ as a solvent for
extracting ω-3 from Dracocephalum kotschyi seed oil. They analyzed the extraction
process too, but using the response surface methodology based on central composite
design. Using this strategy, they studied the effects of pressure, temperature, particle
size and extraction time, on the amount of ω-3 extracted. As a result, they proposed a
quadratic polynomial model for describing their experimental data. They also claim
an excellent quality of the extracted oil by this technique.

We carried out an analysis of research works available in the literature to verify
the tendency of the field. We first focused on the broader subject of supercritical
extraction, finding almost 15,000 reported works since 1976. We then moved on to
including the elements of optimization and properties, which reduced the number of
documents to almost 500. Nonetheless, there has been an evident growing interest
of the research community in both subjects (Fig. 3.2). Furthermore, a quick search
of these topics on Google Patents returns more than 140,000 patents. Thus, research

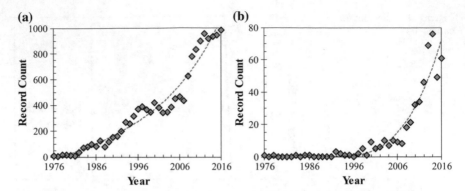

Fig. 3.2 Results analysis for the literature revision carried out. Left: Overview of supercritical extraction. Right: Refinement carried out to include optimization and properties. Dashed lines show data tendency

targeting the fields of supercritical extraction applications and estimation of properties via optimization is worth pursuing. Sadly, due to space restrictions it becomes impossible to cover more of the recent works, but some worthwhile references are [7, 23, 24, 27, 30].

After this brief review, we perceived that while most of the information available deals with the optimization of global parameters such as residence time, pressure levels and so on, there are very few research reports dealing with the thermodynamic aspects of the process itself. In addition, the modeling and simulation of this separation process is demanding due to the high sensitivity of thermodynamic and transport properties with pressure, temperature and composition around the critical properties of the supercritical solvent [10]. The *solvent power* of a gas in such state can experience a huge increase. Consequently, a lab experimental investigation is required, whenever a new process or application is being conceived. This is one of the main reasons why it is so important to select the appropriate phase equilibrium model in order to design a SFE process. Some factors affecting this selection are the nature of the system represented by the solute-solvent interactions and its composition, the operating pressure level, and the thermodynamic data available. In particular, the solubility of the solute in a supercritical fluid is a complicated function of pressure and temperature. Most of the time, there is practically no clue about how it is going to change when pressure or temperature varies. Typically, the design process requires experimental data, which will allow calculating the components distribution between the two phases. Furthermore, it is necessary to know beforehand the solubility limits to select and design the other components of the entire extraction system.

Currently, there are a limited number of alternatives to predict the values of the solubility parameters, separation coefficients, activity coefficients and composition of the experimental systems. When no experimental data are available, it is possible to use the UNIQUAC (as well as UNIFAC) equations for estimating, for example, activity coefficients. These last parameters are fundamental for the modeling and process design of an extraction system. Keep in mind that pressure, temperature and liquid composition depend on those coefficients. Lately, Abdallah et al. [1] proposed a new approach for estimating solubility of some solid drugs in supercritical carbon dioxide using direct and inverse artificial neural networks (ANN), and knowing properties such as the equilibrium temperature and pressure, the critical temperature and pressure, and the acentric factor of the solid solute. They used a hybrid method based on neural networks and particle swarm optimization (PSO), on previously reported data, to develop and validate a model that can predict the solubility of binary systems (solid drugs-sCO$_2$). According to the authors, their model shows high predictive and interpolating abilities at temperatures where no experimental data were found in the literature. They used a set of about seventeen common chemicals for testing and constructing the model. As shown in their work, the sensitivity analysis revealed that all studied inputs had a strong effect on the solubility parameter and allowed the estimation of some solid properties from the solubility data such as critical pressures and temperatures with good accuracy, without using the above mentioned group contribution methods. In the same way, Aminian [4] describes a model for predicting solubility of several solutes in supercritical carbon dioxide but in this case using artificial neural networks. With this particular model, it is possible (according to its author) to get more accurate values, at least for the solutes used for the assessment. Another interesting outcome of this work relates to the superior results in accuracy against the semi-empirical models and the equations of state conventionally used for the prediction of solubilities in SFE, as described in his paper. An additional recent application of neural networks in SFE is presented by Zecovic et al. as described in [36].

3.2   Fundamentals

Following is a brief description of two main topics related to this work: the thermo-dynamics behind SFE processes and the optimization algorithm used.

3.2.1   Some Thermodynamic Aspects of a SFE Process

Phase equilibrium relationships and solubility data are of paramount importance for any supercritical extraction process [6, 11]. These two main characteristics rule the operational and the economic viability of such extraction operations. Currently, there are several ways for correlating and predicting the solubility in supercritical CO$_2$ as a function of pressure and temperature. Among them, there is the Chrastil model, Bartle and Mendez-Santiago-Teja models, and the volume cubic equations of state such as the Peng-Robinson equation, and the Soave-Redlich-Kwong equation, for example [19, 31]. As explained by Hozhabr in [13], there is a new empirical expression useful to predict the solubility of about twenty-four different kinds of solutes in the sCO$_2$ solvent. From the several network arrangements available, he reported that a feedforward network of one hidden layer comprising seven nodes offered the best predicted experimental values. On the other hand, from the classical thermodynamics of SFE processes, there exists the following model to envisage the solute solubility as a function of the mixture or extracting density, pressure and temperature, [16, 34, 35]:

$$ln(Py_2) = \frac{A_0}{T} + A_1\rho + A_2 \tag{3.1}$$

being P the absolute pressure [bar], y_2 the solubility of a solute in sCO$_2$ (solute mole fraction in the gas phase), T the absolute temperature [K], ρ the extracting density, and (A_0; A_1; A_2) the model constants (to be estimated for each particular system). The advantage of this approach is its simplicity, but it requires a numerical technique, as wells as experimental data. Other method, as mentioned before, is the use of equations of state. With those equations, and because of the high pressure (at or above the critical point) the solubility of the solute in sCO$_2$ can be formulated with the assistance of the Poynting correction factor (an exponential term) as,

$$y_2 = \frac{P_2^{sb}}{P}\frac{1}{\phi_2}P_{oy} \therefore P_{oy} = e^{\int_{P_2^{sb}}^{P}\frac{v_2}{RT}dP} \cong e^{\frac{v_2}{RT}(P-P_2^{sb})} \tag{3.2}$$

in which the new terms are the solute sublimation pressure (P_2^{sb}), the solute molar volume (v_2), and the solute fugacity coefficient (ϕ_2), [22].

This last coefficient involves the non-ideality of the fluid phase at high pressure, which can be evaluated with the help of an equation of state. Bear in mind that a cornerstone in the thermodynamic analysis of a SFE is that the fugacity of the solute in the liquid phase must be equal to that in the supercritical phase at the equilibrium. Nevertheless, it is more common in real applications to use the fugacity coefficient in the following way,

$$\phi_2^L x_2 P = \phi_2^{sp} y_2 P \tag{3.3}$$

where x_2 is the composition of the solute in the liquid phase, and y_2 is the composition of solute in the supercritical phase (sp). It is worth mentioning that one can use either the equality of chemical potentials or fugacities as the condition for equilibrium. In a typical SFE, a gas is used as a solvent, which is compressed isothermally above its critical point and exhibits enhanced *solvent power*. In this way, the supercritical fluid acquires desirable transport properties that modify (most of the times rising) its characteristics as a solvent. Besides, the supercritical fluid density gets closer to that of liquids, while its viscosity remains very low comparable to that of gases. The fractionation in this process is due to phase splitting caused by the gas above its critical properties. For this reason, it is paramount to know the equilibrium distribution coefficient of the components between the two phases. This is described by the ratio

$$K = \frac{c^F y}{c^L x} \tag{3.4}$$

where $(y; x)$ are the mass fraction of the mixture in the fluid and liquid phases of a given component, respectively. c^F is the concentration of a specific component in the fluid phase, and c^L is the same but for the liquid phase. Now, using the ratio of these distribution coefficients, it is possible to define the selectivity for a SFE [25, 26].

The feasibility of a separation process is clearly dependent of the knowledge of K. If we want to separate two components, for example A and B, their ratio K_A/K_B should be greater than about 1.1 to ensure a sufficient driving force for fractionation. Although there are limited alternatives for modelling the vapor-liquid equilibrium in a SFE process, the use of empirical equations of state is probably the most common approach. The Peng-Robinson state equation, which is cubic in volume, is a clear example, as it is illustrated next. This state equation is extensively used for thermodynamic properties calculations, and it is given by,

$$Z = \frac{Pv}{RT} = \frac{v}{v-b} - \frac{\frac{a}{RT} + d - 2\sqrt{\frac{ad}{RT}}}{(v+b) + \frac{b}{v}(v-b)} \tag{3.5}$$

where for pure components,

$$a = a(T_c)(1+\kappa)^2$$
$$b = \frac{0.0778 R T_c}{P_c}$$
$$a(T_c) = \frac{0.45724 R^2 T_c^2}{P_c} \tag{3.6}$$
$$\kappa = 0.37464 + 1.54226\omega - 0.26992\omega^2$$
$$d = \frac{a(T_c)\kappa^2}{RT_c}$$

Because the SFE thermodynamic system is a mixture, we require the use of a mixing rule. In this case we selected the Wong-Sandler mixing rule, along with the Van Laar model for calculating the activity coefficient,

$$a = \frac{DQRT}{1-D}$$
$$b = \frac{Q}{1-D}$$
$$D = \sum_i x_i \frac{a_i}{b_i RT} - \left\{ \frac{\sqrt{2}}{ln(1+\sqrt{2})} \right\} \left\{ \frac{G^E}{RT} \right\}$$

$$\frac{G^E}{RT} = \sum_i x_i ln \gamma_i \tag{3.7}$$

$$Q = \sum_i \sum_j x_i x_j \left[\frac{b_i + b_j}{2} - \frac{\sqrt{a_i a_j}}{RT} (1 - k_{ij}) \right]$$

$$ln \gamma_i = \frac{\sum_j z_j a_{ij}}{1 - z_i} \left[1 - \frac{z_i \sum_j z_j a_{ij}}{z_i \sum_j z_j a_{ij} + (1 - z_i) \sum_j z_j a_{ji}} \right]^2$$

Now, for the formulation of the required objective function, we can apply either the ordinary least squares method (OLS) or the maximum likelihood criterion. The former strategy was used in this work. With the help of the above equation, the objective function (OF) was structured as,

$$OF = \sum_i^N \left[P_i^{exp} - P_i^{cal} \right]^2 \tag{3.8}$$

where N is the number of experimental data, P_i^{exp} is the experimental value, and P_i^{cal} is the corresponding pressure at the bubble point (evaluated using the Peng-Robinson equation). For the simulation results presented below, the adjustable parameters were $[a_{12}; a_{21}; k_{12}]$. Nevertheless, Wong et al. describe an alternative way to calculate them [33].

After reviewing some fundamental aspects related to a SFE process, we will briefly describe the use of global optimization algorithms in the modeling of its vapor-liquid equilibrium. We use a particular solute-supercritical solvent system and the harmony search algorithm for illustrative purposes. A few years ago, Bonilla [5] studied the performance of the conventional harmony search algorithm (HS) as a non-linear parameter estimator in vapor-liquid equilibrium modeling. In that work, he used experimental data of binary systems and the ordinary least squares and error-in-variable formulations. He compared his results against other algorithms, such as simulated annealing (SA), differential evolution (DE), particle swarm optimization (PSO) and genetic algorithm (GA), using the least squares method. He found a better performance when using two HS variants known as Improved Harmony Search (IHS) and Global-Best Harmony Search (GBHS). Nevertheless, the conventional HS algorithm was not apt for modeling vapor-liquid equilibrium data using the error-in-variable formulation, as was described in his paper.

3.2.2 The Self-regulated Fretwidth Harmony Search (SFHS) Algorithm

The Harmony Search (HS) algorithm was presented by Geem et al. during 2001 [12]. This algorithm is a simple, yet powerful, approach for dealing with optimization problems, making it an easy to code and versatile tool. In HS, a set of solutions (called Harmony Memory, HM) is preserved and updated continuously with the best candidate solutions. Thus, HM corresponds to a matrix of size ($HMS \times N$), where HMS is the number of solutions to preserve (known as Harmony Memory Size) and N is the number of dimensions of the optimization problem. An advantage of HS is that it balances exploration and exploitation, since it is able to create new candidate solutions by mixing totally random values (in some dimensions) with the values of solutions stored in the HM (for the remaining dimensions). Furthermore, values selected from the memory (HM) can be perturbed as to explore the vicinity of known solutions. This procedure is carried out by using the Pitch Adjusting Rate (PAR) and Eq. (3.9), where $temp$ is the value selected from memory, $rand$ is a uniformly distributed random number between zero and one, and FW is the current fretwidth value. Details about this procedure are given below, when explaining the SFHS algorithm.

$$temp = temp + (rand - 0.5) \cdot FW \tag{3.9}$$

Different variants have appeared throughout the lifespan of HS, as an attempt to improve key components of the algorithm. A recently proposed modification was presented in [3]. This variant of the algorithm was inspired on the good performance of a previously reported approach [9]. The authors proposed varying the fretwidth on each j iteration (i.e. FW becomes $FW(j)$), based on three possible scenarios:

1. Algorithm starting up. In this case, set the fretwidth to an initial fixed value (FW_{ini}).
2. Algorithm finds an improved best solution. Thus, store the fretwidth value so upcoming iterations can search with similar values.
3. Algorithm does not find better solutions after FW_{sat} iterations. In this scenario, reset the fretwidth and switch to an exponential decay.

The first case requires no further comments. The third one is quite similar to the one used in [3] so details will be spared. However, it must be mentioned that the exponential decay follows the behavior shown in Eq. (3.10), where DC_{max} is the maximum number of iterations for arriving at FW_{min}. The remaining case (i.e. scenario two) is the core proposal of SFHS so a bit more explanation will be given. The idea is to create fretwidth values close to the one that allowed finding an improved solution. To do so, Eq. (3.11) is used, where $rand$ is a random number uniformly distributed between zero and one, C_{FW} is an amplitude constant, and A_j is the value that allowed finding the improved solution. The amplitude constant, C_{FW}, is included as a way of controlling the level of adjustment, and it was set to one for the current work. This stochastic behavior is maintained until the third scenario is repeated.

$$FW(j) = FW_{max} e^{\left(\frac{DC}{DC_{max}} ln\left(\frac{FW_{min}}{FW_{max}}\right)\right)} \tag{3.10}$$

$$FW(j) = A_j + (rand - 0.5)C_{FW}A_j \tag{3.11}$$

For this algorithm, the Harmony Memory Size (HMS) is an integer indicating how many solutions are stored at any given time. The Harmony Memory Considering Rate ($HMCR$) and Pitch Adjusting Rate (PAR) are values between zero and one. Based on previously reported recommendations we defined $FW_{ini} = 0.5$, $FW_{max} = 2.0$, $FW_{min} =$, $FW_{sat} = 1000$, and $Sat_{HS} = 10000$. The remaining parameters, i.e. HMS, PAR, and $HMCR$, were defined for each simulation [3, 9]. The overall logic of SFHS is described in algorithm 1.

Algorithm 1: Self-regulated Fretwidth Harmony Search (SFHS) algorithm

---

Require: Memory size (HMS), memory considering rate ($HMCR$), pitch adjusting rate (PAR), initial Fretwidth (FW_{ini}), maximum Fretwidth (FW_{max}), amplitude constant (C_{FW}), and saturation limit (FW_{sat}). Also, define the number of iterations (NI)

HM ← matrix of uniformly distributed random number with size $HMS \times N$, where N is the number of dimensions of the problem

for $idi = 1 : NI$ **do**
    for $idy = 1 : N$ **do**
        if $random(0, 1) < HMCR$ **then**
            $row \leftarrow random(1, HMS)$
            $temp \leftarrow HM(row, idy)$
            if $random(0, 1) < PAR$ **then**
                Adjust $temp$ using eq. 3.9
            end if
        else
            $temp \leftarrow random(searchDomain)$
        end if
        $candidate(idy) \leftarrow temp$
    end for
    if $evaluate(candidate) < worst(HM)$ **then**
        replace $worst(HM)$ with $candidate$
        $A_j \leftarrow FW(j)$
        $Sat_{Ct} \leftarrow 0$
        $Decay \leftarrow false$
        $DC \leftarrow 0$
        Update $FW(j)$ using eq. 11
    else
        if $Decay = false$ **then**
            $Sat_{Ct} \leftarrow Sat_{Ct} + 1$
            if $Sat_{Ct} \geq FW_{sat}$ **then**
                $Decay \leftarrow true$
                $DC \leftarrow 0$
                Update $FW(j)$ using eq. 10
            else
                Update $FW(j)$ using eq. 11
            end if
        else
            $DC \leftarrow DC + 1$
            Update $FW(j)$ using eq. 10
        end if
    end if
end for
return $best(HM)$

Table 3.1 Critical properties of the mixture components [21]

	Tc (K)	Pc (MPa)	Vc (cm³/mole)	Acentric factor
CO2	304.200	7.375	94.000	0.239
α-pineno	630.000	2.890	484.500	0.313

3.3 Methodology

The first part of this work was dedicated to verifying the correct implementation of the SFHS algorithm. Later on, and after using it for calculating the optimum value for some benchmark functions, we ran simulations for the prediction of the thermo-dynamic parameters of the selected system sCO_2:α-pineno. In order to generate the vapor-liquid equilibrium data, we used the state equation and the fugacity coefficients values for the binary mixture. Besides, the fundamental principle of equality of the fugacity coefficients (isofugacity condition) for a component present in both phases was used. Finally, we compiled (for comparison purposes) the experimental data reported in [21]. Table 3.1 includes the critical properties of the two components of the mixture (sCO_2:α-pineno).

3.4 Results

3.4.1 Test Results for Some Optimization Algorithms

Here we include some results dealing with the performance of the following global optimization algorithms such as SFHS, ABHS (Adjustable Bandwidth Harmony Search), IHS (Improved Harmony Search) and firefly. Table 3.2 shows a summary of test data generated for six different test functions in two dimensions, using their standard search domains, and while demanding a fitness of 1×10^{-10}. As it is shown, all three approaches exhibited similar performance (in terms of convergence rate). However, SFHS was the cheapest one (in terms of the number of iterations required to achieve the desired fitness level) in most cases.

Table 3.3 presents a comparison of the algorithm used in this work (i.e. SFHS) against a different variant of the HS algorithm (i.e. ABHS) and against a completely different approach (i.e. the Firefly algorithm), while considering a convergence criteria of 1×10^{-10}. More information about the test functions used for this comparison, as well as about the other tested approaches, can be found in [3]. Data show that, even if the ABHS variant exhibits a similar performance for the problems in the fewest number of dimensions, it degrades as the problem escalates. Moreover, the Firefly algorithm seems like a good approach for some functions, since the number of required iterations only increases slightly. However, its convergence level degrades

Table 3.2 Comparison of the HS variant used in this work (i.e. SFHS) against two previously reported ones (i.e. ABHS and IHS), for six different standard test functions (STF) in two dimensions, and using their standard search range. Data focuses on the average number of required iterations for reaching a fitness of 1×10^{-10}, and on the convergence rate (CR), after 33 repetitions of each test

STF		SFHS iterations	CR (%)	ABHS iterations	CR (%)	IHS iterations	CR (%)
Jong	Avg	3675.73	100	5058.46	100	11016.67	100
	SD	1768.85		839.61		1706.89	
	Max	7840.00		6751.00		14182.00	
	Min	1383.00		2907.00		7184.00	
Rastrigin	Avg	4980.03	97	6601.33	100	16096.33	100
	SD	1672.78		1183.56		1496.68	
	Max	11572.00		8951.00		18497.00	
	Min	2865.00		4437.00		11924.00	
Rosenbrock	Avg	77137.20	15	71456.13	45	16175.00	18
	SD	18145.43		24025.53		1777.04	
	Max	97792.00		98978.00		18585.00	
	Min	50695.00		28686.00		14685.00	
Schwefel	Avg	3282.67	27	4203.90	30	8426.55	33
	SD	844.98		950.09		1195.81	
	Max	4688.00		5962.00		10256.00	
	Min	2183.00		2961.00		6014.00	
Ackley	Avg	8010.00	100	11824.03	100	34565.61	100
	SD	1110.68		1816.08		1380.09	
	Max	9958.00		17050.00		36916.00	
	Min	5841.00		8692.00		31412.00	
Steps	Avg	300.30	100	341.88	100	197.73	100
	SD	160.75		155.18		97.65	
	Max	646.00		675.00		409.00	
	Min	60.00		79.00		43.00	

quickly as the problem grows, and it is unable to reach the requested level of precision for any of the tested number of dimensions with the Rastrigin function.

3.4.2 Interaction Parameters Estimation Using SFHS

Some of the parameters required by SFHS were presented in Sect. 3.2.2. For the remaining ones, however, we used $HMS = 5.0$, $PAR = 0.5$, $HMCR = 0.9$, $C_{FW} = 1.0$ and $\delta \leq 1 \times 10^{-8}$. Figure 3.3 shows an example of the variation of the iterations number, as well as of the convergence time, for the shifted Jong function when using SFHS with the aforementioned parameters.

Table 3.3 Comparison of the approach used in this work (i.e. SFHS) against two reported approaches (i.e. ABHS and the Firefly algorithm), for three different standard test functions (STF) at three different number of dimensions (Dim). Data focuses on the average number of required iterations for reaching a fitness of 1×10^{-10}, and on the convergence rate (CR), after 33 repetitions of each test

STF	Dim	SFHS iterations	CR (%)	ABHS iterations	CR (%)	Firefly iterations	CR (%)
Jong	5	10774	100	10913	100	1876	100
	15	68058	100	244858	67	1951	100
	30	448040	100	877775	64	1954	100
Rastrigin	5	8895	100	12025	100	–	0
	15	46560	100	179616	88	–	0
	30	245894	100	838277	79	–	0
Shifted Jong	5	9698	100	12108	100	1871	100
	15	62076	100	346255	91	1933	64
	30	428798	100	957179	12	–	0

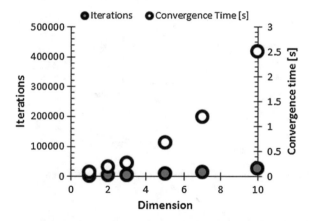

Fig. 3.3 The iterations number and convergence time for the shifted Jong function

As observed, the convergence time is highly dependent of the number of unknown parameters, but this is not the case for the iterations number. For the upcoming simulations, we modified the value of PAR and fixed it into $PAR = 0.8$. The maximum number of iterations (NI) was one million, and we repeated each simulation 30 times. A slight variation of the results was detected. We believe that one of the possible reasons for this behavior, is the apparent random nature of the optimization algorithm, as well as error propagation.

Figure 3.4 shows the vapor (supercritical phase)-liquid equilibrium, for three temperatures. In it, the model prediction for the experimental values P-(x;y) were adjusted. Using the Peng-Robinson state equation, along with the Wong-Sandler

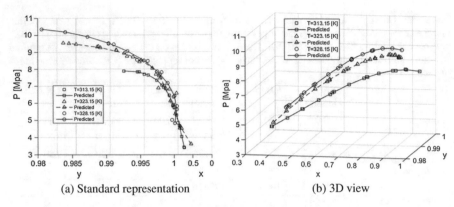

(a) Standard representation   (b) 3D view

Fig. 3.4 V-L equilibrium at three temperatures. Left: Standard representation for this kind of variables, P(x;y). Right: 3D view of the data

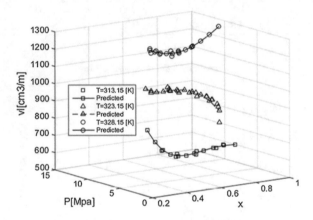

Fig. 3.5 Liquid molar volume variation at three temperatures. A 3D view

mixing rule and the Van Laar model, seems to be a useful path for modeling the vapor-liquid equilibrium at high pressures above critical conditions. However, this type of equation of state (which includes mixing rules for supercritical fluid mixtures), becomes very sensitive to the value of the interaction parameters. Keep in mind that there is some theoretical basis to the cubic equations of state, but this does not hold for such adjustable parameters.

In order to know the molar volume of the liquid phase for the mixture, we recalculated it using the Peng-Robinson state equation along with the parameters found by simulation. Figure 3.5 shows the expected tendency for the volume variation, as a function of the pressure and temperature increment.

Now, we present the way in which the parameters vary after several runs of the estimation procedure. To achieve that, we summarize the data of 33 different runs, at five temperature levels, in three boxplots (one for each estimated parameter). Figure 3.6 shows that, as temperature rises, data for a_{12} become more concentrated,

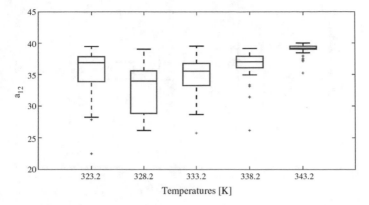

Fig. 3.6 Variation of parameter a_{12} after 33 runs at different temperature levels

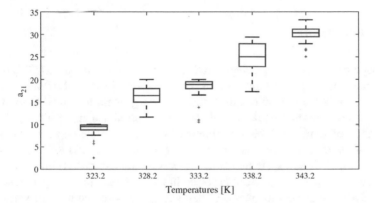

Fig. 3.7 Variation of parameter a_{21} after 33 runs at different temperature levels

increasing the quality of the estimation. Similarly, the median value of this parameter increases by about 8%. Even so, and due to the stochastic nature of the optimization procedure, most temperature levels yielded outliers. In the case of parameter a_{21}, increasing the temperature leads to higher values of interaction (Fig. 3.7). Even so, and as in the previous case, the stochastic nature of the optimizer generates outliers. However, this parameter exhibited an interesting behaviour, as its median value at the highest temperature is thrice the value at the lowest one. Moreover, predictions were, in general, less spread at all temperature values than for a_{12}.

Finally, and in a similar fashion as with the previous parameters, Fig. 3.8 shows that increasing the temperature leads to better predictions of k_{12} (except for the case when $T = 328.2K$), while outliers are still present (for the same reason that was previously commented). However, this time around there was almost no increment in the interaction parameter (when comparing the median value at the highest and lowest temperatures).

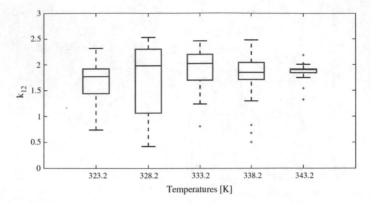

Fig. 3.8 Variation of parameter k_{12} after 33 runs at different temperature levels

3.5 Conclusions

From our literature review it is clear that Supercritical Fluid Extraction (SFE) is a very important industrial process that requires knowledge of process optimization, but also of the system thermodynamic equilibrium. Currently, there is a trend to re-evaluate the use of conventional equations of state near and above the critical point, and the creation of semi-empirical correlations for specific solute-supercritical solvent systems. Experimental investigation on phase equilibrium is still needed to assess the prospect of SFE for new systems. On the other hand, we contrasted simulation results with previous experimental work reported in the literature for the system sCO_2:α-pineno, finding a good agreement. Nevertheless, we observed a small variation of the interaction parameters values. We think that it is due to the use of EOS near the critical point and numerical errors in the optimization solution. Further work is needed for improving the simulation methodology (including new thermodynamic models, and the minimization of the Gibbs free energy), and on the development of approaches to enhance the reliability and reproducibility of this new variant of the Harmony Search algorithm (i.e. SFHS), and of other algorithms.

3.6 Nomenclature

P = Pressure (atm)
v = Molar volume (l/g-mol)
T = Temperature (K)
R = Gas constant (0.08206 (atm-liters/g-mol K))
T_c = Critical temperature (K)
P_c = Critical pressure (atm)
γ = Activity coefficient
ω = Acentric factor

Acknowledgements The authors gratefully acknowledge the financial support provided by Universidad Industrial de Santander and Colciencias.

References

1. Abdallah el hadj, A., Laidi, M., Si-Moussa, C., & Hanini, S. A., (2017). Novel approach for estimating solubility of solid drugs in supercritical carbon dioxide and critical properties using direct and inverse artificial neural network (ANN). *Neural Computing and Applications, 28,* 87–99.
2. Ahmadian-Kouchaksaraie, Z., & Niazmand, R. (2017). Supercritical carbon dioxide extraction of antioxidants from Crocus sativus petals of saffron industry residues: Optimization using response surface methodology. *Journal of Supercritical Fluids, 121,* 19–31.
3. Amaya, I., Cruz, J., & Correa, R. (2015). Harmony Search algorithm: A variant with Self-regulated Fretwidth. *Applied Mathematics and Computation, 266,* 1127–1152. https://doi.org/10.1016/j.amc.2015.06.040.
4. Aminian, A. (2017). Estimating the solubility of different solutes in supercritical CO2 covering a wide range of operating conditions by using neural network models. *Journal of Supercritical Fluids, 125,* 79–87.
5. Bonilla-Petriciolet, A. (2012). On the capabilities and limitations of harmony search for parameter estimation in vapor-liquid equilibrium modeling. *Fluid Phase Equilibria, 332,* 7–20.
6. Byun, H. S., & Rhee, S. Y. (2017). Phase equilibria measurement of binary mixtures for triethylene glycol dimethacrylate and triethylene glycol diacrylate in supercritical CO2. *Korean Journal of Chemical Engineering, 34*(4), 1170–1176. https://doi.org/10.1007/s11814-016-0332-y.
7. Calvo, A., Devenyi, D., Bence, K., Sanz, S., Oelbermann, A., Maier, M., et al. (2017). Controlling concentration of bioactive components in cats claw based products with hybrid separation process. *Journal of Supercritical Fluids, 125,* 50–55.
8. Campos-Dominguez, C., & Gamse, T. (2017). Utilization of micro-mixers for supercritical fluid fractionation: Influence of the residence time. *Journal of Supercritical Fluids.*
9. Contreras, J., Amaya, I., & Correa, R. (2014). An improved variant of the conventional Harmony Search algorithm. *Applied Mathematics and Computation, 227,* 821–830. https://doi.org/10.1016/j.amc.2013.11.050.
10. Garcez, J. J., Barros, F., Lucas, A. M., Xavier, V. B., Fianco, A. L., Cassel, E., et al. (2017). Evaluation and mathematical modeling of processing variables for a supercritical fluid extraction of aromatic compounds from anethum graveolens. *Industrial Crops and Products, 95,* 733–741. https://doi.org/10.1016/j.indcrop.2016.11.042, http://www.sciencedirect.com/science/article/pii/S0926669016308056.
11. García-Pérez, J. S., Robledo-Padilla, F., Cuellar-Bermudez, S. P., Arévalo-Gallegos, A., Parra-Saldivar, R., Zavala-Yoe, R., et al. (2017). Thermodynamics and statistical correlation between supercritical-CO2 fluid extraction and bioactivity profile of locally available Mexican plants extracts. *The Journal of Supercritical Fluids, 122,* 27–34. https://doi.org/10.1016/j.supflu.2016.12.002, http://www.sciencedirect.com/science/article/pii/S0896844616305277.
12. Geem, Z., Kim, J., & Loganathan, G. (2001). A new heuristic optimization algorithm: Harmony search. *Simulation, 76,* 60–68.
13. Hozhabr, S. B., Mazloumi, S. H., & Sargolzaei, J. (2014). Correlation of solute solubility in supercritical carbon dioxide using a new empirical equation. *Chemical Engineering Research and Design, 92*(11), 2734–2739. https://doi.org/10.1016/j.cherd.2014.01.026, http://www.sciencedirect.com/science/article/pii/S0263876214000653.
14. Kashif, A., Hafiz, M. S., & Joong-Ho, K. (2017). Green extraction methods for polyphenols from plant matrices and their byproducts: A review. *Comprehensive Reviews in Food Science and Food Safety, 16,* 295–315.

15. Lindy, J. (2014). *Supercritical fluid extraction: Technology, applications and limitations*. New York: Nova Science Publishers, Incorporated.
16. Luque de Castro, M., Valcarcel, M., & Tena, M. (1994). *Analytical supercritical fluid extraction*. Berlin: Springer.
17. Martin, A., & Aydin, K. S. (1997). *Supercritical fluids: Extraction and pollution prevention* (Vol. 670). ACS symposium series. American Chemical Society.
18. Mchuge, M., & Krukonis, V. (1994). *Supercritical fluid extraction* (2nd ed.). Boston: Butterworth-Heinemann.
19. Mehdi, G., Mahdi, A., & Lashanizadegan, A. (2017). A comparison between Peng-Robinson and Soave-Redlich-Kwong cubic equations of state from modification perspective. *Cryogenics, 84*, 1319.
20. Meireles, A., & Kiran, E. (2013). *Workshop on supercritical fluids and energy*. Campinas SP, Brasil.
21. Pavlicek, J., & Ritcher, M. (1993). High pressure vapor-liquid equilibrium in carbon dioxide- α-pinine system. *Fluid Phase Equilibria, 90*, 125–133. https://doi.org/10.1016/0378-3812(93)85007-9.
22. Prausnitz, J., Lichtenthaler, R., & de Azevedo, E. (1986). *Molecular thermodynamics of fluid-phase equilibria* (2nd ed.). Englewood Cliffs: Prentice-Hall.
23. Prieto, C., Calvo, L., & Duarte, C. M. M. (2017). Continuous supercritical fluid extraction of emulsions to produce nanocapsules of vitamin E in polycaprolactone. *Journal of Supercritical Fluids, 124*, 72–79.
24. Rai, A., Mohanty, B., & Bhargava, R. (2016). Fitting of broken and intact cell model to supercritical extraction (SFE) of sunflower oil. *Innovative Food Science and Emerging Technologies, 38*, 32–40.
25. Rizvi, S. (1994). *Supercritical fluid processing of food and biomaterials*. London: Blackie Academic & Professional.
26. Seader, J., Henley, E., & Keith Roper, D. (2011). *Separation process principles, chemical and biochemical operations* (3rd ed.). New Jersey: Wiley.
27. Sinclair, L., Baek, D., Thompson, J., Tester, J., & Fox, R. (2017). Rare earth element extraction from pretreated bastnasite in supercritical carbon dioxide. *Journal of Supercritical Fluids, 124*, 20–29.
28. Sodeifian, G., Sajadian, A., & Ardestani, S. (2017). Supercritical fluid extraction of omega-3 from Dracocephalum kotschyi seed oil: Process optimization and oil properties. *Journal of Supercritical Fluids, 119*, 139–149.
29. Subroto, E., Widjojokusumo, E., Veriansyah, B., Tjandrawinata, R. R. (2017). Supercritical $CO2$ extraction of candlenut oil: Process optimization using Taguchi orthogonal array and physicochemical properties of the oil. *Journal of Food Science and Technology*, 1–7.
30. Tao, W., Zhang, H., Xue, W., Ren, L., Xia, B., Zhou, X., et al. (2014). Optimization of supercritical fluid extraction of oil from the fruit of Gardenia jasminoides and its antidepressant activity. *Molecules, 19*, 19350–19360.
31. Venturi, F., Sanmartin, C., Taglieri, I., Andrich, G., & Zinnai, A. (2017). Simplified method to estimate Sc-$CO2$ extraction of bioactive compounds from different matrices: Chili Pepper vs. Tomato by-products. *Applied Sciences, 7*, 361–373.
32. Wankat, P. (2014). *Separation process engineering. Includes mass transfer analysis*. Boston: Prentice Hall.
33. Wong, D., Orbey, H., & Sandler, S. (1992). *Industrial and Engineering Chemistry Research, 31*, 2033–2039.
34. Yasuhiko, A., Takeshi, S., & Takebayashi, Y. (2002). *Transport properties of supercritical fluids* (pp. 127–206). Springer series in materials processing.
35. Yazdizadeh, M., Eslamimanesh, A., & Feridun, E. (2012). Applications of cubic equations of state for determination of the solubilities of industrial solid compounds in supercritical carbon dioxide: A comparative study. *Chemical Engineering Series, 71*, 283–299.
36. Zekovic, Z., Bera, O., Durovic, S., & Pavlic, B. (2017). Supercritical fluid extraction of coriander seeds: Kinetics modelling and ANN optimization. *Journal of Supercritical Fluids, 125*, 88–95.

Chapter 4
Intrusion Detection System Based on a Behavioral Approach

Mehdi Moukhafi, Seddik Bri and Khalid El Yassini

Abstract Intrusion Detection System (IDS) can be defined as a group of tools, methods and resources that help us to predict or identify any unauthorized activity in a network. Current IDSs are mainly based on techniques constructed on heuristic rules, named signatures, in order to detect intrusions in a network environment. The drawbacks of these approach is that it could only detect a known attacks and referenced above. Contrastively, Intrusion Detection behavioral, or anomaly, assume that attacks causes an abnormal use of resources or manifest a strange behavior on the part of the user, by studying the behavior of the different types of network traffic it can identify the known and unknown attacks using the artificial learning algorithm. This study proposes a new behavioral approach of intrusion detection based on combination APSO (Accelerated Particle Swarm Optimization)-SVM (Support Vector Machine) to develop a model for IDS. The simulation results show a significant amelioration in performances, all tests were realized with the NSL-KDD data set. In comparison with other methods based on the same dataset, the proposed model shows a high detection performance.

4.1 Introduction

The Information and Communications Technology (ICT) have a central role in the daily life of people within their societies, due to the International Telecommunications Union report [1], which is carried out in 2016 and claim that 53% of the worlds

M. Moukhafi (✉)
Faculty of Sciences, Moulay Ismail University, Meknes, Morocco
e-mail: mehdi.moukhafi@gmail.com

S. Bri
MIN Group, Superior School of Technology: ESTM, Moulay Ismail University, Meknes, Morocco
e-mail: briseddik@gmail.com

K. El Yassini
Faculty of Sciences, Moulay Ismail University, Meknes, Morocco
e-mail: emailkhalid.elyassini@gmail.com

© Springer International Publishing AG, part of Springer Nature 2019
E.-G. Talbi and A. Nakib (eds.), *Bioinspired Heuristics for Optimization*,
Studies in Computational Intelligence 774,
https://doi.org/10.1007/978-3-319-95104-1_4

population uses the Internet. Indeed, the daily life activities, such as the sending of mail or the online paying, in addition to other strategic areas as the banking sector or the military communications, are mainly based on the ICTs. In this context, the attacks carried out by malicious users exploit the vulnerabilities of those systems which are more and more frequent; especially with the easy access to security test tools that is more accessible to professionals as well as hackers.

As a result of such attacks would touch the credibility of the owner of the information system or cause a significant financial damages. Therefore, the problem of security becomes a key issue for both the users and administrators of those information systems. It is necessary for the specialists of the information technologies field to offer some adequate mechanisms to better secure it.

For security, Intrusion Detection Systems (IDS) are used to monitor and analyze the events in the information system which is presented for the first time by Anderson in 1980 [2] and later formalized by Denning [3], the IDS can be used in a global policy of security, which includes other tools of protection, such as firewalls and anti-virus; where it is essential to take advantage of the collaboration of these tools and their complementarity. The Intrusion Detection System may exist alone, and in this case it is very important to optimize its processes. Nevertheless, in all cases, it is essential to improve the IDS performance.

To deal with the problem mentioned above, we propose a behavioral intrusion detection system, a hybrid approach that combine multi-class support vector machine with accelerated particle swarm optimization, in order to improve the intrusion detection rate. This hybrid technique is used to classify the NSL-KDD datasets. A 10-fold cross-validation is used to extract a best SVM parameters. the system performance intrusion detection is significantly promoted by our method. Then, we compare our results.

The remainder of the paper is organized as follow: In Sect. 4.2, we itemize a global architecture of intrusion detection system. Section 4.3 compare intrusion detection approaches and introduce the problem of behavioral detection. The Sect. 4.4.1 illustrates background of the framework. In the Sect. 4.4.2, we introduce a general framework on behavioral detection problem and present an algorithm for building a behavioral model based on APSO-SVM. In Sect. 4.5, we give an application of the model to the problem of intrusion detection, give proof of concept results and compare our method with other state of the art methods. Finally, Sect. 4.6 draws our conclusions and plan for future work.

4.2   Intrusion Detection System

Intrusion Detection System (IDS) could be defined as any tool, methods and resources that help us to predict or identify any unauthorized activity in a network [2]. In the architecture proposed by the group IDWG5 of the IETF [4, 5], there is three modules: Sensors Module, Analyzing Module and Managing Module, as shown in Fig. 4.1. In this architecture, the main was the definition of a standard for communication

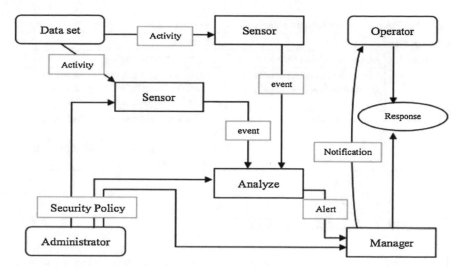

Fig. 4.1 Architecture of an IDS

between the components of the intrusion detection system. This architecture defines a format for the exchange message of the IDS: Intrusion Detection Message Exchange Format (IDMEF), which implicitly contains a data model.

- The intrusion detection features:

- Monitoring and Analysis of the user's activities and system;
- system analysis parameters;
- Identification of typical patterns of attacks;
- the models of abnormal activities analysis.

- Intrusion detection systems tend to ensure the five following properties [6]:

- The accuracy of detection: means a perfect detection of attacks with a minimum risk of false positives;
- Performance: A fast detection of intrusions throughout the analysis of essential events to conduct effective detection;
- Completeness: A detection of known and unknown attacks;
- Fault Tolerance: intrusion detection systems must resist attacks as well as their consequences;
- Speed: A fast analysis of the data which allows to undertake the necessary measures to stop the attack and protect the resources of the network and the intrusion detection system.

Table 4.1 Comparison between behavioral and signature approach

Approaches	Advantages	Drawbacks
Behavioral	–Does not need a base attack	–Top rate of false positive
	–Detection of possible unknown intrusions: few false negatives	–A malicious user can change slowly his behaviour to familiarize the system of a intrusive behaviour: risk of false negatives
Signature	Few false positives	–Difficult to build and maintaining: risk of false negatives
		–No detection of unknown attacks: risk of false negatives

4.3 Type of Approaches of IDS

The intrusion detection systems are generally based on two approaches: behavioral and by signature. The techniques of intrusion detection based on the behavioral approach suppose that the intrusion can be detected by the observation of the behavioral abnormality in relation to the normal behavior or prediction of the system or of the users. By contrast, the signature approach is based on the accumulated knowledge of specific attacks and vulnerabilities in the system (Table 4.1).

4.3.1 Signature Approach

This approach is based on the accumulated knowledge on attacks denoted and vulnerabilities of systems. The Intrusion system detection contains the information about the vulnerabilities then look for any attempt to exploit them. If the IDS detect such attempt, an alarm would react. In other words, any action that is not explicitly known as an attack is considered legitimate, consequently, the precision of IDS based on the signature approach is good. usually there is a number of ways to exploit the same vulnerability, but it is difficult to develop a compact signatures that detect all the variations of the attack and at the same time do not incur in false positives. Lastly, many intrusions are performed by a legitimate user who abuse their privileges, or external attackers who use a variant attacks that are not included in the signature database. As a result, no attack against known vulnerabilities is performed. However, this precision depends always on the updated knowledge on the attacks which must be regular [6], because the IDSs based signature cannot adequately treat a new types of attacks where constantly the environments were changing.

The major drawback of approaches based signature is that they only detect known attacks, which implies a frequent updating of the rules database and the time for the implement.

4.3.2   Behavioral Approach

The techniques of intrusion detection based on the behavioral approach [7] assume that the intrusion can be detected by the observation of the deviation of the normal behavior expected by the system or by the users. firstly, the model of the normal behavior is extracted from the information references collected by various means. Then, the intrusion detection system compares this model with the current activity, if a deviation is detected, an alert will be triggered. generally, we can say that this approach considers any behavior that is not previously recorded as an intrusion. Therefore, this approach may be complete, but the accuracy remains his greatest concern. Intrusion Detection behavioral or by anomaly is based on the assumption that an attack causes an abnormal use of resources or manifest a strange behavior on the part of the user. Consequently, the different approaches have been proposed to learn the normal behavior, in order to be able to detect any significant deviation.

4.3.2.1   Immunological Approach

In this approach, the normal behavior of a process is characterized by a set of execution traces recorded during the learning phase. These traces contain in an orderly manner the whole system calls made by the process during the execution. Concerning the Discovery phase, the last n system calls made by the process is compared to the whole of sub-sequences of N elements contained in the Traces which characterize the basis of reference of this process. Practically, the detection is done from the list of system calls sequences fixed size n statically constructed from traces of reference. Forrest in [7] was the first who use the immunological approach to model the process on a machine. His approach is to describe the normal behavior or the "self" via a finite sequence of calls systems. The sequences called N-gram are used as a basis for comparing the calls of systems process at a monitoring phase. Wespi and al. [8] consider a more general case by analyzing the examination events, they generate sequences a varying sizes of events to model the normal state of the system, then a pattern is selected if there are reasons that follows directly the score of anomaly which is incremented by one and an alarm is triggered when the score exceeds the threshold tolerated.

4.3.2.2   Machine Learning Approach

Learning algorithms can play an important role in detecting attacks (known or unknown). Additionally, the IDSs performances are considerably augmented at the network level.

Govindarajan and Chandrasekaran [9] propose a hybrid system, that combined a prediction of various classifiers. To manufacture their IDS they implemented three modules, the first, based on the RBF neural network, in charge of detection of U2R attacks, the second one is based on SVM for DOS attacks, the third is an RBF-SVM hybrid to recognize a normal traffics, and the latter is an ensemble classifier, RBF-SVM, that detects a Probe and R2L attacks. In addition, the authors implemented a best-first search (BFS), for feature selection.

Pervez and Farid[10] present a hybrid approach based on feature selection and classification for multiple class NSL-KDD, using support vector machines in a one against rest multi-class configuration (OAR-SVM). The feature selection was implemented following the LOO method.

Kanakarajan and Muniasamy[11] have applied a novel tree ensemble technique called GAR-Forest for both binary and multi-class classification intrusion detections. It is a solution based on on a greedy randomized adaptive search procedure with annealed randomness classifier. The GRASP was deployed to generate a set of randomized adaptive decision trees.

Gaikwad and Thool[12] implemented homogenous ensemble classifier to solve the intrusion detection problem. The bagging classifier is constructed with partial decision tree classifier and the Genetic Algorithm is used to select the relevant features from the NSL-KDD99 dataset.

Kevric et al.[13] present a combination of random tree and NBTree algorithms based on the sum rule scheme, they tested several algorithms tree based and they have selected the two best individual classifiers may lead to the best overall performing combination

Aygun and Yavuz[14] propose two deep learning based anomaly detection models using autoencoder and denoising autoencoder respectively. The key factor that directly affects the accuracy of the proposed models is the threshold value which was determined using a stochastic. The AE was trained with only normal data to produce a binary ids (normal/abnormal).

4.4 Proposed Approach

4.4.1 Classification Methods

In the process of intrusion detection, each connection can be classified into either normal traffic or intrusion, and hence it can be considered as a classification problem. In recent years, intrusion detection problem has beneficed from the hybrid machine learning to improve a detection rate of abnormal behaviour. we present a hybrid approach based on a multi-class SVM whose parameters and subset are optimized by APSO.

• **Accelerated Particle Swarm Optimization**

Particle Swarm Optimization (PSO) is a stochastic optimization method, for the nonlinear functions, inspired by the social behavior of insect colonies, bird flocks, fish schools and other animal societies, PSO was invented by Russell Eberhart and James

Kennedy[15] in 1995. Originally, the two began developing software simulations birds flocking around food sources, later after realizing that their algorithm solve optimization problems, they present[16] a discrete binary PSO algorithm developed from the previous PSO and operating in continuous variables.

A particle (candidate solution) that may move to the optimal position by updating its position and its speed. The speed of movement of a particle can be updated by the weight of inertia, cognitive learning factor, and the values of social learning factors.

Each particle is an overall potential of the optimum function f (x) on a given area D is considered a point in D-dimensional space and represented as Xi = [xi1, Xi2,..., XID] and the velocity vector of the particle is V th = [vi1, vi2,..., VD]. In additionally, the best previous position will be replaced by a best fitness value for the particle is PBi th = [pbi1, PbI2,..., pBID] and the best position to date in the area is GBI = [gb1, gb2,...,GBD]. The speed and position of the ith particle is updated according to the Eqs. (4.1) and (4.2):

$$v_{id}^{(k+1)} = wv_{id}^k + \alpha\varepsilon_1(pBest_{id} - x_{id}^k) + \beta\varepsilon_2(gBest - x_{id}^k) \qquad (4.1)$$

$$x_{id}^{k+1} = x_{id} + v_{id}^{k+1}, d = 1, 2, ..., D, id = 1, 2, ..., N \qquad (4.2)$$

Where v_{id}^k is the velocity of particle id at iteration k, The inertial mass represented by w has a value between 0.4 and 0.9, and α, β are the acceleration coefficients (cognitive and social coefficients), $epsilon_1$ and $epsilon_2$ are the random numbers between 0 and 1, x_{id}^k is the current position of particle id at the k iteration, pBest is the best previous position of the id particle, gBest is the position of the best particle in the swarm.

The inertia is calculated by wv_{id}^k and $c_1 r_1(pBest_{id} - x_{id}^k)$ represents a memory (the particle is attracted to the best point in its trajectory) and $c_2 r_2(gBest - x_{id}^k)$ represents the cooperation or information exchange.

The standard particle swarm optimization uses both the current Gbest and Pbest. The goal of using the individual best is primarily to improve the diversity in the quality solutions, however, this diversity can be simulated using some randomness. Subsequently, there is no compelling reason for using the individual best, unless the optimization problem of interest is highly nonlinear and multimodal. In this paper we used a APSO variance which could accelerate the convergence of the algorithm is to use the global best only. Thus, in the accelerated particle swarm optimization (APSO)[17, 18], the velocity vector is generated by a simpler formula:

$$v_{id}^{(k+1)} = wv_{id}^k + \alpha\varepsilon_N + \beta(gBest - x_{id}^k) \qquad (4.3)$$

where ε_N is drawn from N(0, 1) to replace the second term. The update of the position is similar to the standard APSO using formula (2)

• **Support Vector Machine**

Support Vector Machine (SVM) [19] is one of the most popular supervised machine learning algorithms. This is a classification model by evaluating data and identify patterns that retains excellent long generalization capabilities with an integrated resistance to overtraining. This generalization is based on solid theoretical foundations introduced by Vapnik [20]. The basic concept of the SVM regression is to map the non-linearity of data x in a space of high characteristic dimension, and solving a linear regression problem in this feature space.

In the classification of support vector, the separation function is a linear combination of grains as given in Eq. (4.3) and are in contact with the support vector,

$$f(x) = \Sigma_{i \varepsilon S} \mu_i y_i x_i^t x + b \qquad (4.4)$$

Where μ is a Lagrange factor, x_i is training models, y_i $(+1, -1)$ is the corresponding class labels and S denotes the set of support vectors.

Kernel based learning methods consists of a kernel function to generate a kernel matrix for all patterns. A kernel function is a function k(x, y) with Characteristic:

$$k(x, y) = < \phi(x).\phi(y) > \qquad (4.5)$$

Though new kernels are being proposed, the most commonly used are the following four basic kernels:

$$linear : k(x, y) = x.y \qquad (4.6)$$

$$polynomial : k(x, y) = (\gamma x^T y + r)^d, \gamma > 0 \qquad (4.7)$$

$$radial basis function (RBF) : k(x, y) = exp(-\gamma \parallel x - y \parallel^2), \gamma > 0 \qquad (4.8)$$

$$sigmoid : k(x, y) = tanh(\gamma xy + r) \qquad (4.9)$$

4.4.2 Architecture of Proposed IDS

This section is about the description of the proposed system for classification, which aims to improve the accuracy of our classification model by detecting the optimal parameters and features subset for SVM multi-class, in order to achieve a good using of the APSO based optimized framework.

Particularly APSO is attractive for features selection, in this particle swarms will discover the best features combination as they fly with-in the problem space. From an artificial intelligence perspective, create a classifier means creating a template for data, and it is perfect for a model to be as simple as possible.

APSO features selection improve the performances classification (learning time, speed and power of generalization). Furthermore, it increases the comprehensibility of data. This data selection means to select an optimum subset of relevant variables from a set of original variables.

The NSL-KDD has 41 features: it is, relatively, a large number to be processed by the classifier, which cannot complete execution within a reasonable time in the learning phase, then the selection can reduce the features space. We used APSO to select the optimum features, that have the most impact on the prediction of the model.

To implement our proposed approach, the RBF kernel function is used for the multi-class SVM classifier, because this kernel nonlinearly maps samples into a higher dimensional data so it, unlike the linear kernel, can handle the case when the relation between class labels and attributes is nonlinear. Furthermore, The second reason is the number of hyperparameters which influence the complexity of model selection [21].

There are two parameters for an RBF kernel:

- C (penalty parameter): This parameter, common to all SVM kernels, trades off misclassification of training examples against simplicity of the decision surface.
- γ (gamma parameter): It is a specific parameter to RBF kernel function, gamma defines how much influence a single training example has.

The choice of a value for C influences on the classification outcome depends in two majors situation:
–If C is too large, then the classification accuracy rate will be very high in the training phase, but very low in the testing phase.
–If C is too small, then the classification accuracy rate will be lower.
Parameter γ has a notable influence on classification outcomes than C, because its value affects the partitioning outcome in the feature space. An excessively large value for parameter γ results in over-fitting, while a disproportionately small value leads to under-fit-ting [22]. The used parameters (C and γ) as input attributes must be optimized using APSO; Precisely to establish a APSO-SVM based intrusion detection system, the following main steps (Fig. 4.2) must be proceeded.

The k-fold cross-validation technique is typically used to reduce the resulting errors from random sampling in comparing accuracies of a number of predictive models [23].The study divide randomly the data into 10 subsets, for each iteration, nine subsets are used for training multi-class SVM and one subset to validate a model. To evaluate each particle we have used in function fitness the accuracy rate. If the fitness is bigger than the Pbest, Pbest receive a fitness and If Pbests is better than the Gbest, Gbest receive Pbest. Finally the particles velocity and position are updated using (1) and (2). This process continues 10 times, until APSO find the optimal parameters, calculate an accuracy and evaluate a particle (parameters C,γ). The k-fold cross-validation technique is used only to find the optimal parameters of SVM while the best generated model is not used because we are building another model formed by the full NSL-KDD train set (KDDTrain+) and configured by the best parameters founded earlier.

Fig. 4.2 Architecture of a
proposed IDS

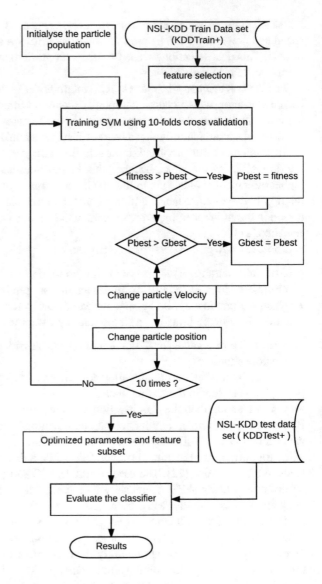

The multi-class SVM classifiers is built by combining several single SVM classifiers, and the method used to make the multi-class SVM model is the One-Against-
All (OVA) [24]. which construct N classifiers, where N is the number of classes. The
resulted hypothesis is given by the formula:

$$f_{ova}(x) = argmax_{i=1...N}(f_i(x)) \tag{4.10}$$

Where $f_{ova}(x)$ is global predict function and $f_i(x)$ is predict function of each classifier, so the ith classifier is trained by labeling all the examples in the ith class as positive and the remainder as negative.

4.5 Results and Discussion

4.5.1 Simulation Tools

All experiments were conducted on a calculation station 24 CPU Intel Core 2.13 GHz, 48 GB RAM, running under Linux CentOS 7, in order to evaluate the performance of the proposed model. Then, the implementation was coded using the Java language.

4.5.2 NSL-KDD Data Set

Many data sets, related to IDS, have been used by researchers to develop solutions based on a behavioral approach, the most popular is certainly the KDD99 [25], the latter is a revised version of the DARPA 98, created by Cybersystems and MIT technology group of Lincoln Laboratory by simulating the LAN US Air Force with a multiple attacks, it was used for the first time in competitions kdd99.

However, the KDD99 has major defects [26, 27], the most important are the huge number of redundant records (for about 78% in the train set), which causes the learning algorithms to be biased towards the frequent records, and thus prevent them from learning infrequent records which are usually more harmful to networks such as U2R and R2L attacks. To solve these issues, a new dataset, NSL-KDD [28] was proposed, which consists of only selected records form the complete KDD99 dataset and does not suffer from any KDD99 shortcomings, that was summarized into record connections where each connection is a single row vector consisting of 41 features. The attack classes presented in this data set are grouped into four categories, namely, denial of service attack, probes, remote to user attack and user to root attack. In this research, We used the NSL-KDD to elaborate our model of prediction which was evaluated by the same data base. We also compared the results with the results of a model trained with the KDD99.

4.5.3 Computational Results and Comparison

As it is mentioned below, the NSL-KDD has a big number of features (41 feature) which is used by the classifier does not offer a best result because it provokes an over-learning of the model. APSO was used to select the optimum features, as a result the number of fields were reduced from 41 to 15 (Table 4.2). Selected features: 2, 3, 4, 5, 6, 8, 14, 23, 26, 29, 30, 35, 36, 37, 38.

Table 4.2 Comparison
between the data set and
subset

Data set	0, 1, 18, 10, 491, 0, 0, 0, 0, 0, 0, 0, 0, 0, 0, 0, 0, 0, 0, 0, 0, 0, 0, 2, 2, 0, 0, 0, 0, 1, 0, 0, 150, 25, 0.17, 0.03, 0.17, 0, 0, 0, 0.05, 0, normal
Subset	1,18, 10, 491, 0, 0, 0, 2, 0, 1, 0, 0.03, 0.17, 0, 0, normal

To evaluate our approach, we have used four performance indicators from intrusion detection research [29]. The accuracy is the correct classification rate of all classes. Additionally, the detection rate is defined as the number of correctly classified normal and intrusion packets divided by the total number of the validation data set, while The false positive rate is defined as the total number of normal connections, which were incorrectly classified as attacks, and divided by the total number of normal connections. the false negative rate is defined as the total number of intrusion connections that were incorrectly classified as normal traffic divided by the total number of intrusion connections. The performance of the proposed method of intrusion detection was evaluated on NSL-KDD test set (KDDTest+), and the train set (KDDTrain+) is used for training the model. Lastly, Tables 4.3 and 4.4 illustrate the confusion matrix, we achieved 90,87% of instances classified correctly and 93,43 % of intrusion detection rate.

The Fig. 4.3, compares a model trained by the NSL-KDD data set and another model trained by the KDD99 data set, both tested on the NSL-KDD test set, it is clear that the model of the KDD99 suffers classification inaccuracy, especially as regards the Probes, R2L and U2R attacks, which proves that the KDD99 suffers from major defects.

The Table 4.5 is a comparison between the performance of the Svm model optimized by APSO and SVM models optimized by Genetic Algorithm and Ant colony. To perform a statistic analysis, we have execute each experiment ten times with each algorithm using a 10-folds cross validation, in the end, we take the model which offers the best experiments for the validated with the totality of the NSL-KDD test set. addi-

Table 4.3 Confusion matrix

	Normal	DOS	Probe	R2L	U2R
Normal	9118	202	210	21	160
Dos	25	7143	195	97	0
Probe	121	291	1991	15	3
R2L	673	15	0	2197	0
U2R	24	0	0	6	37

Table 4.4 Standard metrics
for Intrusion system
evaluation

	Normal	Intrusions
Normal	TN = 9118 93,89 %	FN = 593 6,11 %
Intusion	FP = 843 6,57 %	TP = 11990 93,43 %

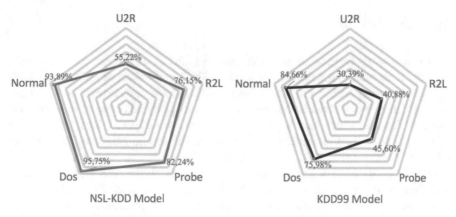

Fig. 4.3 The accuracy rates per attack

Table 4.5 Statistical analyzes of optimization algorithms

Algorithm	Min accu (%)	Max accu (%)	Avg accu (%)	KDDTest+ accu (%)
APSO-SVM	98,77	98,95	98,84	90,87
GA [30]	93,58	95,71	94,39	82,12
ACO [31]	88,41	90,62	89,75	66,84

tionaly we used the recomendations of the parameter settings of Eiben and Smit [30] for the genetic algorithm, and the recommendations of Wong and Komarudin [31] for Ant colony. The results show that APSO gives the best performances and above all a better stability of the results, contrary to the GA and ACO which offer the performances that vary considerably.

To show the validity of the used technique (SVM-APSO), Fig. 4.4 compares the proposed method with approaches of state- of-the-art which detect all attacks and use only the NSL-KDD as a testing dataset, because several researchers use the KDD99 to validate their models, the latter suffers from duplicate records in the test sets; Which give partially false results that do not indicate the real performance of the prediction model. The above results show that our approach improve the performance of the IDS. To conclude APSO-SVM is more reliable than state-of-the-art methods.

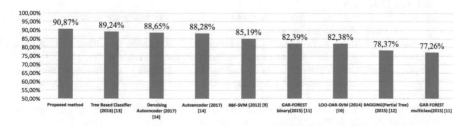

Fig. 4.4 Comparison of proposed model with other methods by accuracy

4.6   Conclusion

In this study, we present an intrusion detection system based on a behavioral approach, in order to develop our solution, we used an SVM multi-class of OVA (one vs all) type optimized by APSO. The main of using APSO in this study is in two-fold. At one hand, we have reduced the SVM feature vectors space by detecting or searching for the most features which have a great impact on the model predictions. The number of input features in a classifier should be limited to ensure good predictive power without an excessively computationally intensive model, as well as reducing the number from 41 to only 15 features. At the other hand, we successfully optimize SVM parameters used (C, gamma). Our training model consists of big datasets with distributed environment that improves the performance of Intrusion detection system. In the experiments the NSL-KDD data set is used to learn and evaluate our model, the performance of proposed IDS is better than that of other existing machine learning approaches tested on the same dataset.

For our future works, our searches will be oriented toward the heterogeneous multi-classifiers, in which we will develop a solution a base for the two levels; the first is composed by a different types of classifiers, the second, is a meta-classifier which will be conceptualized as chair of a committee with the members of the first level.

Acknowledgements This work is supported in part by the High School of Technology, Moulay Ismail University Meknes, which has provided the calculation station where we have executed our experiences.

References

1. Report of the international telecommunications union (ITU). (2016). http://www.itu.int/en/ITU-D/Statistics/Pages/facts/default.aspx. Accessed 13 April 2017.
2. Anderson, J. P. (1980). *Computer security threat monitoring and surveillance*. Technical report.
3. Denning, D. E. (1987). An intrusion-detection model. *IEEE Transactions on Software - Engineering, 13*(2), 222–232.
4. Curry, D., & Debar, H. (2006). Intrusion detection message exchange format data model and extensible markup language (xml) document type definition. https://tools.ietf.org/html/draft-ietf-idwg-idmef-xml-16. Accessed 13 April 2017.
5. Intrusion detection message exchange format (IDMEF). www.ietf.org/rfc/rfc4765.txt. Accessed 13 April 2017.
6. Debar, I. H., Dacier, M., & Wespi, A. (2000). A revised taxonomy for intrusion-detection systems. *Annales des Telecommunications., 55*(7), 361–378.
7. Forrest, I. S., Hofmeyr, S. A., Somayaji, A., Longstaff, T. A. (1996). A sense of self for unix processes. In *The 12th IEEE Symposium on Research in Security and Privacy* (pp. 120–128).
8. Wespi, A., Dacier, M., & Debar, H. (2000). Intrusion detection using variable-length audit trail patterns. In *The 3rd International Workshop on the Recent Advances in Intrusion Detection*, (pp. 110–129).
9. Govindarajan, M., & Chandrasekaran, R. (2012). Intrusion detection using an ensemble of classification methods. *World Congress on Engineering and Computer Science, 1*, 1–6.

10. Pervez, M. S., Farid, D. M. (2014). Feature selection and intrusion classification in NSL-KDD cup 99 dataset employing SVMs. In *The 8th International Conference on Software, Knowledge, Information,Management and Applications* (pp. 1–6).
11. Kanakarajan, N. K., Muniasamy, K. (2015). Improving the accuracy of intrusion detection using gar-forest with feature selection. In *The 4th International Conference on Frontiers in Intelligent Computing: Theory and Applications (FICTA)* (pp. 539–547). Berlin: Springer.
12. Gaikwad, D., & ThoolR, C. (2015). Intrusion detection system using bagging with partial decision treebase classifier. *Procedia Computer Science, 49*, 92–98.
13. Kevric, J., Jukic, S., & Subasi, A. (2016). An effective combining classifier approach using tree algorithms for network intrusion detection. *Neural Computing and Applications*, 1–8. online available.
14. Aygun, R. C., Gokhan, A. Y. (2017), Network anomaly detection with stochastically improved autoencoder based models. In *The IEEE 4th International Conference on Cyber Security and Cloud Computing* (pp. 193–198).
15. Kennedy, J., Eberhart, R. C.(1995). Particle swarm optimization. In *The 4th IEEE International Conference on Neural Networks* (pp. 1942–1948).
16. Kennedy, J., Eberhart, R. C. (1997). A discrete binary version of the particle swarm algorithm. In *The IEEE International Conference on Systems, Man, and Cybernetics* (pp. 4104–4108). Orlando.
17. Yang, X. S. (2008). *Nature-Inspired Metaheuristic Algorithms*. Luniver Press.
18. Yang, X. S. (2010). *Engineering Optimization: An Introduction with Metaheuristic Applications*. Chichester: Wiley.
19. Burges, C. (1998). A tutorial on support vector machines for pattern recognition. *Data Mining and Knowledge Discovery, 2*(2), 121–167.
20. Vapnik, V., & Cortes, C. (1995). Support vector networks. *Machine Learning, 20*(3), 273–297.
21. Chang, C. (2005). *Practical guide to support vector classification.* Technical report. 10.1.1.224.4115.
22. Pardo, M., & Sberveglieri, G. (2005). Classification of electronic nose data with support vector machines. *Sensors and Actuators B Chemical, 107*, 730–737.
23. Salzberg, S. L. (1997). On comparing classifiers: pitfalls to avoid and a recommended approach. *Data Mining and Knowledge Discovery, 1*, 317–327.
24. Christopher, M. B. (2006). *Pattern Recognition and Machine Learning* (1st ed.). Berlin: Springer.
25. Data set KDD99. http://kdd.ics.uci.edu/databases/kddcup99. Accessed 15 August 2017.
26. zgr, A., & Erdem, H. (2016). A review of KDD99 dataset usage in intrusion detection and machine learning between 2010 and 2015. *Peer J Preprints, 4*, e1954.
27. Mahoney, M. V., & Chan, P. K. (2003). An analysis of the 1999 DARPA/Lincoln laboratory evaluation data for network anomaly detection. *Proceedings of the Sixth International Symposium on Recent Advances in Intrusion Detection.* Lecture Notes in Computer Science (pp. 220–237). Verlag: Springer.
28. Data set NSL-KDD99. http://www.unb.ca/cic/research/datasets/nsl.html. Accessed 15 August 2017.
29. Hassan, M. (2013). Current studies on intrusion detection system, genetic algorithm and fuzzy logic. *International Journal of Distributed and Parallel Systems, 4*(2), 35–47.
30. Eiben, A. E., & Smit, S. K. (2011). Evolutionary algorithm parameters and methods to tune them. In E. M. Y. Hamadi & F. Saubion (Eds.), *Autonomous Search*. Berlin: Springer.
31. Wong, K. Y., Komarudin. (2008). Parameter tuning for ant colony optimization: a review. In *Proceedings of the International Conference on Computer and Communication Engineering (ICCCE08): Global Links for human development.* Kuala Lumpa, Malaysia, 13–15 May.

Chapter 5
A New Hybrid Method to Solve the Multi-objective Optimization Problem for a Composite Hat-Stiffened Panel

Ahmad El Samrout, Oussama Braydi, Rafic Younes, Francois Trouchu and Pascal Lafon

Abstract In this paper we present a new hybrid meta heuristic by combining Multi-objective bat algorithm (MOBA) and variable neighborhood search (VNS). The hybrid meta heuristic is coupled with response surface methodology (or meta modeling) to solve the mechanical multi-objective optimization problem of hat stiffened composite panel. The optimization criteria were the weight and the rigidity of the panel. Experimental results show that our suggested approach is quite effective, as it provides solutions that are competitive with the results obtained by using MOBA alone and other state of the art optimization algorithms.

Keywords Hat stiffened panel · Multi-objective optimization · Meta modeling
Multi-objective bat algorithm · Variable neighborhood search

5.1 Introduction

The design of industrial structures in aeronautics is a constantly evolving field, due to the perpetual need to gain weight and space in this domain. A lighter aircraft means fewer loads to compensate, and thus less fuel consumption, which decreases the greenhouse effect. Therefore, a lighter aircraft is beneficial at both economic and environmental levels.

A. El Samrout (✉) · O. Braydi · P. Lafon
University of Technology of Troyes, Institut Charles Delaunay, Troyes, France
e-mail: ahmad.el_samrout@utt.fr

O. Braydi · R. Younes
Faculty of engineering, Lebanese University, Hadath, Lebanon

F. Trouchu
Montreal Polytechnic, Quebec, Montreal, Canada

© Springer International Publishing AG, part of Springer Nature 2019
E.-G. Talbi and A. Nakib (eds.), *Bioinspired Heuristics for Optimization*,
Studies in Computational Intelligence 774,
https://doi.org/10.1007/978-3-319-95104-1_5

Composite material has been used in aerospace construction because of their high stiffness to weight ratio and their resistance to fatigue and corrosion. In particular, stiffened composite panels are widely used in aircraft's fuselage, as well as in wings and tail sections. However, such materials induce an additional weight to the overall structure. Therefore, a rigidity-weight optimization of stiffened panels becomes a necessity in aerospace industry.

In [1] a new method for manufacturing self-stiffened composite panels using flexible injection was presented. Some mechanical properties of the new panels, such as stiffness, were then evaluated using a three points bending test. The author studied the influence of the stiffener on the mechanical properties of the composite.

The aim of our study is to improve the hat stiffened panels produced in [1] using meta heuristics and response surface methodology. Two objective functions are considered; the weight minimization and the rigidity maximization.

The kind of problem is well known in panel optimization literature. While some researchers try to optimize the geometrical shape of the panel [2–4], others rely on the stacking sequence of the composite panel as their optimization variable [5–8].

The use of meta modeling and multi objective meta heuristics is also well known in these optimization problems, for instance [9] presented an optimization procedure for a geometric design of a composite material stiffened panel with conventional stacking sequence using static analysis and hygrothermal effects. The procedure is based on a global approach strategy, composed by two steps: first, the response of the panel is obtained by a neural network system using the results of finite element analyses and, in a second step, a multi-objective optimization problem is solved using a genetic algorithm. In [10] a process to compare three genetic algorithms (GAs) for the solution of multiobjective optimization problem of a T-shape composite stringer under compression loads has been presented.

The major contribution of this work is the hybridization between two meta heuristics that have not been coupled before to the best of our knowledge, in an attempt to discover new hybridization potentials. The work is based on the rational observation that the nature of the two candidate hybrids are compatible.

This paper is organized as follows. In Sect. 5.2 the geometrical dimensions, physical proprieties, boundary conditions and finite element model of the panel are illustrated. In Sect. 5.3 the formulation of the optimization problem along with the meta model and the hybrid meta heuristic are introduced. It is also devoted for the presentation and the interpretation of the results. Finally, conclusion is given in Sect. 5.4.

5.2 Model Presentation and Validation

The author in [1] developed a new procedure to produce composite plates of size $400 \times 140 \times 3$ mm reinforced with a centrally located Omega feature, then tested these panels using three points bending tests that he called omega test and inverse omega test.

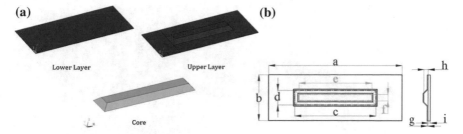

Fig. 5.1 a Panel composition , **b** Panel geometry

Table 5.1 Variable initial values

Variables	c	d	e	f	g	h
Values (mm)	260	46	240	20	1.5	14

5.2.1 Panel Composition and Geometry

The sandwich hat-stiffened composite panel consists of three components: an upper and a lower composite layers, separated by a foam core (see Fig. 5.1a)

The geometric parameters of this panel are presented in Fig. 5.1b. Eight variables can represent the panel in a sufficient manner, but since we are trying to study the influence of foam core, it is safe to assume that $a = 406$ mm and $b = 140$ mm are constant. Also for the sake of simplicity we will assume that i is equal to g. Table 5.1 shows the initial values of the panel's variables.

5.2.2 Material Proprieties

The core is made from Foam $K20(E = 1.5$ GPa; $v = 0.3)$. The upper and lower layers and the contour are made from glass fiber-epoxy. The ply lay-up is $[90, 0, 90]$ with a total number of 3 plies, each has a thickness of 0.47 mm.

5.2.3 Meshing

The meshing (see Fig. 5.2) is divided into two steps, the first step is to generate the mesh of the core with tetrahedral elements. The second step is to generate the skins of mesh (lower and upper) with hexahedral elements in each element with 8 nodes and consists of 3 sections 'Shell' (90/0/90) with respect to the main axis X. Orientation of the elements was taken into account.

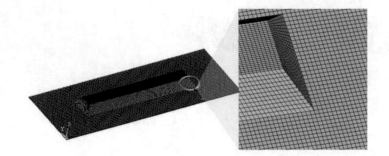

Fig. 5.2 Meshing

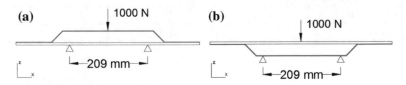

Fig. 5.3 Loading **a** Omega loading, **b** inverse omega

5.2.4 Loading and Boundary Conditions

Our model will treat only the elastic behavior of the panel. According to [1], the composite panel will act as an elastic panel while the bending force is under 2000 N. In our finite element model a static force of 1000 N will be applied to mid plan of the panel (upper layer in the case of omega test, and lower layer in the case of inverse omega). While two supports distant by 209 mm hold the panel on the opposite side (see Fig. 5.3).

The displacement along axis X, Y, and Z will be blocked for the supports. To represent the force, the displacement along axis X and Y will be blocked along the central support and the loading is distributed across all nodes in the Z direction ($1000/50 = 20$ N per node for omega and $1000/72 = 13.88$ N per node for inverse omega).

5.2.5 Result and Verification

The results of the finite element simulation are shown in Fig. 5.4.

The comparison of stiffness values for the composite panel, in the elastic range, between the experimental results and the finite elements results, for both omega and inverse omega is given in Table 5.2. This small range of error justifies our model.

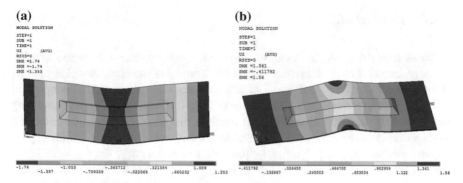

Fig. 5.4 Finite element results tests: **a** omega, **b** inverse omega

Table 5.2 Comparison between theoretical and experimental results

	Experimental results (mm)	Finite elements results (mm)	Error %
OMEGA	580.75	574.71	1.05
INV OMEGA	584.43	640.61	8.77

Table 5.3 Upper and lower bounds for each variable

Variables	c	d	e	f	g	h
Lower bound X_l (mm)	234	44	200	18	1.24	11.4
Upper bound X_u (mm)	266	52	249	26	1.56	14.6

5.3   Multi Objective Optimization

The two objectives of the optimization problem are to maximize the rigidity R and to minimize the weight W. The design variables are the dimensions of the panel $X = \{c, d, e, f, g, h\}$. The values of the lower bound X_l and upper bound X_u of X are shown in Table 5.3.

Therefore, the multi-objective optimization problem can be formulated as follows:

$$Minimize\ F_{obj}\left(W(X), \frac{1}{R(X)}\right), \tag{5.1}$$

$$Subject\ to:$$

$$X_l < X < X_u \tag{5.2}$$

5.3.1 Meta Modeling

Finite element analysis usually costs a huge computational time. A common solution to this problem is to use Meta modeling. A Meta model or surrogate model is a model of the model, i.e. a simplified model of an actual model. In our case, the Meta model will try represent the mechanical behavior of the panel with response to the bending experiences. It should be able to predict the deformation of the plate, and hence it's rigidity simply by knowing the geometrical dimensions of the plate. Various types of Meta models include polynomial equations, neural network, Kriging, etc. These types often share the same steps; first a design of experiments is established where a certain number of experiments is done, each time with a different set of inputs. Then a systematic method to determine the relationship between inputs affecting the process and the output of that process is applied.

In our example the cubic face centered design was adopted as design of experiments, then the finite element experience was repeated 80 times. The Meta model used in our example to find the correlation between the deformation of the panel and its geometrical parameters is Kriging which is a method of interpolation for which the interpolated values are modeled by a Gaussian process governed by prior covariances. It is widely used in the domain of spatial analysis and computer experiments.

By calculating the error percentage, i.e. the difference between the response of the finite element model and the response of the meta model, one can determine the degree of accuracy of our meta model in both omega and inverse omega tests. In the case of omega, the error is smaller than $4 \times 10^{-2}\%$, while in the case of inverse omega the error is around $3 \times 10^{-6}\%$.

5.3.2 Hybrid Algorithm

In this section, we attempt to optimize the meta model obtained in Sect. 5.3.1 w.r.t the problem formulated in Eq. 5.1 using a new hybrid based on MOBA and VNS.

5.3.2.1 Multi-objective Bat Algorithm

The idea behind multi-objective bat algorithm (MOBA) was first introduced in [11]. It is a meta heuristic that imitates the bat's echolocation system. It can be summarized as follows: Each virtual bat flies randomly with a velocity v_i at position (solution) x_i with a varying frequency or wavelength $\varphi \in [\varphi_{min}, \varphi_{max}]$ and loudness A_i and where $\beta \in [0, 1]$ is a random vector drawn from a uniform distribution and x_* is the current global best location (solution) which is located after comparing all the solutions among all the n bats at each iteration t. As a bat searches and finds its prey, it updated its position and velocity and changes frequency, loudness and pulse emission rate r. Search is intensified by a local random walk. Selection of the best continues until a stop criteria is met.

The standard MOBA routine is presented in Algorithm 2.

Algorithm 2: MOBA

Objective functions $f_1(x), …, f_k(x)$

Initialize the bat population x_i $(i = 1, 2, …, n)$ and v_i

for $j = 1 \rightarrow N$*(points on Pareto fronts)* **do**

    Generate K weights $w_k \geq 0$ so that $\sum_{k=1}^{K} w_k = 1$

    Form a single objective $f = \sum_{k=1}^{K} w_k f_k$

    while $t <$ *Max number of iterations* **do**

        $\varphi_i = \varphi_{min} + (\varphi_{max} - \varphi_{min})\beta$

        $v_i^{t+1} = v_i^t + (x_i^t - x_*)\varphi_i$

        $x_i^{t+1} = x_i^t + v_i^t$

        if *rand* $> r_i$ **then**

            Random walk around a selected best solution

        end

        Generate a new solution by flying randomly

        if $(rand < A_i)$ & $(f(x_i) < f(x_*))$ **then**

            Accept the new solutions

            increase r_i and reduce A_i

        end

        Rank the bats and find the current best x_*

    end

end

5.3.2.2   Variable Neighborhood Search

The idea behind variable neighborhood search (VNS) was initially introduced by [12]. It is a meta heuristic that explores distant neighborhoods of the current incumbent solution, and moves from there to a new one if and only if an improvement was made. The local search method is applied repeatedly to get from solutions in the neighborhood to local optima. The standard VNS routine is presented in Algorithm 3.

Algorithm 3: VNS

$k = 1$

while $(k \leq k_{max})$ & *(Max number of iterations is not reached)* **do**

  Shaking: *generate a point x' at random from the kth neighborhood of x*

  Local search: *apply a local search method with x' as initial solution;*

  denote with x'' the so obtained local optimum

  if $(x''$ *is better than x)* **then**

    $x \leftarrow x''$

    $k \leftarrow 1$

  end

  else

    $k \leftarrow k + 1$

  end

end

5.3.2.3 Hybrid Algorithm

The choice MOBA and VNS as components for the hybrid algorithm is justified by the fact that these two meta heuristics are complementary to each other; MOBA is a population-based method that is capable of exploring the search space , while VNS is a trajectory-based method that is known for intensifying the search.

In our hybrid algorithm, the solutions obtained by MOBA are taken as initial solutions for VNS. From these "good" solutions, VNS divides the search space into sub-structures and then guides the search aggressively towards better solutions. The algorithm is clarified in Algorithm 4. We implemented the algorithms in MATLAB 2015a language and ran it on a PC with 2.2 GHz and 8 GB RAM memory, with a bat population size of 20 bat and an initial loudness parameter of 0.25 and a pulse rate of 0.5. The frequency range was [0,2].

5.3.3 Results

Figure 5.5 shows the Pareto frontier (PF) for both omega and inverse omega cases, obtained using MOBA and Hybrid MOBA (MOBA + VNS). All the points on PF are equally "good", and each point represent a set of dimension that guaranties an optimum panel.

In order to evaluate the solutions obtained using MOBA + VNS, and compare them with those obtained using only MOBA, the metrics proposed by [13] are used:

1. MID (mean ideal distance): The closeness between Pareto solution and ideal point (0, 0). The lower value of MID, the better the quality of the solution is.

Algorithm 4: Hybrid MOBA

Procedure MOBA:

Objective functions $f_1(x), ..., f_k(x)$

Initialize the bat population x_i ($i = 1, 2, ..., n$) and v_i

for *$j = 1 \to N$(points on Pareto fronts)* **do**

    Generate K weights $w_k \geq 0$ so that $\sum_{k=1}^{K} w_k = 1$

    Form a single objective $f = \sum_{k=1}^{K} w_k f_k$

    while *$t < $ Max number of iterations* **do**

        $\varphi_i = \varphi_{min} + (\varphi_{max} - \varphi_{min})\beta$

        $v_i^{t+1} = v_i^t + (x_i^t - x_*)\varphi_i$

        $x_i^{t+1} = x_i^t + v_i^t$

        if *rand $> r_i$* **then**

            | *Random walk around a selected best solution*

        end

        Generate a new solution by flying randomly

        if *(rand $< A_i$) & ($f(x_i) < f(x_*)$)* **then**

            Accept the new solutions

            increase r_i and reduce A_i

        end

        Rank the bats and find the current best x_*

    end

    Record x_* as a non-dominated solution

end

VNS procedure:

Divide the set of non dominated solutions obtained in MOBA into k_{max} structures

for *every solution x obtained in MOBA* **do**

    $k = 1$

    while *($k \leq k_{max}$) & (Max number of iterations is not reached)* **do**

        Shaking: *generate a point x' at random from the kth neighborhood of x*

        Local search: *apply a local search method with x' as initial solution; denote with x'' the so obtained local optimum*

        if *(x'' is better than x)* **then**

            $x \leftarrow x''$

            $k \leftarrow 1$

        end

        else

            | $k \leftarrow k + 1$

        end

    end

end

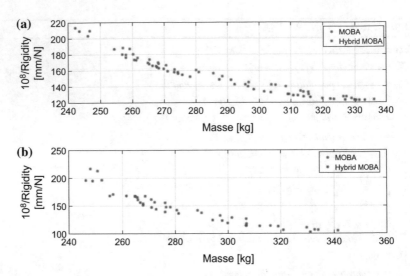

Fig. 5.5 Pareto frontier using MOBA and Hybrid MOBA **a** omega, **b** inverse omega

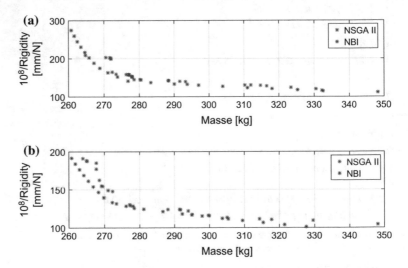

Fig. 5.6 Pareto frontier using NSGA-II and NBI **a** omega, **b** inverse omega

2. SNS: The spread of non-dominance solution. The higher value of SNS, the better the quality of the solution is.

Table 5.4 gives a comparison between MOBA and hybrid MOBA in omega and inverse omega cases using MID and SNS metrics. It shows clearly that our hybrid model is superior to the original algorithm except for the spread (SNS) in the case of inverse omega test.

Table 5.4 Comparison between MOBA and hybrid MOBA

	Hybrid MOBA omega	MOBA omega	Hybrid MOBA inverse omega	MOBA inverse omega
MID	326.8727	330.5729	318.1531	326.7924
SNS	13.6755	13.2973	11.3325	13.7764

Table 5.5 Comparison between NSGA II and hybrid NBI

	NSGAII omega	NBI omega	NSGAII inverse omega	NBI inverse omega
MID	322.3037	337.3789	317.3037	319.5815
SNS	9.3521	20.1274	9.3245	15.4160

Furthermore, the results were extended by solving the problem using normal boundary intersection (NBI) [14] and non dominated sorting genetic algorithm (NSGA-II) [15] which are two state of the art optimization algorithms. NSGA-II was applied with a population of 200 individuals, a crossover percentage of 70% and a Mutation percentage of 40%. While for NBI, 20 points constitute the Pareto front. The optimization results are shown in Fig. 5.6.

Table 5.5 shows that in both cases of omega and inverse omega, NBI has the upper hand over NSGA-II in terms of the spread, while being lower in terms of the closeness to the PF.

By comparing Tables 5.4 and 5.5, hybrid MOBA proves to be a compromise between NSGA-II and NBI, as it is better than NBI and worst than NSGA-II in terms of the closeness to the PF and better than NSGA-II and worst than NBI in terms of the spread.

5.4   Conclusion

In this paper we presented a new hybrid meta heuristic based on the combination of MOBA and VNS, because of their complementary strengths. The meta modeling technique is also used to make the optimization more suitable. The hybrid algorithm is tested on a multi-objective optimization problem of hat stiffened composite panel. The optimization criteria were the weight and the rigidity of the panel. Experimental results show that our suggested approach is quite effective, as it provides solutions that are competitive with the results obtained by using NSGA-II, NBI and MOBA alone.

References

1. Rifay, M. (2010). *Fabrication de panneaux auto-raidis par injection flexible*. Masters thesis, cole Polytechnique de Montral
2. Kaufmann, M., Zenkert, D., & Mattei, C. (2008). Cost optimization of composite aircraft structures including variable laminate qualities. *Composite Science Technology, 68*(13), 2748–2754.
3. Rikards, R., Abramovich, H., Kalnins, K., & Auzins, J. (2006). Surrogate modeling in design optimization of stiffened composite shells. *Composite Structures, 73*(2), 244–51.
4. Badran, S. F., Nassef, A. O., & Metwalli, S. M. (2009). Y-stiffened panel multi-objective optimization using genetic algorithm. *Thin Wall Structures, 47*(11), 1331–42.
5. Irisarri, F. X., Bassir, D. H., Carrere, N., & Maire, J. F. (2009). Multiobjective stacking sequence optimization for laminated composite structures. *Composites Science and Technology, 69*(7–8), 983–90.
6. Irisarri, F. X., Laurin, F., Leroy, F. H., & Maire, J. F. (2011). Computational strategy for multiobjective optimization of composite stiffened panels. *Composite Structures, 93*(3), 1158–67.
7. Todoroki, A., & Ishikawa, T. (2004). Design of experiments for stacking sequence optimizations with genetic algorithm using response surface approximation. *Composite Structures, 64*(3–4), 349–357.
8. Todoroki, A., & Sekishiro, M. (2008). Stacking sequence optimization to maximize the buckling load of blade-stiffened panels with strength constraints using the iterative fractal branch and bound method. *Composites Part B: Engineering, 39*(5), 842–50.
9. Marín, L., Trias, D., Badalló, P., Rus, G., & Mayugo, J. A. (2012). Optimization of composite stiffened panels under mechanical and hygrothermal loads using neural networks and genetic algorithms. *Composite Structures, 94*(11), 3321–3326.
10. Badalló, P., Trias, D., Marín, L., & Mayugo, J. A. (2013). A comparative study of genetic algorithms for the multi-objective optimization of composite stringers under compression loads. *Composites Part B: Engineering, 47*, 130–136.
11. Yang, X. (2011). Bat Algorithm for the multi-objective optimization. *International Journal of Bio-Inspired Computation, 3*, 267–274.
12. Mladenović, N., & Hansen, P. (1997). Variable neighbourhood search. *Computers and Operations Research, 24*(11), 1097–1100.
13. Behnamian, J., Fatemi Ghomi, S. M. T., & Zandieh, M. (2009). A multi-phase covering Pareto-optimal front method to multi-objective scheduling in a realistic hybrid flowshop using a hybrid metaheuristic. *Expert Systems with Applications, 36*(8), 11057–11069.
14. Das, I., & Dennis, J. E. (1998). Normal-boundary intersection: a new method for generating the Pareto surface in nonlinear multicriteria optimization problems. *SIAM Journal on Optimization, 8*(3), 631–657.
15. Deb, K., Pratap, A., Agarwal, S., & Meyarivan, T. (2002). A fast and elitist multiobjective genetic algorithm: NSGA-II. *Transactions on Evolutionary Computation, 6*(2), 182–197.

Chapter 6
Storage Yard Management: Modelling and Solving

Gustavo Campos Menezes, Geraldo Robson Mateus
and Martín Gómez Ravetti

Abstract The volume of cargo transported and stored around the world grows every year. To ensure competitiveness, productivity and cost savings, companies continually invest in process improvement and the development of IT tools. This chapter discusses various optimization problems existing in the process of storing and transporting loads. In this work, an integrated problem of planning and scheduling is defined, a mathematical programming model is investigated, and solutions are obtained through the use of heuristics and a commercial optimization package. The computational experiments (based in real cases) showed that the method is more efficient in producing a feasible solution than the solver.

6.1 Introduction

The storage yard optimization problem, can be defined, in general, as follows: There are a set of supply nodes, where products are available for transportation, storage nodes where the products are stocked and demand nodes or delivery subsystem

G. C. Menezes (✉)
Departamento de Eletroeletrônica e Computação, Centro Federal de Educação,
Tecnológica de Minas Gerais, Avenida Doutor Antônio Chagas Diniz 655, CEP,
Cidade Industrial, Contagem, MG 32210-160, Brazil
e-mail: gustavo@cefetmg.br

G. R. Mateus
Departamento de Ciência da Computação, Universidade Federal de Minas Gerais,
Av. Antônio Carlos 6627, CEP, Belo Horizonte, MG 31270-010, Brazil
e-mail: mateus@dcc.ufmg.br

M. G. Ravetti
Departamento de Engenharia de Produção, Universidade Federal de Minas Gerais,
Av. Antônio Carlos 6627, CEP, Belo Horizonte, MG 31270-901, Brazil
e-mail: martin.ravetti@dep.ufmg.br

© Springer International Publishing AG, part of Springer Nature 2019
E.-G. Talbi and A. Nakib (eds.), *Bioinspired Heuristics for Optimization*,
Studies in Computational Intelligence 774,
https://doi.org/10.1007/978-3-319-95104-1_6

for shipping products. Specialized equipment with predefined capacities is used to transport the products within the network. An equipment route between nodes has a given capacity and handle one product at a time.

The defined problem can be adapted for different applications, such as: in mining industry, where a supply node can be the ore mining, a yard (mineral storage yard) and a demand nodes a railway. In a terminal port, where a supply nodes are the arrival points of products (wagon turner), a storage yard, the place where the products are stored in the terminal (whether bulk or container) and the point of demand a berth where the ships await loading. In the agroindustry, where products handled can be grains, such as: soy, corn, wheat, among others. In goods delivery companies, in which specific equipment receives the products of transport vehicles, store these products in storage areas to deliver in final destinations.

The equipment used to transport these cargoes can be diverse, such as: conveyor belts, wagon turners, automotive vehicles, stackers and reclaimers, ship loaders, cranes (for handling containers), among others. To transport these products, a sequence of equipment is used to carry the transport between a point of origin (supply or yard) to the destination (yard or demand node). This sequence of equipment is called a route. Routes can be direct (when they depart from a point of supply and reach the demand), or indirect (when departing from the storage yard to the demand).

The main contributions of this research are related to a brief literature review about storage yard problems and the use of a hierarchical framework to solve a production planning, stockyard allocation and scheduling problems for the delivery of products.

The remainder of this article is structured as follows: Sect. 6.2 presents the literature review. Section 6.3 defines the problem (based on iron ore storage terminal). Section 6.4 presents the mathematical formulation. Section 6.5 discusses the solution strategy applied. Section 6.6 is dedicated to computational experiments and the paper ends with conclusions.

6.2   Storage Yard Problems

Minimizing the transportation costs, through the intelligent use of resources (labor, equipment and space) is the great competitive differential in the present day. Various studies in this line of research, such as those by [19, 21, 25], stress the need for integrated solutions that take into account the problems associated with the three levels: strategic, tactical, and operational. There are several challenges involved in storage yards management. Figure 6.1 highlights the three major nodes related to this problem.

The offer node represents the arrival or supply of products to the system. This offer can be programmed when the arrival of products is known and distributed in a planning horizon, or not programmed, when only the volume of products that will be manipulated is known, and not their distribution in the time intervals.

The stock node, responsible for temporary storage. This storage can be done in sheds, outdoors (in the case of iron ore) or in silos (in the case of grains). Stocked

Fig. 6.1 Offer, stock and demand nodes

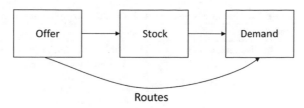

Fig. 6.2 Storage yard management: integrated problems

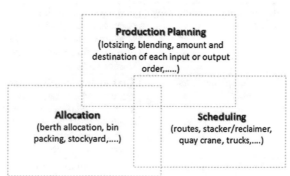

products may undergo some processing before being shipped to the destination. This processing may be done by plants, which process the product prior to transportation, or simpler processes such as blending. In blending, products with different characteristics can be combined into a single storage area to ensure a quality metric. In situations involving the transport of iron ore, blending ensures a minimum quality of iron in the final product.

Finally, the demand node, which represents the destination of the products. This destination can be the end customer, industries or other cargo terminal. Trucks, conveyor belts, ships and trains can be used to transport products between these nodes. As an equipment can only transport one product at a time, it is necessary to scheduling the equipment to ensure that all products are transported between the three nodes.

Figure 6.2, highlights the main modules of the integrated storage yard management problem: Production Planning, Scheduling and Allocation. These three modules can be subdivided into other subproblems: lotsizing, blending, berth allocation, bin packing, stockyard allocation, route scheduling, equipment scheduling, yard truck scheduling, among many others. These problems when solved individually already have high combinatorial characteristics and complexity. When solved in an integrated way, the challenges are even greater and require a great research effort. Table 6.1 highlights some of these problems, as well as solution methods employed to solve them.

The problems evaluated in Table 6.1 can be solved in several ways: exact or heuristic approaches, as well as modeled as deterministic, stochastic or integrated problems. In general, most of these problems are recently studied in an integrated way.

Table 6.1 Operations research and storage yard problems

Problem	Definition	Solution methods
Berth allocation problem	Determine when and where the ships should moor along a quay. The objective is to minimize the total time spent by the ships at the port	Exact algorithms [23, 26] and Heuristics [2, 22]
Stacker/reclaimer scheduling problem	Given a set of handling operations in a yard, the objective is to find an optimal operation scheduling on each stacker-reclaimer	Formulations and algorithms [1, 11]
Production planning and scheduling	The problem is to define the amount and destination of each input or output order in a bulk cargo terminal and establishing a set of feasible routes to transport the cargo	Branch and price [17], hierarchical approach [16] and metaheuristic [15]
Yard truck scheduling	The problem is to assing a fleet of trucks to transport containers to minimize the makespan	Formulation and algorithms [12, 13, 20]
Stockyard Allocation	How to arrange and holding of cargo at a shipping point for future transportation	Algorithms and model [7, 24]
Mine production planning	The problem is to schedule multiple sequential acitivities that are interrelated, such as: mining, blending, processing and transporting	Deterministc and stochastic formulations [10, 14, 18]

6.3 Case Study: Iron Ore Storage Terminal

The port complex can be represented as a set of three subsystems: Reception, stockyards and berths. Iron ore from the mines reach the Reception subsystem through the rail system. At the Reception, products can be transferred directly to moored ships or to stockyards. When the ship docks at one of the berths, a demand is generated that needs to be met. The flow of products between reception, stockyards and berths is performed using specific equipment: conveyor belts, trucks, ore stackers, and reclaimers, among others. Figure 6.3 provides a schematic representation of the problem.

The number of routes is limited and they may share equipment. Thus, if two different products are assigned to routes sharing equipment, these routes must be active at non-overlapping intervals. Figure 6.4 shows a case where two routes (routes 1 and 2) share the same equipment.

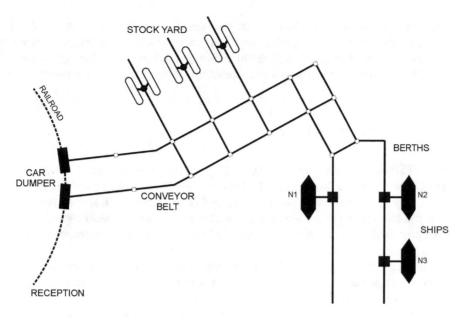

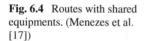

Fig. 6.3 Reception, stockyards and berths. (Menezes et al. [17])

Fig. 6.4 Routes with shared equipments. (Menezes et al. [17])

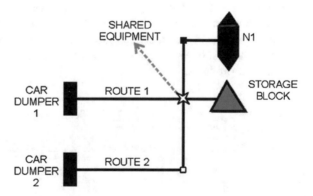

The problem can be defined as follows: lets consider a variety of products arriving at a port terminal (supply), they need to be transferred to meet the demand in a set of pier or to a local storage area. To make this transfer, products need a feasible route. The production planning and scheduling problem defines the amount and destination of each product and simultaneously establishes a set of feasible routes to guarantee that the ships will load on schedule. In the remainder of this text, this problem will be called the Product Flow Planning and Scheduling Problem, or $PFPSP$.

Some related work to $PFPSP$ can be: Bilgen and Ozkarahan [4], study the problem of blending and allocating ships for grain transportation. Conradie et al. [8] address the optimization of the flow of products (in this case coal) between mines and factory. Barros et al. [2] develop an integer linear programming model for the problem of allocating berths in conjunction with the storage conditions of the stockyard. Robenek et al. [23] proposes an integrated model for the integrated berth allocation and yard assignment problem in bulk ports. Finally, [22] Investigate the berth allocation in an ore terminal with demurrage, despatch and maintenance.

6.4 PFPSP Formulation

The PFPSP mathematical model was initially proposed in [17]. All production is planned for a given time horizon, divided into T periods. Product supply is related to the arrival of a train, and demand is related to the arrival of a ship. Routes are classified into three types: routes x that transport products from the Reception to the Stockyard, routes y from the Reception to the Ships, and routes z from the Stockyard to the Ships.

Table 6.2 defines the sets used to model PFPSP and gives the input parameters of the model, which define the capacity limits for periods, storage subareas, equipment, routes, and costs associated with the objective function.

Table 6.3 shows all decision variables used in PFPSP modeling. These variables are associated with the allocation of storage subareas and the allocation and scheduling of routes and for returning which demands were not met in each period.

The objective function (17.1) seeks to minimize the penalty of not meeting the supply of products at the Reception subsystem, the penalty of not meeting the demand of ships, the cost of product allocation in the stockyard and the cost of using routes to transport products.

$$\min f(z) = \sum_{p \in P} \sum_{t \in T} \alpha_{pt} IR_{pt} + \sum_{n \in N} \sum_{p \in P} \sum_{t \in T} \beta_{npt} IP_{npt}$$
$$+ \sum_{s \in S} \sum_{p \in P} \sum_{p' \in (P \cup 0)} \sum_{t \in T} \gamma^s_{pp't} Sf^s_{pp't} + \sum_{p \in P} \sum_{p' \in P} \sum_{t \in T} \lambda_{pp'} \left(\sum_{r \in R^y} c^{ry} y^r_{pp't} + \sum_{r \in R^z} c^{rz} z^r_{pp't} \right)$$
$$+ \sum_{p \in P} \sum_{t \in T} \sum_{r \in R^x} \sigma^r (c^{rx} x^r_{pt}) + \sum_{p \in P} \sum_{p' \in P} \sum_{t \in T} \sum_{r \in R^y} \sigma^r (c^{ry} y^r_{pp't}) + \sum_{p \in P} \sum_{p' \in P} \sum_{t \in T} \sum_{r \in R^z} \sigma^r (c^{rz} z^r_{pp't})$$

$$(6.1)$$

$$\sum_{r \in R^x} c^{rx} x^r_{pt} + \sum_{r \in R^y} c^{ry} \left(\sum_{p' \in P} y^r_{p'pt} \right) - IR_{p(t-1)} + IR_{pt} = a_{pt}, \qquad \forall p \in P, \forall t \in T.$$

$$(6.2)$$

$$\sum_{r \in R^z} c^{rz} \left(\sum_{p' \in P} z^r_{pp't} \right) + \sum_{r \in R^y} c^{ry} \left(\sum_{p' \in P} y^r_{pp't} \right) - IP_{np(t-1)} + IP_{npt} = d_{npt}, \qquad \forall n \in N, \forall p \in P, \forall t \in T.$$

$$(6.3)$$

Table 6.2 Set definition and input parameters for the PFPSP model

Set	Description	Parameter	Description
T	Set of periods	a_{pt}	Supply of product p at the beginning of period t
P	Set of products	d_{npt}	Demand of product p at a ship moored at berth n at the beginning of period t
S	Set of storage sub-areas	K	High-value constant
R^x	Set of routes (Reception/stockyard)	l^s_{pt}	Storage capacity of subarea s for product p in period t
R^x_s	Subset of routes x that reach subarea s	b^m	Capacity of equipment m (in ton/hour)
R^y	Set of routes (Reception/Piers)	j^m_t	Available time (in hours) for the use of equipment m in period t
R^z	Set of routes (Stockyard/Pier)	H_t	Total time in period t (in hours)
R^z_s	Subset of routes z from subarea s		
M	Set of equipment	c^{rx}	Capacity (in tons/hour) of route $r \in R^x$
R^x_m	Subset of routes x that use equipment m	c^{ry}	Capacity (in tons/hour) of route $r \in R^y$
R^y_m	Subset of routes y that use equipment m	c^{rz}	Capacity (in tons/hour) of route $r \in R^z$
R^z_m	Subset of routes z that use equipment m	β_{npt}	Penalty for not meeting the demand of a ship moored at berth n with product p in period t
R	Set of all available routes $R = R^x \cup R^y \cup R^z$	α_{pt}	Penalty for not meeting the supply at the reception of product p in period t
N	Set of available mooring berths	$\gamma^s_{pp't}$	Preparation cost associated with replacing product p by product p' in subarea s at period t
E	Pairs of routes that share at least one piece of equipment to transport products	$\lambda_{pp'}$	Cost associated with the loss of income by replacing product p by product p' to meet the demand of product p. When $p = p'$, $\lambda_{pp'} = 0$

$$e^s_{p(t+1)} = e^s_{pt} + \sum_{r \in R^x_s} c^{rx} x^r_{pt} - \sum_{r \in R^z_s} c^{rz} (\sum_{p' \in P} z^r_{p'pt}), \qquad \forall s \in S, \forall p \in P, \forall t \in T.$$

$$(6.4)$$

$$e^s_{pt} \leq l^s_{pt}, \qquad \forall s \in S, \forall p \in P, \forall t \in T. \qquad (6.5)$$

$$\sum_{p \in P} (\sum_{r \in R^x_m} c^{rx} x^r_{pt} + \sum_{r \in R^z_m} c^{rz} (\sum_{p' \in P} z^r_{pp't}) + \sum_{r \in R^y_m} c^{ry} (\sum_{p' \in P} y^r_{pp't})) \leq j^m_t b^m, \qquad \forall m \in M, \forall t \in T.$$

$$(6.6)$$

Table 6.3 Variable Definition

Variable	Description
$f_{pt}^s, \forall s \in S, \forall p \in P, \forall t \in T$	Has unit value when subarea s is allocated for product p in period t
$Sf_{pp't}^s, \forall s \in S, \forall p \in P, \forall p' \in (P \cup 0), p \neq p', \forall t \in T$	Has a value of 1 when product p has been replaced with product p' at period t. This replacement can occur only when the amount of product p in subarea s has been exhausted in the preceding period $t - 1$
$x_{pt}^r, \forall r \in R^x, \forall p \in P, \forall t \in T$	Time taken in period t to transport product p from the reception to the stockyard using route $r \in R^x$
$y_{pp't}^r$, $\forall r \in R^y, z_{pp't}^r \forall r \in R^z, \forall p, p' \in P, \forall t \in T$	Time taken to transport product p' to meet the demand of product p in period t using route r from sets R^y and R^z. When p' is equal to p, the product delivered is the same as was requested
$e_{pt}^s, \forall s \in S, \forall p \in P, \forall t \in T$	Amount of product p stored at subarea s in period t
$IR_{pt}, \forall p \in P, \forall t \in T$	Represents the amount of product p in the Reception subsystem that was not delivered at the end of period t
$IP_{npt}, \forall n \in N, \forall p \in P, \forall t \in T$	Represents the amount of product p that was not delivered at mooring berth n at the end of period t
$t_{pp't}^r, \forall r \in R, \forall p, p' \in P, \forall t \in T$	Start time for the processing operation using route $r \in R$ in period t. For each variable x_{pt}^r, $y_{pp't}^r$ and $z_{pp't}^r$, there is one start time ($t_{pp't}^r$). When $t_{pp't}^r$ is associated with x_{pt}^r, p and p' are the same product
$u_{pp't}^r, \forall r \in R, \forall p, p' \in P, \forall t \in T$	Binary variable. It has a value of 1 if the product p' used to meet the demand of product p uses the route r from set R in period t. For all $r \in R^x$, $p = p'$
$\theta_{pp'\hat{p}\hat{p}'t}^{rr'}, \forall p, p', \hat{p}, \hat{p}' \in P, \forall t \in T$	Binary variable. It has a value of 1 if the product p or p' (used to meet p) precedes the product $\hat{p}$ or $\hat{p}'$ (used to meet $\hat{p}$) in the conflicting routes r, r' $\in E$ in period t

$$\sum_{p \in (P \cup 0)} f_{pt}^s = 1, \qquad \forall s \in S, \forall t \in T. \tag{6.7}$$

$$Sf_{pp't}^s \geq f_{p,(t-1)}^s + f_{p't}^s - 1, \qquad \forall s \in S, \forall t \in T, \forall p \in P, \forall p' \in (P \cup 0), p \neq p'. \tag{6.8}$$

$$l_{pt}^s \qquad f_{pt}^s - e_{pt}^s \geq 0, \qquad \forall s \in S, \forall p \in P, \forall t \in T. \tag{6.9}$$

$$l_{pt}^s \qquad f_{pt}^s - \sum_{r \in R_s^x} c^{rx} x_{pt}^r \geq 0, \qquad \forall s \in S, \forall p \in P, \forall t \in T \qquad (6.10)$$

$$x_{pt}^r \leq H_t \quad u_{pp't}^r, \qquad \forall p, p' \in P, p = p', \forall t \in T, \forall r \in R^x \qquad (6.11)$$

$$y_{pp't}^r \leq H_t \quad u_{pp't}^r, \qquad \forall p, p' \in P, \forall t \in T, \forall r \in R^y \qquad (6.12)$$

$$z_{pp't}^r \leq H_t \quad u_{pp't}^r, \qquad \forall p, p' \in P, \forall t \in T, \forall r \in R^z \qquad (6.13)$$

$$H_t(1 - u_{pp't}^r) + H_t(1 - u_{\hat{p}\hat{p}'t}^{r'}) + H_t(1 - \theta_{pp'\hat{p}\hat{p}'t}^{rr'}) + t_{\hat{p}\hat{p}'t}^{r'} \geq t_{pp't}^r + x_{pt}^r \qquad (6.14)$$

$$H_t(1 - u_{pp't}^r) + H_t(1 - u_{\hat{p}\hat{p}'t}^{r'}) + H_t(\theta_{pp'\hat{p}\hat{p}'t}^{rr'}) + t_{pp't}^r \geq t_{\hat{p}\hat{p}'t}^{r'} + x_{\hat{p}t}^{r'} \qquad (6.15)$$

$$\forall (r, r') \in E, (r, r') \in R^x, \forall p = p' \in P, \forall \hat{p} = \hat{p}' \in P, \forall t \in T$$

$$H_t(1 - u_{pp't}^r) + H_t(1 - u_{\hat{p}\hat{p}'t}^{r'}) + H_t(1 - \theta_{pp'\hat{p}\hat{p}'t}^{rr'}) + t_{\hat{p}\hat{p}'t}^{r'} \geq t_{pp't}^r + x_{pt}^r \qquad (6.16)$$

$$H_t(1 - u_{pp't}^r) + H_t(1 - u_{\hat{p}\hat{p}'t}^{r'}) + H_t(\theta_{pp'\hat{p}\hat{p}'t}^{rr'}) + t_{pp't}^r \geq t_{\hat{p}\hat{p}'t}^{r'} + y_{\hat{p}\hat{p}'t}^{r'} \qquad (6.17)$$

$$\forall (r, r') \in E, r \in R^x, r' \in R^y, \forall p = p' \in P, \forall \hat{p}, \hat{p}' \in P, \forall t \in T$$

$$H_t(1 - u_{pp't}^r) + H_t(1 - u_{\hat{p}\hat{p}'t}^{r'}) + H_t(1 - \theta_{pp'\hat{p}\hat{p}'t}^{rr'}) + t_{\hat{p}\hat{p}'t}^{r'} \geq t_{pp't}^r + x_{pt}^r \qquad (6.18)$$

$$H_t(1 - u_{pp't}^r) + H_t(1 - u_{\hat{p}\hat{p}'t}^{r'}) + H_t(\theta_{pp'\hat{p}\hat{p}'t}^{rr'}) + t_{pp't}^r \geq t_{\hat{p}\hat{p}'t}^{r'} + z_{\hat{p}\hat{p}'t}^{r'} \qquad (6.19)$$

$$\forall (r, r') \in E, r \in R^x, r' \in R^z, \forall p = p' \in P, \forall \hat{p}, \hat{p}' \in P, \forall t \in T$$

$$H_t(1 - u_{pp't}^{r'}) + H_t(1 - u_{\hat{p}\hat{p}'t}^{r'}) + H_t(1 - \theta_{pp'\hat{p}\hat{p}'t}^{rr'}) + t_{\hat{p}\hat{p}'t}^{r'} \geq t_{pp't}^r + y_{pp't}^r \qquad (6.20)$$

$$H_t(1 - u_{pp't}^{r'}) + H_t(1 - u_{\hat{p}\hat{p}'t}^{r'}) + H_t(\theta_{pp'\hat{p}\hat{p}'t}^{rr'}) + t_{pp't}^r \geq t_{\hat{p}\hat{p}'t}^{r'} + y_{\hat{p}\hat{p}'t}^{r'} \qquad (6.21)$$

$$\forall (r, r') \in E, (r, r') \in R^y, \forall p, p', \hat{p}, \hat{p}' \in P, \forall t \in T$$

$$H_t(1 - u_{pp't}^{r'}) + H_t(1 - u_{\hat{p}\hat{p}'t}^{r'}) + H_t(1 - \theta_{pp'\hat{p}\hat{p}'t}^{rr'}) + t_{\hat{p}\hat{p}'t}^{r'} \geq t_{pp't}^r + z_{pp't}^r \qquad (6.22)$$

$$H_t(1 - u_{pp't}^{r'}) + H_t(1 - u_{\hat{p}\hat{p}'t}^{r'}) + H_t(\theta_{pp'\hat{p}\hat{p}'t}^{rr'}) + t_{pp't}^r \geq t_{\hat{p}\hat{p}'t}^{r'} + z_{\hat{p}\hat{p}'t}^{r'} \qquad (6.23)$$

$$\forall (r, r') \in E, (r, r') \in R^z, \forall p, p', \hat{p}, \hat{p}' \in P, \forall t \in T$$

$$H_t(1 - u_{pp't}^{r'}) + H_t(1 - u_{\hat{p}\hat{p}'t}^{r'}) + H_t(1 - \theta_{pp'\hat{p}\hat{p}'t}^{rr'}) + t_{\hat{p}\hat{p}'t}^{r'} \geq t_{pp't}^r + z_{pp't}^r \qquad (6.24)$$

$$H_t(1 - u_{pp't}^{r'}) + H_t(1 - u_{\hat{p}\hat{p}'t}^r) + H_t(\theta_{pp'\hat{p}\hat{p}'t}^{rr'}) + t_{pp't}^{r'} \geq t_{\hat{p}\hat{p}'t}^{r'} + y_{\hat{p}\hat{p}'t}^{r'} \qquad (6.25)$$

$$\forall (r, r') \in E, r \in R^z, r' \in R^y, \forall p, p', \hat{p}, \hat{p}' \in P, \forall t \in T$$

$$t^r_{pp't} + x^r_{pt} \leq \sum_{i=1}^{t} H_i \qquad (6.26)$$

$$t^r_{pp't} \geq \sum_{i=1}^{t-1} H_i \qquad (6.27)$$

$$\forall r \in R^x, \forall p = p' \in P, \forall t \in T$$

$$t^r_{pp't} + y^r_{pp't} \leq \sum_{i=1}^{t} H_i \qquad (6.28)$$

$$t^r_{pp't} \geq \sum_{i=1}^{t-1} H_i \qquad (6.29)$$

$$\forall r \in R^y, \forall p, p' \in P, \forall t \in T$$

$$t^r_{pp't} + z^r_{pp't} \leq \sum_{i=1}^{t} H_i \qquad (6.30)$$

$$t^r_{pp't} \geq \sum_{i=1}^{t-1} H_i \qquad (6.31)$$

$$\forall r \in R^z, \forall p, p' \in P, \forall t \in T$$

Constraints (17.2) formulate the meeting of the supply at the Reception subsystem. As previously stated, meeting the supply at this subsystem consists of unloading the trains, and therefore, these constraints guarantee that unmet supplies are updated. The unmet supplies IR_{p0} at period zero are an input data of the problem. Meeting the demand at the Pier subsystem, i.e., loading cargo onto the ships, is imposed by constraints (6.3). The unmet demands (IP_{np0}) at period zero are an input data of the problem.

Constraints (6.4) guarantee that stocks are kept up-to-date at each subarea. Constraints (6.5) define the storage capacity of each subarea. Constraints (6.6) ensure that no equipment will have its capacity exceeded. Constraints (6.6) allow two routes that share the same equipment to be used simultaneously as long as they do not exceed its capacity. Constraints (6.7) control the stockyard allocation. Once a subarea is allocated for a product, it cannot be used for any other product in the same period. When a subarea is empty, the product 0 is allocated to it, i.e., $f^s_{0t} = 1$. Constraints (6.8) control the replacement of products in a subarea. If $Sf^s_{pp't} = 1$, then product p has been replaced with p' at period t. The requirement that there can be only one product p stockpiled at a subarea s in period t if the stockyard allocation decision variable is valued 1 is enforced by constraints (6.9) and (6.10). If a route r at period

t is used ($u^r_{pp't} = 1$) to carry a product p or p', constraints (6.11)–(6.13) guarantee that its availability and capacity (measured in hours) are met.

Equations (6.14) and (6.15) define disjunctive constraints for each pair of conflicting routes $(r, r' \in E)$ and $(r, r') \in R^x$. They also establish the order of products p and $\hat{p}$ sharing equipment. If $\theta^{rr'}_{pp'\hat{p}\hat{p}t} = 1$, (6.15) is redundant, and (6.14) ensures that product p or (p') precedes $\hat{p}$ or $(\hat{p}')$ and that the start time of $\hat{p}$ is greater than the start time of p; if $\theta^{rr'}_{pp'\hat{p}\hat{p}t} = 0$, $\hat{p}$ precedes p. The same is true for all the other pairs of conflicting routes $(r, r' \in E)$ such that $r \in R^x$ and $r' \in R^y$ constraints (6.16), (6.17); $r \in R^x$ and $r' \in R^z$ constraints (6.18), (6.19); $r, r' \in R^y$ constraints (6.20), (6.21); $r, r' \in R^z$ constraints (6.22), (6.23), and $r \in R^z$ and $r' \in R^y$ constraints (6.24), (6.25). Constraints (6.26)–(6.31) ensure that a product cannot be introduced before or after the period in which it is established in the production plan.

6.5 Solution Approach

In the PFPSP formulation, the time required to transport the products is limited by the duration of one period, and the scheduling is guaranteed through constraints (6.14)–(6.31). These constraints impose great difficulty in solving the PFPSP in an integrated manner as their number is combinatorial and generates large memory consumption and processing. One alternative to overcome this difficulty, was to adopt a hierarchical approach where production planning and scheduling are solved separately. Figure 6.5 presents the solution strategy.

Two heuristic methods were used to solve the PFPSP: The first approach, based on a feedback procedure, published in [16] and a second up-down solution, which will be described below.

In the up-down approach, production planning and scheduling are solved period by period. In the production phase, a relaxed version of the PFPSP is solved through a commercial solver. In this version, scheduling and integrality constraints are relaxed. In the remainder of this article, this relaxed problem will be called relaxed PFPSP. The production variables (x^r_{pt}, $y^r_{pp't}$ and $z^r_{pp't}$) are sent to the scheduling phase to select and schedule the routes for a given period t. The scheduling phase defines the start time and the end time for each task, considering the sharing of equipment among the selected routes.

If a feasible schedule is not found, we transfer activities for the next period (backlog). This transfer is carried out as follows: If only part of the task is unfeasible in the period t, a new task is created in $t + 1$, considering its length only part unprocessed in t. If the task completely violated the period t, the task is completely sent to the period $t + 1$. In case of finding a feasible schedule, we just move towards the next period. The Fig. 6.6 illustrates this procedure: Task 6 has been completed over the time period. Therefore, a part of your processing time will be transferred to the following period and scheduled with other tasks.

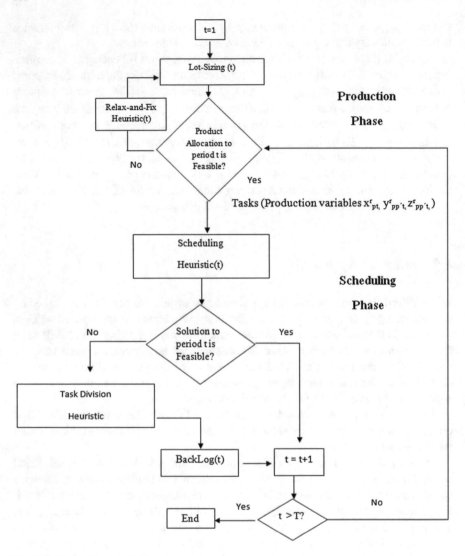

Fig. 6.5 Hierarchical strategy

6.5.1 Production Phase

At this phase, the method solves a capacitated lot sizing. The problem allocates products into sub-areas, with the aim of selecting the best sub-area for each product. While the lot sizing is easily solved by a commercial solver, the allocation of products considers several integer variables, making the problem harder to be solved. Therefore, we use a relax-and-fix strategy to efficiently solve the allocation problem.

Fig. 6.6 Task transfer
between periods

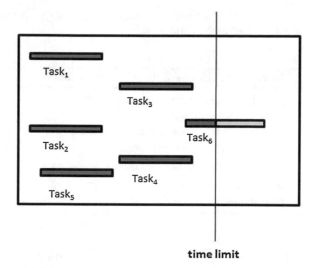

time limit

The relax-and-fix heuristic works by fixing decision variables in a sequence until reaching a feasible solution. In our case, we first deal with the number of products allocated in each sub-area, variable f^s_{pt}, selecting the variable with more allocated products. The relaxed PFPSP is solved again, and the process is repeated until reaching a feasible set of variables f^s_{pt}.

6.5.2 Scheduling Phase

The PFPSP production variables define the product type, the quantity and the route used to transport the products. The scheduling problem consists of establishing the start and end times for these tasks, considering incompatibility restrictions. Preemption is not allowed and the objective is to minimize the makespan. Hereafter, this scheduling problem is called the scheduling problem with incompatibility jobs (SPIJ). The SPIJ is not a new problem, many solutions have been proposed in the literature, such as scheduling with incompatible jobs [6], scheduling jobs using an agreement graph [3], and multi-coloring and job-scheduling [5].

6.5.3 SPIJ Heuristic

To efficiently find good solutions, a greedy randomized search procedure (GRASP) was implemented. GRASP is an iterative algorithm first proposed to [9], that basically consists of two phases: greedy construction and local search. The greedy construction phase builds a feasible solution s, whereas the local search investigates the neighborhood of s, searching for better solutions. The main phases of the heuristic are described in the following.

Algorithm 5: GRASP-Scheduling procedure

---

jobs = Production variables (relaxed PFPSP)

for i = 1 : MaxIteration **do**
  Solution = GreedyRandomizedConstruction(Seed)
  Solution = LocalSearch(Solution)
  Solution = UpdateSolution(Solution, BestSolution)
end
BestSolution

Algorithm 6: GreedyRandomizedConstruction(Seed) procedure

Solution = $\emptyset$

Evaluate the incremental costs of the candidate jobs

while Solution is not a complete solution **do**
  Build the restricted candidate list (RCL)
  Select an element i from the RCL at random
  $Solution = Solution \cup i$
  Reevaluate the incremental costs
end
Return Solution

At each iteration of procedure *GreedyRandomizedConstruction*, the algorithm considers the jobs extracted from planning not yet scheduled, as the list of candidate elements. A greedy solution for the SPIJ is constructed as follows: Select randomly a job i from list of candidate elements at random. Next, define the lowest start time for the job, keeping already scheduled jobs that conflict with i without overlapping. Once all jobs are scheduled, a feasible solution for the SPIJ is provided. The local search consists in changing the order of jobs found in the greedy construction phase.

6.6 Computational Experiments

The experiments are performed based on an real product flow problem in an iron ore port terminal in Brazil, recognized as one of the largest worldwide. The basic parameters are the number of periods, the products and the routes. In general, they work with seven periods of one day or fourteen periods of twelve hours. Table 6.4

Table 6.4 Data used to generate the instances

Parameter	Description
Stockyard	The product storage area is divided into four stockyards
Delivery	Two berths: two ships can be loaded simultaneously at berth 1
Equipment	Five car dumpers, four ore reclaimers, three stackers/reclaimers (equipment that performs both tasks), and eight stackers
α_{pt}	2 (two monetary units)
β_{npt}	10 (ten monetary units for the berth number one), 50 (fifty monetary units for the berth number two)
$\gamma^s_{p,p',t}$	10 (ten monetary units)
$\lambda_{pp'}$	Based on the following formula: 0.01 (monetary unit) $*|p - p'|$, where $|p - p'|$ represents the quality deviation between the product p and p'
σ^r	Based on the following formula: 0.01 (monetary unit) $*$ length of route r

highlights the main parameters used to create the instances. The parameters α_{pt}, β_{npt}, $\gamma^s_{p,p',t}$, $\lambda_{pp'}$ and σ^r are part of the PFPSP formulation.

In the iron ore port terminal, the demand nodes are those where the ship moors to receive the products. For the experiments, three demand nodes are considered: two moored ships in berth 1 and one ship in berth 2. Likewise, the supply nodes represent the points where the wagons unload the products. These points are related to the car dumpers. For the experiments, five car dumpers were considered. In this system, various products and quantities can arrive at different periods.

The priority in a port terminal is to meet the demand, so the penalty of not meeting it β_{npt}, is usually higher than the non-fulfillment of the supply (α_{pt} parameter). Even among the berths, the penalty is differentiated (β_{npt} value in Table 6.4). In our particular case, berth 2 meets larger ships so the priority is always to meet the ships of this berth. In the experiments, the cost of exchange products in the stockyard $\gamma^s_{pp't}$ is the same for any pair.

6.6.1 Instances

In this experiment, instances are created considering the following features: initial empty stock and initial stock with 30% of capacity, products supplied equal to those demanded and instances where the supplied and demanded products are different (when the product type is switched to meet the demand) and the cost of the exchange is calculated based on the parameter $\lambda_{pp'}$ from Table 6.4.

The first, second and third columns of the Tables, contain the number of the instance, the type and the name. For example, instance $8P5Prod$ corresponds to a planning horizon of eight periods and five different products being handled. Column Z_{LB} provides the lower limit for the PFPSP obtained by its linear relax-

Table 6.5 110 routes-capacity ranging from 8,000 to 12,000 t/h-supply equal demand

Number	Type	Instance	CPLEX							Heuristic approach	
			Z_{LB}	t_{LB}	Z_{UB}	t_{UB}	Z_{BB}	Z_{UB_1}	T_{UB_1}		
1		4P5Prod	138.16	5	142.14	205	142.13	143.6	12		
2		4P10Prod	140.4	34	143.25	785	143.25	143.6	13		
3	Empty stock	8P5Prod	276.32	10	281.62	683	281.62	287.2	36		
4	Products	8P10Prod	279.7	1065	–	–	–	287.2	42		
5	Equal	10P5Prod	345.4	12	351.96	1159	351.931	359	58		
6		10P10Prod	348.67	2089	–	–	–	359	63		
7		4P5Prod	838.16	5	842.14	168	842.14	843.6	12		
8		4P10Prod	2688.16	5674	–	–	–	2693.6	17		
9	Empty stock	8P5Prod	1476.32	13	1481.67	1202	1481.56	1487.2	38		
10	Different	8P10Prod	6226.32	9871	–	–	–	6237.2	60		
11	Products	10P5Prod	1945.4	33	1952.96	1977	1952.78	1959	60		
12		10P10Prod	7345.4	2381	–	–	–	7359	83		
13		4P5Prod	197.35	5.00	200.74	289.00	200.74	237.99	22		
14		4P10Prod	204.22	1236.00	206.64	11.003.00	206.64	237.48	25		
15	Stock in 30%	8P5Prod	–	–	–	–	–	286.76	43		
16	Different	8P10Prod	–	–	–	–	–	300.30	63		
17	Products	10P5Prod	–	–	–	–	–	376.2	61		
18		10P10Prod	–	–	–	–	–	370.4	65		

Table 6.6 110 routes - capacity ranging from 8,000 to 12,000 t/h - supply greater than demand

Number	Type	Instance	CPLEX						Heuristic approach	
			Z_{LB}	t_{LB}	Z_{UB}	t_{UB}	Z_{BB}		Z_{UB_1}	T_{UB_1}
19		4P5Prod	148.3	4	–	–	–		163.39	51
20		4P10Prod	150.4	34	–	–	–		154.53	36
21	Empty stock	8P5Prod	297.50	9	–	–	–		316.75	116
22	Products	8P10Prod	305.46	879	–	–	–		323.45	91
23	Equal	10P5Prod	372.82	12	–	–	–		391.65	170
24		10P10Prod	380.9	1683	–	–	–		419.44	118
25		4P5Prod	651.66	5	–	–	–		675.40	78
26		4P10Prod	1876.34	854	–	–	–		2063.2	57
27	Empty stock	8P5Prod	710.14	10	–	–	–		773.48	168
28	Different	8P10Prod	2541.04	9422	–	–	–		2635.61	163
29	Products	10P5Prod	860.48	12	–	–	–		882.46	200
30		10P10Prod	–	–	–	–	–		3121.53	208
31		4P5Prod	–	–	–	–	–		151.47	49
32		4P10Prod	–	–	–	–	–		152.68	39
33	Stock in 30%	8P5Prod	–	–	–	–	–		335.83	91
34	Different	8P10Prod	–	–	–	–	–		310.55	89
35	Products	10P5Prod	–	–	–	–	–		398.93	134
36		10P10Prod	–	–	–	–	–		388.99	133

ation. Columns Z_{UB} and Z_{BB} provide the upper and lower bounds (best bounds) obtained with the branch-and-cut algorithm of the CPLEX solver. Regarding the hierarchical procedure, column Z_{UB_1} provide the upper bound obtained for the hierarchical solution. Finally, t_{LB}, t_{UB} and t_{UB1} are the elapsed computational time to obtain the values of Z_{LB}, Z_{UB}, Z_{UB_1} respectively, expressed in seconds.

The results shown in Table 6.5 indicate that solving the PFPSP in optimization packages is not feasible. The solver was able to produce solutions only for half (nine) of them. In the rest, because insufficient memory, it was not possible to obtain even an upper bound. With the heuristic, it is possible to obtain solutions for all instances, all supplies and demands were met, and all tasks are scheduled respecting the duration of each period. In cases where the scheduling was unfeasible, the tasks were transferred to the next period (backlog) and scheduled.

In the experiments described in Table 6.6, it was necessary to manipulate more products and routes (as excess supply must remain at the stockyard); therefore, the optimization package was not able to find a feasible solution for any case. Similar to the results found in Table 6.5, with the hierarchical approach (Table 6.6), it is possible to obtain solutions for all instances,

6.7 Conclusions

In this chapter, we investigated: optimization problems related to the storage yards and an integrated problem of planning and scheduling. It was presented a mathematical model and a heuristic capable of solving it. The computational experiments showed that the method is more efficient in producing a feasible solution than the solver. Furthermore, it is possible to solve medium and large instances, that with optimization packages is computationally unfeasible. The solution approach, can be used to represent various problems related to the flow of products in bulk cargo (iron ore, coal and grains). Future works includes adaptation of these methods, as well as new methods. Efforts are also concentrated in to consider uncertainty surrounding supply, demand and equipment failure.

Acknowledgements This research is supported by the following institutions: Instituto Tecnolgico Vale (ITV), Fundação de Amparo à Pesquisa do Estado de Minas Gerais (FAPEMIG) and Conselho Nacional de Desenvolvimento Científico e Tecnológico (CNPq).

References

1. Angelelli, E., Kalinowski, T., Kapoor, R., & Savelsbergh, M. W. (2016). A reclaimer scheduling problem arising in coal stockyard management. *Journal of Scheduling*, *19*(5), 563–582.
2. Barros, V. H., Costa, T. S., Oliveira, A. C. M., & Lorena, L. A. N. (2011). Model and heuristic for berth allocation in tidal bulk ports with stock level constraints. *Computers & Industrial Engineering, 60*, 606–613.

3. Bendraouche, M., & Boudhar, M. (2012). Scheduling jobs on identical machines with agreement graph. *Computers & Operations Research, 39*(2), 382–390.
4. Bilgen, B., & Ozkarahan, I. (2007). A mixed-integer linear programming model for bulk grain blending and shipping. *International Journal of Production Economics, 107*(2), 555–571.
5. Blchliger, I., & Zufferey, N. (2013). Multi-coloring and job-scheduling with assignment and incompatibility costs. *Annals of Operations Research, 211*(1), 83–101.
6. Bodlaender, H. L., Jansen, K., & Woeginger, G. J. (1994). Scheduling with incompatible jobs. *Discrete Applied Mathematics, 55*(3), 219–232.
7. Boland, N., Gulezynski, D., & Savelsbergh, M. (2012). A stockyard planning problem. *EURO Journal on Transportation and Logistics, 1*(3), 197–236.
8. Conradie, D., Morison, L. E., & Joubert, J. W. (2008). Scheduling at coal handling facilities using simulated annealing. *Mathematical Methods of Operations Research, 68*(2), 277–293.
9. Feo, T. A., & Resende, M. G. (1989). A probabilistic heuristic for a computationally difficult set covering problem. *Operations Research Letters, 8*(2), 67–71.
10. Goodfellow, R., & Dimitrakopoulos, R. (2017). Simultaneous stochastic optimization of mining complexes and mineral value chains. *Mathematical Geosciences, 49*(3), 341–360.
11. Hu, D., & Yao, Z. (2010). Stacker-reclaimer scheduling for raw material yard operation. In *Third International Workshop on Advanced Computational Intelligence*, pp. 432–436.
12. Kaveshgar, N., & Huynh, N. (2015). Integrated quay crane and yard truck scheduling for unloading inbound containers. *International Journal of Production Economics, 159*, 168–177.
13. Lee, D. H., Cao, J. X., Shi, Q., & Chen, J. H. (2009). A heuristic algorithm for yard truck scheduling and storage allocation problems. *Transportation Research Part E: Logistics and Transportation Review, 45*(5), 810–820.
14. Matamoros, M. E. V., & Dimitrakopoulos, R. (2016). Stochastic short-term mine production schedule accounting for fleet allocation, operational considerations and blending restrictions. *European Journal of Operational Research, 255*(3), 911–921.
15. Menezes, G. C., Mateus, G. R., & Ravetti, M. G. (2015). Scheduling with incompatible jobs: model and algorithms. In *Proceedings of the 7th Multidisciplinary International Conference on Scheduling : Theory and Applications (MISTA)*, 25–28 Aug, Prague, Czech Republic, (pp. 776–781).
16. Menezes, G. C., Mateus, G. R., & Ravetti, M. G. (2016). A hierarchical approach to solve a production planning and scheduling problem in bulk cargo terminal. *Computers & Industrial Engineering, 97*, 1–14.
17. Menezes, G. C., Mateus, G. R., & Ravetti, M. G. (2017). A branch and price algorithm to solve the integrated production planning and scheduling in bulk ports. *European Journal of Operational Research, 258*(3), 926–937.
18. Montiel, L., & Dimitrakopoulos, R. (2015). Optimizing mining complexes with multiple processing and transportation alternatives: An uncertainty-based approach. *European Journal of Operational Research, 247*(1), 166–178.
19. Newman, A. M., Rubio, R., Caro, R., Weintraub, A., & Eurek, K. (2010). A review of operations research in mine planning. *Interfaces, 40*(3), 222–245.
20. Niu, B., Xie, T., Tan, L., Bi, Y., & Wang, Z. (2016). Swarm intelligence algorithms for yard truck scheduling and storage allocation problems. Neurocomputing *188*, 284–293 (2016): In *Advanced Intelligent Computing Methodologies and Applications Selected papers from the Tenth International Conference on Intelligent Computing (ICIC 2014)*.
21. Pimentel, B. S., Mateus, G. R., & Almeida, F. A. (2013). Stochastic capacity planning and dynamic network design. *International Journal of Production Economics, 145*(1), 139–149.
22. Ribeiro, G. M., Mauri, G. R., de Castro Beluco, S., & Antonio, L. (2016). Berth allocation in an ore terminal with demurrage, despatch and maintenance. *Computers & Industrial Engineering, 96*, 8–15.
23. Robenek, T., Umang, N., & Bierlaire, M. (2014). A branch-and-price algorithm to solve the integrated berth allocation and yard assignment problem in bulk ports. *European Journal of Operational Research, 235*(2), 399–411.

24. Savelsbergh, M. (2015). Cargo assembly planning. *EURO Journal on Transportation and Logistics*, *4*(3), 321–354.
25. Singh, G., Sier, D., Ernst, A. T., Gavriliouk, O., Oyston, R., Giles, T., et al. (2012). A mixed integer programming model for long term capacity expansion planning: A case study from the hunter valley coal chain. *European Journal of Operational Research*, *220*(1), 210–224.
26. Vacca, I., Salani, M., & Bierlaire, M. (2013). An exact algorithm for the integrated planning of berth allocation and quay crane assignment. *Transportation Science*, *47*(2), 148–161.

Chapter 7
Multi-capacitated Location Problem: A New Resolution Method Combining Exact and Heuristic Approaches Based on Set Partitioning

Mohammed El Amrani, Youssef Benadada and Bernard Gendron

Abstract In this paper, we present one generalization of the famous capacitated p-median location problem, called *budget constraint multi-capacitated location problem* (MCLP). This new generalization is characterized by allowing each facility to be used with different capacity levels. The MCLP solution can be represented as a set of disjoint clusters (pair of one facility and a subset of customers). Creating these clusters satisfies implicitly some constraints of the problem. In this work, we present the new formulation of the MCLP based on set partitioning, then we suggest an adapted solving method, which will be called NFF (Nearest Facility First). This new method can be used in two ways: as a heuristic by taking only the first solution found or exact approach when waiting finished the execution. Computational results are presented at the end using instances that we have created under some criteria of difficulties or adapted from those of p-median problems available in literature. The NFF method provides very good results for low and medium difficulty instances, but it is less effective for the more complex ones. To remedy this problem, the method will be supplemented by column generation approach.

Keywords Location · p-median · Set partitioning · Heuristic · Column generation

M. El Amrani (✉) · Y. Benadada
ENSIAS, Mohammed V University in Rabat, Madinat Al Irfane, BP 713,
Agdal Rabat, Morocco
e-mail: melamrani86@gmail.com

Y. Benadada
e-mail: youssef.benadada@um5.ac.ma

B. Gendron
CIRRELT Universite de Montreal, CP 6128, Succursale Centre-ville Montreal,
Montreal, Quebec H3C 3J7, Canada
e-mail: bernard.gendron@cirrelt.ca

© Springer International Publishing AG, part of Springer Nature 2019
E.-G. Talbi and A. Nakib (eds.), *Bioinspired Heuristics for Optimization*,
Studies in Computational Intelligence 774,
https://doi.org/10.1007/978-3-319-95104-1_7

7.1   Introduction

Locating facilities is one of the main problems when it comes to making strategic or tactical decisions. The objective of this kind of problems is usually to minimize a cost function that can include the cost for the assignment as well as the opening cost of facilities. In many location variants, the facilities opening cost is initially fixed by a budget constraint; in such case, only the assignment cost is minimized [1].

Facility location has been the subject of a large number of publications in the fields of supply chain optimization and operational research. The p-median is the most famous location problem that we can find hugely in literature. The Capacitated P-Median location Problem CPMP is a variant of this well-known problem subject to the capacity.

Unlike the p-median problem, widely discussed in the literature, the MCLP is a new problem that we did not find any existing study, only in [2]. MCLP problem is an NP-complete problem because it represents a generalization of CPMP [3]. Variants of the latter appeared in [4–8], Set Partitioning formulation is applied by [7], dynamic location problems in [9–11] and network problems in [12, 13]. To solve the CPMP several approaches have been proposed: [6] used a cutting planes algorithm based on Fenchel cuts [4, 14, 15] proposed the application of Branch and Price and branch and Bound methods based on Lagrangian relaxation, [5] proposed a resolution with column generation and [16–18] used other methods.

This paper is organized as follows: After the introduction, we present the generalization of the capacitated p-median location problem. After that, we discuss in section three the new formulation of the multi-capacitated location problem based on set partitioning. The fourth section is devoted to the solving methods, namely the Branch and cut used by CPLEX and a new suitable method, called NFF. Computational results are presented in the fifth section before the conclusion.

7.2   Problem Context and Generalization of CPMP

In this section, we underline the capacitated p-median location problem and its mathematical formulation as well as the generalization proposed to obtain the new multi-capacitated location problem.

7.2.1   Capacitated p-Median Location Problem

Given a bipartite graph G(V, U, E) where V and U are respectively the sets of customers and facilities nodes and E is the set of edges. V and U are each independent set such that every edge connects a vertex in U to one in V (Fig. 7.1).

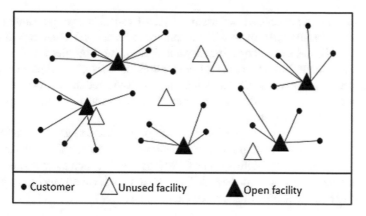

Fig. 7.1 Graphical representation of location problem

The location problem graph is composed of many connected sub-graphs; each one is also made up of either one facility solely (white triangle) or a facility with a partition of customers (black triangle, points).

Let $E = \{(i, j) : i \in V, j \in U\}$ and $c_{i,j}$ the assignment cost of the customer i to the facility j, the customer demand d_i and the facility capacity u_j are associated respectively with node $i \in V$ and node $j \in U$.

Let $x_{i,j}$ be the binary variable associated with the edge $(i, j) \in E$ ($x_{i,j} = 1$ if customer i is assigned to facility j, 0 otherwise) and let y_j be the binary variable associated with the median node $j \in E$ ($y_j = 1$ if facility j is used, 0 otherwise). The integer linear mathematical formulation is:

$$Min \quad \sum_{i \in N} \sum_{j \in M} d_i c_{i,j} x_{ij} \tag{7.1}$$

$$\sum_{j \in M} x_{ij} = 1, \quad i \in N \tag{7.2}$$

$$\sum_{i \in N} d_i x_{ij} \leq u_j y_j, \quad j \in M \tag{7.3}$$

$$x_{ij} \leq y_j, \quad i \in N, j \in M \tag{7.4}$$

$$\sum_{j \in M} y_j = p \tag{7.5}$$

$$x_{ij} \in \{0, 1\}, \quad i \in N, j \in M \tag{7.6}$$

$$y_j \in \{0, 1\}, \quad j \in M \tag{7.7}$$

The objective (1) is only to minimize the assignment costs, which may occur as transport costs by a unit of distance. In this variant, we are not looking to minimize the opening costs of the medians. The demand constraints (2) force each customer to be assigned to one and only one median. Constraints (3) impose that the capacity of each median must not be exceeded, we assume that all facilities have the same capacity u. Constraints (4) are used to restrict customers to be assigned to a closed facility. It can be noted that in the capacitated p-median location problem, this constraint becomes redundant with the capacity constraints; however, the experience shows that this constraint represents a valid high inequality to reduce the execution time. Constraint (5) is to specify that the number of medians must be equal to p. (6) and (7) are the integrality constraints.

The search for p-median nodes in a network or graph is a classical location problem. The purpose is to locate p facilities (medians) in order to minimize the assignment cost of customers to facilities. The CPMP defines for each median one capacity that characterizes the service provided. The total demands for all customers assigned to one facility cannot exceed its service capacity.

7.2.2 Generalization of the Capacitated p-Median Location Problem

In field of industry, service costs increase with the capacity used. The application of a single CPMP problem is wasteful in terms of resources, the capacity the presented service can considerably exceed the customers' requirements. In order to generalize the CPMP for more complex situations faced in industry, [2] propose a new model called MCLP (budget constraint Multi-Capacitated Location Problem), which defines several capacity levels for each facility. Thus, a facility can be open at only one capacity level, the total demands of the assigned customers define the capacity level to use and each opened level has a corresponding cost.

Several applications in the industry use the MCLP concept such as telecommunications, energy management, and many others. This can explain the fact that this is one of the most important known problems having impact on the strategic decisions.

The aim of this problem is to optimize the related transport cost of assigning customers to facilities. Each customer has a fixed demand served by a single open facility. The facilities can be used in one of many pre-set levels of capacity; no one can be open for more than one level at the same time. By assigning customers to facilities, we have to check that the total demands of customers served by each facility is less than its level capacity used. The sum of facility opening costs is bounded by a limit budget.

In this study of proposing new formulation of MCLP based on set partitioning, we will test the new model by using resolution methods, namely the Branch and Cut used by CPLEX solver and a new resolution method called NFF.

The NFF proposed to solve MCLP consists on assigning each customer to the nearest possible facility, to do that we will sort all facilities by the assignment cost. At the first iteration, all customers are assigned to the nearest facility. Then we compare the sum of the demands assigned to each facility by its capacity level used. If this iteration provides one solution, so the solution found is optimal, otherwise we allow customers to be assigned to the next nearest facilities until arriving at a feasible solution. The NFF algorithm can be used in two ways, when generating clusters, the formulation applied starts to provide some very good solutions (according to the NFF philosophy). If we intend to look for the optimal solution we let the program finish the execution, otherwise some earlier solutions can be enough as a heuristic approach in very reasonable execution time.

7.3   Formulation

The MCLP problem is a location problem with capacity where facilities can be used at several levels, each level is characterized by a certain capacity to respect and the facility can open only one level at once. In the mathematical formulation, we will need to create additional variables and notations and modify the constraints of the previous problem such that each facility must respect the maximum capacity of the selected level. The mathematical formulation is as follows:

$$Min \quad \sum_{i \in N} \sum_{j \in M} d_i c_{i,j} x_{ij} \tag{7.8}$$

$$\sum_{j \in M} x_{ij} = 1, \quad i \in N \tag{7.9}$$

$$\sum_{i \in N} d_i x_{ij} \leq \sum_{k \in K} u^k y_j^k, \quad j \in M \tag{7.10}$$

$$\sum_{k \in K} y_j^k \leq 1, \quad j \in M \tag{7.11}$$

$$x_{ij} \leq \sum_{k \in K} y_j^k, \quad i \in N, j \in M \tag{7.12}$$

$$\sum_{j \in M} \sum_{k \in K} f^k y_j^k \leq B \tag{7.13}$$

$$x_{ij} \in \{0, 1\}, \quad i \in N, j \in M \tag{7.14}$$

$$y_j^k \in \{0, 1\}, \quad i \in N, k \in K \tag{7.15}$$

Where K denotes the set of levels, u^k represent capacity of level k, f^k is the opening cost associated with level k, B is the limit budget on the total opening costs. y_j^k is decision binary variable that is one if and only if the facility j is opened and used at the level k.

This model generalizes the capacitated p-median problem including the capacity levels concept. This new data appears in the capacity constraints (10) and at the budget constraint (13), because the facilities have different opening costs. (12) are also additional constraints which represent valid inequalities that cut the feasible

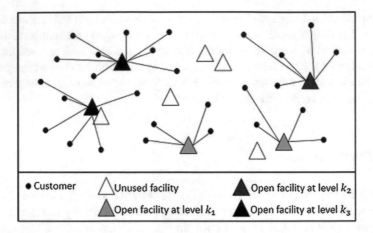

Fig. 7.2 Graphical representation of MCLP problem

region. (10) are capacity constraints and (11) force the facility to be open at one level at most. (14) and (15) are the integrality constraints.

In this section, we choose to re-formulate our MCLP as a set partitioning problem. Given a bipartite graph $G(V, U, E)$ where V and U are respectively the sets of customers and facilities nodes and E is the set of edges. V and U are independent sets such that every edge connects a vertex in U to one in V (Fig. 7.2).

The MCLP problem graph is composed of many connected sub-graphs, the facility (triangle) can be operated in many levels: white triangle for closed facility, light gray one for the level k_1, dark gray for the level k_2 or black for a level with the biggest capacity (k_3). The capacity level is chosen according to customers' demands. We can easily see that the final solution will be in form of a set of pair (facility, customers' subset), this pair will be called "cluster". All of these clusters will build the feasible region of this new model. This new formulation bellow will be called MCLP based on set-partitioning.

$$Min \quad \sum_{j=1}^{m}\sum_{p=1}^{l} c_p^j x_p^j \tag{7.16}$$

$$\sum_{j=1}^{m}\sum_{p=1}^{l} a_{ip} x_p^j = 1, \quad i \in N \tag{7.17}$$

$$\sum_{p=1}^{l} b_{jp} x_p^j \leq 1, \quad j \in M \tag{7.18}$$

$$\sum_{j=1}^{m}\sum_{p=1}^{l} f_p x_p^j \leq B \tag{7.19}$$

$$x_p^j \in \{0, 1\}, \qquad p \in \{1, \ldots, l\}, \ j \in M \tag{7.20}$$

$S = \{S_1, S_2, , S_l\}$ indicates the set of subset of N, and $S \subseteq P(N)$.
c_p^j is the total assignment cost of the customer subset S_p to the facility j.
$c_p^j = \sum_{i \in S_p} d_i c_{ij}$

$$A = [a_{ip}]_{n \times l} \quad with \quad a_{ip} = \begin{cases} 1 & if \ \ i \in S_p \\ 0 & otherwise \end{cases} \quad satisfying \quad \sum_{i \in N} d_i a_{ip} \leq max_k\{u^k\}$$

$$B = [b_{jp}]_{m \times l} \quad with \quad b_{jp} = \begin{cases} 1 & if \ \ j \ \ is \ assigned \ to \ \ S_p \\ & 0 \ \ otherwise \end{cases}$$

$f_p = Min_k\{f^k \sum_{i \in N} d_i a_{ip} \leq u^k\}$

x_p^j is a decision binary variable that is equal to one if and only if cluster build by facility j and customer subset S_p belongs to the solution.

It may be noted that in this formulation, the constraints (8), (9), (13) are respectively equivalent to (16), (17), (19) and the constraints (10), (11), (12) appeared explicitly in the initial formulation become implicitly considered. Constraints (18), which is not declared in the first model explicitly, prevents us to have, in the feasible solution, the same facility in more than one cluster.

If S is the set of all subsets of N, the formulation can give an optimal solution to the MCLP. However, the number of subsets may be very huge, and an exact resolution with a solver such as CPLEX becomes impossible in a reasonable time. Therefore, we must think of a way to solve this problem with a reduced computational time. To circumvent this situation, we propose in the next section a method to prohibit the assignment of customers to facilities supposed too far, limiting the number of produced clusters.

7.4 Solving Methods

To proceed with solving the problem, we will suggest to test the new formulation with an exact method well known, that of Branch and Cut, although this method presents difficulties for large instances, it will serve to validate the formulation. The new adapted method for the present formulation will be detailed afterwards.

7.4.1 Branch and Cut

The Branch and cut is a combination of two algorithms into one, namely the Branch and Bound and cutting plans.

Algorithm:

1. At each node of the resolution tree, a linear relaxation is solved by cutting plane approach.
2. If the solution is feasible (or if it is above a known solution), in case of minimization, the node is declared pruned.
3. If all nodes are pruned, STOP, the best solution found is optimal.
4. Otherwise, choose a non-pruned node, select a fractional variable x_i, and consider two subproblems by fixing x_i at 1 and x_i at 0 (Branching phase).
5. Solving every problem by generating new unsatisfied constraints (cutting phase). Go to step 3.

Although the Branch and Bound is a reliable method, it may be too slow in some cases, while the cutting methods are much faster, but they do not ensure the reliability of the solution. The Branch and Cut is a hybridization of both, it has the advantage of being both reliable and fast

7.4.2 Nearest Facility First

In this section, we present the NFF resolution method, which consists of banning the creation of clusters with a too high total assignment cost.

To solve this problem, we start by generating clusters; Firstly, we classify, for each customer, the facilities in ascending order of assignment costs. Then we assign to each customer the least expensive facility. After this initial assignment, and for all iterations that follow, we compare for each facility the sum of customer demands assigned with the different capacity levels. If the total of these demands exceeds the highest level, then the cluster is rejected by unfeasibility. Otherwise, we create the clusters with the lowest level satisfying all demands. Thus, the opening cost of the cluster is that of the selected level, and the assignment cost of the cluster is the sum of all assignment.

The solution found in this step, if it exists, can only be optimal, because it presents an ideal case, where each customer is assigned to the nearest facility, which means the least costly. Otherwise, we regenerate the clusters by adding the second closest facility for each customer in terms of assignment cost, and so on until achieve a feasible solution or arrive at a number of clusters considered too high for reasonable computational time. If necessary, we use the column generation approach.

To explain more the logic of this method, we detail its steps in the following algorithm:

Algorithm:

Generate Clusters

- If the set CL (clusters set) does not exceed a maximum number (5000 clusters).

  1. Search all assignments already made or clusters already created, each facility with its customers associated.
     Let's n the number of customers, for each pair (facility, part of customers).
  2. If the sum of all demands exceeds the maximum capacity.
     a. The cluster with n customers is rejected by unfeasibility.
     b. If $n > 0$ we create n clusters each one with $n - 1$ customers. Otherwise $STOP$.
     c. Come back to 2.
  3. Otherwise
     $i = 0$;
     a. If the sum of demands does not exceed the capacity i, we add this cluster in the set L with the capacity i and the corresponding opening cost and total assignment costs.
     b. Otherwise $i \leftarrow i + 1$, come back to **a**.
     c. Come back to 2.

- Otherwise we move to a resolution with columns Generation.

EndGenerate Clusters

1. Initially there are a set N of customers, a set M of facilities and an empty set CL.
2. For each customer, sort the facilities in ascending order in terms of assignment.
3. Assign to each customer the nearest facility.
   Repeat $|N| * (|M| - 1)$ times
4. Generate the clusters.
5. If solution is found, it can only be optimal, $STOP$. (the algorithm stops by optimality).
6. Otherwise,
   add the best assignment (customers, facility) and come back to 4.
   EndRepeat
   (if no solution found)
   Repeat $|CL|$ times, and let $p = 1$
7. Develop the cluster p with all possible combinations.
8. If one solution is found, $STOP$. Otherwise, $p \leftarrow p + 1$, come back to 6.

This algorithm always finishes by finding the solution, if such a solution exists (Fig. 7.3). The following diagram describe the sequence of the algorithm:

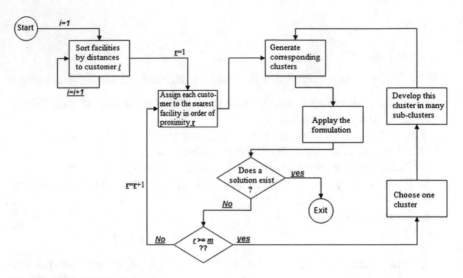

Fig. 7.3 NFF Algorithm Diagram

7.5 Computational Results

We did not find any reference in the literature on MCLP. This model therefore has, as we know, no existing instances for testing and comparison. For this, we will create, in the context of this work, instances using semi-random values based on a justified and appropriate choice. We will also use some p-median instances to supplement the computational tests, since the MCLP is a generalization of the latter.

We turn the two algorithms using the CPLEX solver. We use the version 7 of Java and version 12 of Cplex and we run the program on a machine i7-2600 CPU @ 3.40 GHz.

Our data set consists of five instance classes which have five levels of difficulty (easy, medium, difficult, very difficult and complex). The difficulty of these instances is based on the size of the problem, which is often measured by the number of customers. In contrast, the number of facilities and the number of levels used have a low impact on the size of the problem. Each difficulty level contains several instances. These instances also represent difficulty sub-levels. They are characterized by the dispersion of the points (customers compared to plants) and also by the amount of available resources. Experience shows that the difficulty of the problem varies proportionally to the variance of customer-facility distances and customer demands. At the same time, it varies inversely with the budget allocated to the opening of facilities, and to their capacity levels. Thus, increasing the difficulty, while keeping the feasibility, we multiply the number of iterations needed to find the optimal solution.

This algorithm is composed of two parts: the first concerns the reformulation of the mathematical model and the adaptation of data in the form of clusters by manipulation with JAVA. In the second we solved the new model using the CPLEX solver.

The parameters of these instances are described in the following table:

Inc: Instances

DL: Difficulty levels (E: easy, M: medium, D: difficult et VD: very difficult, C: complex)

NC: Number of Customer

NF: Number of Facility

NL: Number of capacity Levels

NV: Number of Variables

NCL: Number of Clusters used to find solution.

Obj: Objective value.

CPU: Execution time (in seconds).

For simplification reasons for large size problems, we add at each iteration, a set of clusters instead of adding one. This means that the number of clusters with which we find the feasible solution is not minimum, and the first solution found when running is not necessarily optimal, but very near the optimal one, so we have to wait in such case finishing the calculation.

The following table lists the various instances used and the computational results found:

Inc	NC	NF	NL	Nbr var	B&C		NFF					
							NCL	First Solution Found			Optimal Sol	
					Obj	CPU		Obj	CPU	GAP	Obj	CPU
E1	10	3	2	36	103	0.04	14	103	0.15	0%	103	0.15
E2	10	5	2	60	586	1.26	473	586	0.20	0%	586	0.20
F3	20	5	3	115	387	3.04	489	387	0.15	0%	387	0.15
E4	30	8	3	264	334	4.75	734	334	0.16	0%	334	0.16
M1	50	4	3	212	8 826	0.07	69	8 826	0.55	0%	8 826	0.55
M2	50	6	4	324	2 612	2.40	1 014	2 612	1.52	0%	2 612	1.52
M3	70	6	4	444	7 079	10.55	1 116	7 079	2.43	0%	7 079	2.43
D1	100	10	5	1 050	12 618	0.124	55	12 618	0.45	0%	12 618	0.45
D2	100	15	5	1 575	1 587	0.38	259	1 587	0.52	0%	1 587	0.52
D3	200	15	8	3 120	90 312	0.592	112 845	112 763	8.02	19.91%	90 312	19.36
VD1	300	25	10	7 750	34 175	0.842	1 214	34 584	1.14	1.18%	34 175	1.95
VD2	300	30	10	9 300	25 037	0.967	2 034	25 146	1.26	0.43%	25 037	2.05
VD3	402	30	12	12 420	42 233	0.827	3 248	43 011	2.35	1.80%	42 233	4.45
VD4	402	40	12	16 560	39 805	1.872	10 451	42 614	1.02	6.59%	39 805	12.43
C1	500	50	4	25 200	25 452	11.87	4 365	26 314	5.28	3.27%	25 452	49.66
C2	1 000	100	4	100 400	46 719	143.53	4 813	47 624	12.69	1.90%	46 719	72.46
C3	3 038	600	10	1 828 800	-	-	3 128	68 421	52.2	12.39%	59 942	89.35
C4	3 038	700	10	2 133 600	-	-	2 305	45 126	46.52	-	-	-
C5	3 038	1 000	10	3 048 000	-	-	1 345	97 542	20.14	-	-	-

According to the computational results, the NFF method is based on the concept of attacking the simplest first; we start by assigning each customer to the nearest facility, if a solution is found it will be optimal, if necessary we test the next facility. And so on until arrive at the solution.

The NFF provides two solution approaches:

– Approximate solution by taking the first solution appeared while executing program and this has the advantage of reduced computational time. This can be used when we need only very good (not optimal) solution and the execution time is more important than the solution quality.

– Optimal solution when we let the program finish the instructions until the end. This is used when execution time is not very important.

We notice from the table of results that the method works very well with a very reasonable execution time for easy instances in terms of dispersion points and in terms of resource availability, but it is less effective for difficult instances.

The NFF heuristic approach continue to provide solutions even for complex instances, but unfortunately, these solutions are not optimal. It is for this reason that we will harness the power of our method in such instances and we will complete the method by column generation, if the NFF is no longer successful after the generation of 5000 clusters.

In these results, we have defined the GAB for only the NFF heuristic approach, because the exact approach always gives optimal solutions and the GAB should normally always equal to zero.

7.6 Conclusion

In this paper, we introduced the multi-capacitated location problem that represents a generalization, in terms of capacity, of the famous p-median location problem. This generalization describes real situations existing in the industry where the concept of multi-capacity is present. To implement this problem, we decide to use a new mathematical formulation based on a set partitioning approach. The Branch and Cut algorithm is used to validate the new formulation and we adopt, for resolution, the NFF method in two ways: as an optimal approach for instances in which the difficulty level is medium or difficult, and as a heuristic for more complex instances. As perspective, we propose an optimal alternative for complex instances using column generation. In a future work, we may also improve the formulation by including the multi-service concept. The problem will consider indeed that each facility can offer several services at once and each one is characterized by an suitable capacity.

References

1. Klose, A. (2000). A Lagrangean relax-and-cut approach for the two-stage capacitated facility location problem. *European Journal of Operational Research, 126*, 408–421.

2. El Amrani, M., Benadada, Y., & Gendron, B. (2016). Generalization of capacitated p-median location problem: modeling and resolution. In *International IEEE Conference of Logistics Operations Management* (pp. 1–6). FES: IEEE, SCOPUS.
3. Garey, M. R. & Johnson, D. S. (1979). *Computers and Intractability; A Guide to the Theory of NP-Completeness*.
4. Senne, E. L. F. & Pereira, M. A. (2004). A branch-and-price approach to p-median location problems.
5. Lorena, L. A. N. & Senne, E. L. F. (2003). A column generation approach to capacitated p-median problems. http://www.lac.inpe.br/~lorena/instancias.html.
6. Boccia, M., Sforza, A., Sterle, C. & Vasilyev, I. (2007). A cut and branch approach for the capacitated p-median problem based on fenchel cutting planes.
7. Baldacci, R., Hadjiconstantinou, E., Vittorio, M., & Mingozzi, A. (2001). A new method for solving capacitated location problems based on a set partitioning approach. *Computers & Operations Research, 0*, 365–386.
8. Dantrakul, S., Likasiri, C., & Pongvuthithum, R. (2014). Applied p-median and p-center algorithms for facility location problems.
9. Behmardi, B.. & Shiwoo, L. (2008). Dynamic multi-commodity capacitated facility location problem in supply chain. In *Industrial Engineering Research Conference* (pp. 1914–1919). Corvallis
10. Dias, J., Captivo, M. E., & Climaco, J. (2006). Capacitated Dynamic Location Problems with opening, closure and reopening of facilities. *IMA Journal of Management Mathematics, 0*, 1–32. Advance access published.
11. Current, J., Ratick, S., & ReVelle, C. (1997). Dynamic facility location when the total number of facilities is uncertain: a decision analysis approach. *European Journal of Operational Research, 110*, 597–609.
12. Eberyab, J., Krishnamoorthya, M., Ernsta, A., & Bolandb, N. (2000). The capacitated multiple allocation hub location problem: Formulations and algorithms. *European Journal of Operational Research, 120*, 614–631.
13. Shu, J., Ma, Q., & Li, S. (2010). Integrated location and two-echelon inventory network design under uncertainty. *Annals of Operations Research, 0*, 233–247.
14. Ceselli, A., & Giovanni, R. (2005). A branch-and-price algorithm for the capacitated p-median problem.
15. Ceselli, A. (2003). Two exact algorithms for capacitated p-median problem.
16. Da Gama, F. S., & Captivo, M. E. (1998). A heuristic approach for the discrete dynamic location problem. *Location Science, 6*, 211–223.
17. Holmberg, K. (1998). Exact solution methods for uncapacitated location problems with convex transportation costs. *European Journal of Operational Research, 114*(1999), 127–140.
18. Frantzeskakis, M., & Watson-GANDY, C. D. T. (1989). The use of state space relaxation for the dynamic facility location problem. *Annals of Operations Research, 18*, 189–212.

Chapter 8
Application of Genetic Algorithm for Solving Bilevel Linear Programming Problems

M. Ait Laamim, A. Makrizi and E. H. Essoufi

Abstract Bilevel linear programming problem is a special class of nonconvex optimization problems which involves two levels with a hierarchical organization structure. In this paper, we present a genetic algorithm (GA) based approach to solve the bilevel linear programming (BLP) problem. The efficiency of this approach is confirmed by comparing the results with Kuo and Han's method HGAPSO consisting of a hybrid of GA and particle swarm optimization algorithm (PSO) in Kuo and Han (Applied Mathematical Modelling 35:3905–3917, 2011, [15]) using four problems in the literature and an example of supply chain model. These results show that the proposed approach provides the optimal solution and outperforms HGAPSO for most test problems adopted from the literature.

Keywords Bilevel linear programming · Genetic algorithm · KKT optimality conditions

8.1 Introduction

Bilevel optimization problem was formulated by Von Stackelberg in 1934 [20] in the area of game theory as a program with two hierarchical levels called Stackelberg game [19]. This problem was introduced in the optimization field in 1973 by J. Bracken and J. Mc Gill [8].

In a basic model of bilevel programming, that is a leader-follower game, decision is shared between two decision-makers within a hierarchical structure, in which they try to optimize their individual objectives, subject to a series of interdependent

M. Ait Laamim (✉) · A. Makrizi · E. H. Essoufi
Faculty of Sciences and Techniques, Hassan 1st University, Settat, Morocco
e-mail: marwa.laamim@gmail.com

A. Makrizi
e-mail: a_makrizi@yahoo.fr

E. H. Essoufi
e-mail: e.h.essoufi@gmail.com

© Springer International Publishing AG, part of Springer Nature 2019
E.-G. Talbi and A. Nakib (eds.), *Bioinspired Heuristics for Optimization*,
Studies in Computational Intelligence 774,
https://doi.org/10.1007/978-3-319-95104-1_8

constraints [4]. The decision-maker of the higher level is usually called the leader, and that of the lower level is called the follower. The leader plays first, chooses a strategy which aims at optimizing its objective, then the follower observes and reacts to the leader's decision by a rational feedback. Since the set of the available choices for the two players is interdependent, the decisions taken by the leader influence the set of possible reactions of the follower and vice-versa.

General bilevel programming problems can be formulated as

$$\max_{(x, y^*) \in \Omega} F(x, y^*),$$

$$where \quad y^* \in \arg \max_{y \in \Omega(x)} f(x, y), \tag{8.1}$$

where $x \in \mathbf{R}^{n_1}$ and $y \in \mathbf{R}^{n_2}$ are the decision vectors of the leader and the follower, respectively. $F, f : \mathbf{R}^n \to \mathbf{R}, n = n_1 + n_2; \Omega \subset \mathbf{R}^n$ defines the common constraint region and $\Omega(x) = \{y \in \mathbf{R}^{n_2} : (x, y) \in \Omega\}$.

Due to the nested structure of bilevel problems, they are nonconvex and difficult to be dealt with, even when all involved functions are linear [4, 10]. Indeed, the bilevel linear programming (BLP) problem has been proven to be NP-hard by Bard, Ben-Ayed and Blair [3, 6]. Also, Hansen, Jaumard and Savard [13] have proved that the BLP problem is a strongly NP-hard problem.

A large number of studies have been focused on the bilevel linear problem and on its applications [6, 7, 13]. To date, a variety of approaches have been developed for solving the BLP problem. There are two popular approaches: enumeration and transformation with KKT conditions. Enumeration methods use the fact that an optimal solution of the BLP problem occurs at an extreme points of the feasible region. The transformation approach consists of replacing the follower's level of the BLP problem with its KKT conditions, thus yielding a nonstandard one-level problem.

Despite the considerable progress made in classic optimization for solving bilevel linear optimization problems, most of these methods require knowledge of the search space, which is not satisfied in general. Besides, they are not efficient in terms of computational time, especially for large-scale problems. For these problems, metaheuristics seem to be a good alternative since they have been proven to be robust in finding good solutions to complex optimization problems and solving large-scale problems in a reasonable amount of time. Therefore, many metaheuristic approaches have been proposed for solving the bilevel programming problem, like genetic algorithms, which are the main topic of this paper, tabu search [11], particle swarm optimization [15] and simulated annealing [18].

El-Ghazali Talbi [21] provides good general references on the taxonomy of metaheuristics for bilevel optimization. Genetic algorithms are one of the artificial intelligence stochastic techniques based on genetic processes of biological organisms; it simulates the natural evolution of a population of individuals through operators such as selection, crossover and mutation. Genetic algorithms have been used in a wide variety of applications and proved their robustness in finding good solutions to complex problems in a reasonable computational time [12, 17].

On the subject of BLP problems, when all functions and constraints are linear, Mathieu et al. [16] proposed a genetic algorithm for solving the bilevel linear problem in which each chromosome is $n_1 + n_2$ string of base-10 digits representing a feasible solution and where only mutation is used. The leader's variables are generated while the follower's ones are found by solving the follower's linear problem.

Later, Hejazi et al. [14] firstly transform the BLP problem into a one-level problem by including KKT conditions of the follower's problem as constraints of the leader's problem. Then a genetic algorithm is developed in which chromosomes are $m + n_2$ binary components, each feasible chromosome represents a bilevel feasible extreme point. Crossover and mutation are applied in the usual way.

Calvet et al. [9] developed an approach for solving bilevel linear problems using a combination of genetic algorithm and classical enumeration techniques by associating chromosomes with extreme points of the constraint region.

The main purpose of this paper is to apply genetic algorithm for solving the BLP problems. Using some test problems in the literature and a supply chain model, the results prove that the proposed genetic algorithm based approach is more suitable for such problems, yielding optimal solutions in a reasonable time in comparison with the approach based on HGAPSO algorithm presented by Kuo et al. [15].

The paper is organized as follows. Section 8.2 recalls the general formulation of the BLP problem and introduces basic notations and related definitions to be used in the paper. The applied algorithm to solving BLP problem is presented and discussed in Sect. 8.3, while Sect. 8.4 is devoted to the numerical results that evaluate the performance of the proposed algorithm. Finally, Sect. 8.5 provides a conclusion of the paper with some final remarks and directions for future work.

8.2 Bilevel Linear Optimization

8.2.1 Mathematical Formulation of the BLP Problem

We consider the BLP problem with the following form:

$$
\begin{aligned}
\min_{x_1} \ & F(x_1, x_2) = c_1 x_1 + d_1 x_2 \\
\min_{x_2} \ & f(x_1, x_2) = c_2 x_1 + d_2 x_2 \\
s.t. \quad & A_1 x_1 + A_2 x_2 \leq b, \\
& x_1, \ x_2 \geq 0,
\end{aligned}
\tag{8.2}
$$

where F is the objective function of the leader and f is the objective function of the follower. Also, x_1 is an $n_1 \times 1$ vector, x_2 is an $n_2 \times 1$ vector, c_1 and d_1 are n_1-dimensional row vectors, c_2 and d_2 are n_2-dimensional row vectors, A_1 is an $m \times n_1$-matrix, A_2 is an $m \times n_2$-matrix and b is an m-dimensional column vector. To give a characterization of an optimal solution to the BLP problem, we need the following definitions [2].

8.2.2 Definitions

(1) Constraint region of the BLP problem:

$$\Omega := \{(x_1, \ x_2) : x_1 \geq 0, x_2 \geq 0, A_1 x_1 + A_2 x_2 \leq b\}$$

(2) For each x_1 taken by the leader, the feasible set for the follower:

$$\Omega_{x_2}(x_1) := \{x_2 \geq 0 : A_2 x_2 \leq b - A_1 x_1\}$$

(3) Projection of Ω onto the leader's decision space:

$$P := \{x_1 \in \mathbf{R}^{n_1} : \exists\, x_2 \in \mathbf{R}^{n_2}, (x_1, \ x_2) \in \Omega\}$$

(4) For $x_1 \in P$, $R(x_1)$ defines the follower's rational reaction set:

$$R(x_1) := \{x_2 : x_2 \in argmin\{f(x_1, \ x_2) : x_2 \in \Omega_{x_2}(x_1)\}\}$$

(5) Inducible region:
$$IR := \{(x_1, \ x_2) \in \Omega : x_2 \in R(x_1)\}$$

(6) A point $(x_1, \ x_2)$ is a feasible solution if $(x_1, \ x_2) \in IR$

(7) A point $(x_1^*, \ x_2^*)$ is an optimal solution to the BLP problem if for all feasible pairs $(x_1, \ x_2) \in IR$, $F(x_1^*, \ x_2^*) \leq F(x_1, \ x_2)$

To ensure that the BLP problem is well posed, we assume that the polyhedron Ω is nonempty and compact, for each decision x_1 chosen by the leader, $R(x_1)$ is nonempty and it's a point-to-point map [3]. The inducible region IR defines the set over which the leader may optimize. So according to the above notation, the BLP problem can be expressed as: min $\{F(x_1, \ x_2) : (x_1, \ x_2) \in IR\}$. This allows to give the following fundamental result illustrated in [4].

Corollary 1 *The solution to the BLP problem occurs at a vertex of IR.*

8.2.3 KKT Formulation of the BLP Problem

The basic principle is to transform the BLP problem into a one-level problem by replacing the follower's problem with its KKT conditions and appending the resultant system to the leader's problem.

Let $u \in \mathbf{R}^m$ and $v \in \mathbf{R}^{n_2}$ be the dual variables associated with the constraints $A_1 x_1 + A_2 x_2 \leq b$ and $y \geq 0$, respectively, we have the following proposition.

Proposition 1 ([4]) *A necessary condition that* (x_1^*, x_2^*) *solves the BLP problem* (8.2) *is that there exist vectors* u^* *and* v^* *such that* (x_1^*, x_2^*, u^*, v^*) *solves:*

$$
\begin{aligned}
\min \quad & c_1 x_1 + d_1 x_2 \\
\text{s.t.} \quad & u A_2 - v = -d_2, \\
& u(b - A_1 x_1 - A_2 x_2) + v x_2 = 0, \\
& A_1 x_1 + A_2 x_2 \le b, \\
& x_1 \ge 0, \; x_2 \ge 0, \; u \ge 0, \; v \ge 0.
\end{aligned}
\tag{8.3}
$$

8.3 Genetic Algorithm to Solve the BLP Problem

As mentioned previously, there is an extreme point of IR which solves the BLP problem. The main idea of the proposed genetic algorithm is to associate chromosomes with extreme points of IR which gives optimal solutions in acceptable computational time. After generating the initial population of chromosomes, the algorithm proceeds by crossover, mutation, evaluation and selection. In each generation a population of extreme points of IR is maintained. The algorithm performs the following steps:

8.3.1 Encoding

Let ω be the vector of slack variables controlled by the follower's problem. Thus problem (8.3) can be expressed as:

$$
\begin{aligned}
\min \quad & c_1 x_1 + d_1 x_2 \\
\text{s.t.} \quad & u A_2 - v = -d_2, \\
& A_1 x_1 + A_2 x_2 + \omega = b, \\
& u\omega = 0, \\
& v x_2 = 0, \\
& x_1, x_2, u, v, \omega \ge 0.
\end{aligned}
\tag{8.4}
$$

The genetic algorithm will be applied to solve problem (8.4).

Each chromosome consists of $m + n_2$ string of binary components, whose m first components are associated with the vector u and the remaining components are associated with the vector v. Each chromosome transforms problem (8.4) into problem (8.5) according to the following rule, given in [14].

If the value of the jth component corresponds to u_j is equal to zero, consequently the variable u_j is equal to zero and the variable ω_j is greater than or equal to zero, if not u_j is greater than or equal to zero and ω_j is equal to zero.

Furthermore, if the kth component corresponds to the variable v_k is zero, consequently the value of variable v_k is equal to zero and $(x_2)_k$ is greater than or equal to zero, if not v_k is greater than or equal to zero and $(x_2)_k$ is equal to zero.

Hence, the resulting problem can be written as:

$$
\begin{aligned}
\min \quad & c_1 x_1 + d_1' x_2' \\
s.t. \quad & u' A_2'' - v = -d_2, \\
& A_1 x_1 + A_2' x_2' + \omega' = b, \\
& x_1, \ x_2', \ u', \ v', \ \omega' \geq 0.
\end{aligned}
\tag{8.5}
$$

with x_2', u', v' and ω' are the variables of x_2, u, v and ω that are greater than or equal to zero. d_1' is the component of d_1 associated with x_2'. A_2'' and A_2' are the rows and columns of A_2 which are associated with u' and x_2', respectively.

Thus, for each chromosome, if the optimal solution of problem (8.5) exists, then this solution corresponds to an extreme point of IR. So, if a chromosome represents an extreme point of IR, it will be called a feasible chromosome. Otherwise, it will be a non-feasible chromosome.

8.3.2 Initial Population

The initial population is composed of N bilevel feasible solutions. In the implementation, we get these solutions by solving N times the following linear problem:

$$
\begin{aligned}
\min \quad & z x_1 + d_2 x_2 \\
s.t. \quad & A_1 x_1 + A_2 x_2 \leq b, \\
& x_1, x_2 \geq 0,
\end{aligned}
\tag{8.6}
$$

where z is an $(1 \times n_1)$ random vector which is generated N times. Generated solutions are extreme points of IR [14], they are converted into chromosomes. Since we look for an optimal solution of the BLP problem, we define the fitness of a chromosome as the value of the leader's objective function.

8.3.3 Crossover

In this step, two chromosomes are selected randomly from the population as parents with crossover probability pc, after that we generate a uniform random number $c \in [0, 1]$, if c is lower than pc then we generate children using the following procedure:

Let r and h be the length of the chromosome and an integer random number in the interval $[1, r - 1]$, respectively.

The hth first genes of the children are the same genes as the respective parents.

The rest of genes are obtained according to the following rules:

1. The $(h + j)$th gene of the first child is changed to the $(r - j + h)$th gene of the second parent (for $j = 1, 2,..., r - h$).
2. The $(h + j)$th gene of the second child is changed to the $(r - j + h)$th gene of the first parent (for $j = 1, 2,..., r - h$).

After crossover, children are evaluated and the feasible ones are added to the present population, then the whole population is passed on to the mutation step.

8.3.4   Mutation

To perform the mutation process, one chromosome is selected randomly from the new population with mutation probability pm, after that we generate a uniform random number $c' \in [0, 1]$, if c' is lower than pm then an integer random number l is generated in the interval $[1, r]$ and the lth gene of the new chromosome is changed to 1, if it was initially 0 and 0, if it was initially 1. The evaluation of the new chromosome is carried out and the feasible chromosome is added to the present population.

8.3.5   Selection

Population in this step consists of the present population plus the feasible offspring resulting from crossover and mutation. If the number of generated races is sufficient then we go to the next step. Otherwise, we sort out all chromosomes in ascending order of fitness value then we select the first N chromosomes. This is considered as the new population.

8.3.6   Evaluation

The evaluation of new chromosomes generated after crossover and mutation is obtained using the decomposition of problem (8.5) into two linear problems, which gives us the possibility of rejecting nonfeasible chromosomes [14].

8.3.7   Termination

The algorithm continues by selecting the best extreme point in IR with respect to F and repeating the process until a specified number of generations will be satisfied. In the end, an extreme point of IR with the best fitness value is provided which is the optimal solution to the BLP problem.

8.4 Numerical Results

The computational experiment is divided into two parts. The first part is devoted to proving the quality of the solution provided by the proposed genetic algorithm (GA2). For this purpose, we compare the results obtained with those of the genetic algorithm proposed by Wang et al. [22] (called GA1 from now on), PSO algorithm and HGAPSO method, which integrated the mutation mechanism with PSO and called HGAPSO-3, described in [15], using four test problems and a supply chain distribution model, chosen by Kuo et al., because of different levels of complexities more detailed in [15]. The second part of this section goes on to confirm the efficiency of GA2 also from the computational point of view using other test problems taken from the literature. All test problems are presented in "Appendix".

All simulations were performed on a personal computer with Intel(R) Core(TM)i5 at 2.5 GHz having 4 GB of RAM, under Windows 7 professional (32 bits). The genetic algorithm code was written in MATLAB R2010.

To get the initial population, the vector z in problem (8.6) is a n_1 dimensional random row vector whose components are generated according to a uniform distribution in the interval [1, 50].

In order to compare these methods, we consider the solution obtained from Lingo[1] as the best solution, we take the same parameters values described in [15] and we calculate the error rate expressed for a M method as follows:

$$\text{Error rate} = \frac{|f_l^* - f_M^*|}{f_l^*} \times 100\%,$$

where f_l^* is the optimal solution based on Lingo, f_M^* is the optimal solution obtained from M method. Table 8.1 depicts the corresponding parameter setup for the four BLP problems.

8.4.1 Numerical Comparison of Metaheuristic Algorithms for Solving Standard Test Problems from the Literature

For test problem 1, the best result for each method is listed in Table 8.2. We can find that HGAPSO-3 and the developed genetic algorithm (GA2) give better results than GA1 and PSO. Besides, They have the smallest error rates. Tables 8.3 and 8.4 illustrate the best solutions of test problem 2 and test problem 3, respectively. It shows that GA2, PSO and HGAPSO-3 have better performance. The parameter setup and the best solutions of test problem 4 are described in Tables 8.5 and 8.6, respectively. It also reveals that GA2 has better results than GA1 and PSO. Regarding HGAPSO-3, GA2 has much smaller error rates than those of HGAPSO-3, which are 0.0664 and 0.2113. So GA2 still has the most promising performance.

---

[1] An optimization modeling software.

Table 8.1 Parameter setup

Method	GA1	PSO	HGAPSO-3	GA2
Parameter	Population: 20 Pc: 0.9 Pm: 0.1 Generations: 200	Population: 20 V_{max}: 10 Iw[a]: 1.2 − 0.2 Iterations: 200	Population: 20 V_{max}:10 Iw[a]: 1.2 − 0.2 Iterations: 200 Pm: 0.1	Population: 20 Pc: 0.9 Pm: 0.1 Generations: 200

[a]Inertia weight parameter

Table 8.2 The best solutions of test problem 1

Method	GA1	PSO	HGAPSO-3	GA2	Lingo
x_1	17.4328	17.4417	17.4545	17.4545	17.4545
x_2	10.8657	10.8788	10.9091	10.9091	10.90909
F	84.6578	84.8511	85.0909	85.0909	85.0909
f	−50.0302	−50.0781	−50.1818	−50.1818	−50.18182
F er[a](%)	0.51	0.28	0	0	N/A
f er[a](%)	0.30	0.21	0	0	N/A

[a]Error rate

Table 8.3 The best solutions of test problem 2

Method	GA1	PSO	HGAPSO-3	GA2	Lingo
x_1	15.9984	15.9999	16	16	16
x_2	10.9968	10.9998	11	11	11
F	10.9968	10.9998	11	11	11
f	−10.9968	−10.9998	−11	−11	− 11
F er[a](%)	0.03	0.00	0	0	N/A
f er[a](%)	0.03	0.00	0	0	N/A

[a]Error rate

Table 8.4 The best solutions of test problem 3

Method	GA1	PSO	HGAPSO-3	GA2	Lingo
x_1	3.9994	4	4	4	4
x_2	3.9974	4	4	4	4
F	15.9917	16	16	16	16
f	−3.94636	−4	−4	−4	−4
F er[a](%)	0.05	0	0	0	N/A
f er[a](%)	1.34	0	0	0	N/A

[a]Error rate

Table 8.5 Parameter setup of test problem 4

Method	GA1	PSO	HGAPSO-3	GA2
Parameter	Population: 20 Pc: 0.9 Pm: 0.1 Generations: 150	Population: 20 V_{max}: 10 Iw[a]: 1.2 − 0.2 Iterations: 150	Population: 20 V_{max}:10 Iw[a]: 1.2 − 0.2 Iterations: 150 Pm: 0.1	Population: 20 Pc: 0.9 Pm: 0.1 Generations: 150

[a]Inertia weight parameter

Table 8.6 The best solutions of test problem 4

Range [0, 1]	GA1	PSO	HGAPSO-3	GA2	Lingo
x_1	0.15705	0.02192	0	0	0
x_2	0.86495	0.86693	0.8992	0.9	0.9
y_1	0	0	0	0	0
y_2	0.47192	0.56335	0.5994	0.6	0.6
y_3	0.51592	0.34108	0.3978	0.4	0.4
F	21.52948	24.81256	25.9827	26	26
f	−3.39072	−3.1977	−3.1932	−3.2	−3.2
F er[a](%)	17.19	4.57	0.0664	0	N/A
f er[a](%)	5.96	6.21	0.2113	0	N/A

[a]Error rate

8.4.2 Supply Chain Model Results

This part of the study considers the model of supply chain of two distribution centers and one manufacturer, formulated as a BLP problem in [15]. It is assumed that the supply chain leader is the distribution center, he has the control to determine the quantity and the cost of the inbound inventory, which influences the manufacturer's distribution behavior. X_{ij} represents the product demand of the ith distribution center from the jth manufacturer and Y_i is the total amount of products that the ith distribution center has, where i and j indicate the distribution center number and the manufacturer number, respectively. The supply chain model states as follows:

$$\begin{aligned} \max \quad & 110X_{11} + 120X_{21} - 40Y_1 - 50Y_2 \\ \min \quad & 130X_{11} + 145X_{21} \\ s.t. \quad & X_{11} \leq Y_1, \\ & X_{21} \leq Y_2, \\ & Y_1 \leq 1000, \\ & Y_2 \leq 500, \\ & Y_1 + Y_2 \geq 750, \\ & X_{ij} \geq 0, \; Y_i \geq 0, \quad i = 1, 2 \; and \; j = 1. \end{aligned} \tag{8.7}$$

Table 8.7 Solution of the supply chain model

Range [0, 1000]	GA1	PSO	HGAPSO-3	GA2	Lingo
X_{11}	997.3735	997.7685	999.9299	1000	1000
X_{21}	496.5783	497.1162	499.9319	500	500
Y_1	999.5942	999.3908	999.9323	1000	1000
Y_2	498.0463	498.2986	499.9325	500	500
F	104414.4	104517.9	104990.2	105,000	105,000
f	201662.42	201791.7	202481.01	202.500	202.500
F er^a(%)	0.5577	0.4591	0.0093	0	N/A
f er^a(%)	0.4136	0.3498	0.0094	0	N/A

^{a}Error rate

Table 8.8 Solutions of test problems

Test Problem	Solutions obtained by GA2	T(s)	Solutions in literature
5 [2]	(0.8889, 2.2222)	32.0345	(8/9, 20/9)
2	(16, 11)	39.7550	(16, 11)
1	(17.4545, 10.9091)	55.0114	(17.4545, 10.90909)
6 [5]	(1, 0, 0.5, 1)	63.3822	(1, 0, 0.5, 1)
4	(0, 0.9, 0, 0.6, 0.4)	72.32960	(0, 0.9, 0, 0.6, 0.4)

Table 8.7 shows the results of the supply chain model for all methods. We found that GA2 gives the best solutions and also has the smallest error rates compared with other methods.

8.4.3 Computational Results

Table 8.8 contains the computational results, of the previous four test problems plus two others taken from the literature, corresponding to the following parameters. A maximum number of iterations, $NGMAX = 50$ with population size $N = 20$, $pc = 0.7$ and $pm = 0.05$. The results thus obtained [1] are compatible with the solutions in the literature. So, it has been proven the efficiency of GA2 both from the computational point of view and the quality of solutions obtained.

8.5 Conclusion

In this paper, we have presented a genetic algorithm (GA2) for solving BLP problems. We were also interested in comparison with GA1 proposed by Wang et al. [22], PSO and HGAPSO-3 described in [15]. The experiment results [1] of different test problems taken from the literature indicate that the proposed GA2 outperforms all these methods for most of the problems. However, GA2 can be applied to solve a particular case of BLP problems in which there are no first-level constraint involving second-level variables. On the basis of the promising findings presented in this study, further research is needed to extend the current approach for more general linear bilevel problems.

Appendix

Test problem 1

$$\max_{x_1} \; F = -2x_1 + 11x_2$$
$$where \; x_2 \; solves$$
$$\max_{x_2} \; f = -x_1 - 3x_2$$
$$s.t. \quad x_1 - 2x_2 \le 4$$
$$2x_1 - x_2 \le 24$$
$$3x_1 + 4x_2 \le 96 \tag{8.8}$$
$$x_1 + 7x_2 \le 126$$
$$-4x_1 + 5x_2 \le 65$$
$$x_1 + 4x_2 \ge 8$$
$$x_1, x_2 \ge 0.$$

Test problem 2

$$\max_{x_1} \; F = x_2$$
$$where \; x_2 \; solves$$
$$\max_{x_2} \; f = -x_2$$
$$s.t. \; -x_1 - 2x_2 \le 10$$
$$x_1 - 2x_2 \le 6$$
$$2x_1 - x_2 \le 21 \tag{8.9}$$
$$x_1 + 2x_2 \le 38$$
$$-x_1 + 2x_2 \le 18$$
$$x_1, x_2 \ge 0.$$

Test problem 3

$$\max_{x_1} \ F = x_1 + 3x_2$$
$$where \ x_2 \ solves$$
$$\max_{x_2} \ f = -x_2$$
$$s.t. \ -x_1 + x_2 \leq 3$$
$$x_1 + 2x_2 \leq 12$$
$$4x_1 - x_2 \leq 12$$
$$x_1, x_2 \geq 0.$$

(8.10)

Test problem 4

$$\max_{x} \ F = 8x_1 + 4x_2 - 4y_1 + 40y_2 - 4y_3$$
$$where \ y \ solves$$
$$\max_{y} \ f = -x_1 - 2x_2 - y_1 - y_2 - 2y_3$$
$$s.t. \quad y_1 - y_2 - y_3 \geq -1$$
$$-2x_1 + y_1 - 2y_2 + 0.5y_3 \geq -1$$
$$-2x_2 - 2y_1 + y_2 + 0.5y_3 \geq -1$$
$$x_1, x_2, y_1, y_2, y_3 \geq 0$$

(8.11)

Test problem 5

$$\max_{x_1} \ F = -x_1 - x_2$$
$$where \ x_2 \ solves$$
$$\max_{x_2} \ f = 5x_1 + x_2$$
$$s.t. \quad -x_1 - \frac{x_2}{2} \leq -2$$
$$-\frac{x_1}{4} + x_2 \leq 2$$
$$x_1 + \frac{x_2}{2} \leq 8$$
$$x_1 - 2x_2 \leq 4$$
$$x_1, x_2 \geq 0$$

(8.12)

Test problem 6

$$\max_{x} \ F = 2x_1 - x_2 + 0.5y_1$$
$$where \ y \ solves$$
$$\max_{y} \ f = -x_1 - x_2 + 4y_1 - y_2$$
$$s.t. \quad 2x_1 - y_1 + y_2 \geq 2.5$$
$$-x_1 + 3x_2 - y_2 \geq -2$$
$$-x_1 - x_2 \geq -2$$
$$x, y \geq 0$$

(8.13)

References

1. Ait Laamim, M., Makrizi, A., & Essoufi, E. H. (2017) Genetic algorithm for solving a linear bilevel program. In *5th International Congress of the SM2A, March 16–18, 2017, Meknes, Morocco.*
2. Bard, J. (1984). Optimality conditions for the bilevel programming problem. *Naval Research Logistics, 31,* 13–26.
3. Bard, J. (1991). Some properties of the bilevel programming problem. *Journal of Optimization Theory and Applications, 68,* 371–378.
4. Bard, J. (1998). *Practical bilevel optimization. algorithms and applications.* Boston: Kluwer Academic Publishers.
5. Bard, J. F., & Falk, J. E. (1982). An explicit solution to the multi-level programming problem. *Computers and Operations Research, 9,* 77–100.
6. Ben-Ayed, O., & Blair, C. E. (1990). Computational difficulties of bilevel linear programming. *Operations Research, 38,* 556–560.
7. Ben-Ayed, O., Boyce, D. E., & Blair, C. E. (1988). A general bilevel linear programming formulation of the network design problem. *Transportation Research, 22,* 311–318.
8. Bracken, J., & McGill, J. (1973). Mathematical programs with optimization problems in the constraints. *Operations Research, 21,* 37–44.
9. Calvete, H. I., Galé, C., & Mateo, P. M. (2008). A new approach for solving linear bilevel problems using genetic algorithms. *European Journal of Operational Research, 188,* 14–28.
10. Dempe, S. (2002). *Foundations of bilevel programming.* Boston: Kluwer Academic Publishers.
11. Gendreau, M., Marcotte, P., & Savard, G. (1996). A hybrid tabu-ascent algorithm for the linear bilevel programming problem. *Journal of Global Optimization 8,* 217–233.
12. Goldberg, D. (2002). *The design of innovation: lessons from and for competent genetic algorithms.* Boston: Kluwer Academic Publishers.
13. Hansen, P., Jaumard, B., & Savard, G. (1992). New branch-and-bound rules for linear bilevel programming. *SIAM Journal on Scientific and Statistical Computing, 13,* 1194–1217.
14. Hejazi, S. R., Memarian, A., Jahanshahloo, G., & Sepehri, M. M. (2002). Linear bi-level programming solution by genetic algorithm. *Computers and Operations Research, 29,* 1913–1925.
15. Kuo, R. J., Han, Y. S. (2011). A hybrid of genetic algorithm and particle swarm optimization for solving bi-level linear programming problem - A case study on supply chain model. *Applied Mathematical Modelling 35,* 3905–3917.
16. Mathieu, R., Pittard, L., & Anandalingam, G. (1994). Genetic algorithm based approach to bi-level linear programming. *Operations Research, 28,* 1–21.
17. Michalewicz, Z. (1996). *Genetic algorithms + data structures = evolution programs* (3rd ed.). Berlin: Springer.
18. Sahin, K. H., & Ciric, A. R. (1998). A dual temperature simulated annealing approach for solving bilevel programming problems. *Computers and chemical engineering, 23,* 11–25.
19. Simaan, M., & Cruz, J. P. (1973). On the Stackelberg strategy in nonzero-sum games. *Journal of Optimization Theory and Applications, 11,* 533–555.
20. Stackelberg, H. (1952). *The theory of the market economy.* Oxford: Oxford University Press.
21. Talbi, E. G. (2013). A taxonomy of metaheuristics for bi-level optimization. *Metaheuristics for bi-level optimization* (Vol. 482, pp. 1–39). Berlin: Springer.
22. Wang, G. M., Wang, X. J., Wan, Z. P., & Chen, Y. L. (2005). Genetic algorithms for solving linear bilevel programming. In *Proceedings of the Sixth International Conference on Parallel and Distributed Computing Applications and Technologies (PDCAT'05).* China: IEEE.

Chapter 9
Adapted Bin-Packing Algorithm for the Yard Optimization Problem

Chafik Razouk, Youssef Benadada and Jaouad Boukachour

Abstract Given the importance of the Maritime transportation to move goods between continent, optimizing their processes becomes the objective of many types of research. In this paper, we present a new model of the yard optimization problem which contains three important components: unloading/loading, transfer and storage process. Our proposed method called *Adapted Bin Packing Algorithm for the yard optimization problem* (ABPAYOP) focus on using the approach of the Bin packing algorithm to build Bins (free positions in the yard, subgroup of containers) this will be a generalization of the container stacking problem as it will include the use of yard cranes, quay cranes and internal trucks. the ABPAYOP solutions can be represented as a set of disjoint clusters satisfying of given number of constraints (yard bays, subgroup of containers). In this work, we present the new formulation of the yard optimization problem, then we will apply our heuristic ABPAYOP to solve it. Computational results are presented at the end using instances created and adapted to the ones existing in the literature. Our results illustrate the performance of the applied method for the medium and big instances.

Keywords Maritime transportation · Yard optimization · Adapted Bin Packing
Heuristic · Cranes · Internal trucks

C. Razouk · Y. Benadada
ENSIAS, Mohammed V University in Rabat, Madinat Al Irfane, BP 713,
Agdal Rabat, Morocco
e-mail: chafik.razouk@gmail.com

Y. Benadada
e-mail: youssef.benadada@um5.ac.ma

J. Boukachour (✉)
University le Havre, 25 Street Philippe Lebon, 76600 Le Havre, France
e-mail: jaouad.boukachour@univ-lehavre.fr

© Springer International Publishing AG, part of Springer Nature 2019   137
E.-G. Talbi and A. Nakib (eds.), *Bioinspired Heuristics for Optimization*,
Studies in Computational Intelligence 774,
https://doi.org/10.1007/978-3-319-95104-1_9

9.1 Introduction

The containerization has been used remarkably due to the globalization in the last ten years according to Steenken et al. [1], it has as objective the cost reduction and the minimization of the possibility damages.

According to the last study made by the maritime journal AFP in February 2017, around 3 Millions containers twenty feet equivalent unit (TEU) has been changed during 2016 in the Tanger Med Port1 [2], 90% has been received under the transshipment process and the remaining 10% belong to the import and export operations. So, this is representing an increase of 40% comparing with the last study made in 2012. Tanger Med Port considered as a bridge between the 5 continents, it has more than 170 connection with other ports of 67 countries.

A container terminal is a temporary storage space, where vessels berth in port terminal to unload inbound containers during the import operations and pick up outbound containers during the export operations, otherwise move containers from one vessel to another in case of transshipment using a temporary intermediate storage location. To be more competitive, containers has been standardized in the world wide, the most famous ones are: 20, 40, 45 feet (TEU: twenty feet equivalent units). They can be stacked then on top of each other following the last-in, first-out (LIFO) strategy. The most important terminal port equipment's which can be used are: quay crane, internal trucks, yard cranes and external trucks.

The quay cranes are assigned to the vessel once they are assigned and docked at the quay side, they start the unloading process according to the stowage plan already provided, the very critical points are: the unloading time window, the number of moves and the vessel stability. The unloaded containers are transferred from the seaside to the yard side using internal trucks, they transfer them one by one (it can be two TEU at the same time depending on the port equipment capability) to the assigned blocks in the storage area, the very critical points are: the congestion between the internal trucks, the time limit of each operation and moving the containers in a safe mode. Once arrived at the assigned block, the yard crane picks the container and store it at the assigned container location taking into account the different container characteristics we have: destination, estimated departure time (EDT), weight, size, type. The external trucks moves the containers to the final customer destination. The outbound containers follow the reverse process (Fig. 9.1).

For the transshipment process the containers are moves from one vessel to another using a temporary storage location, this management requires a lot of optimization studies as it contains the main activity of the port.

Some containers can be stored in the yard blocks at the bottom of the stack following the LIFO strategy, in order to search them some reshuffles are needed. It occurs since other vessel have been unloaded later or containers have been stacked in the wrong order due to lack of accurate information. This reduces the productivity of the cranes and create additional tasks (reshuffles) with no value added. So the yard management become the first bottleneck which impact the efficiency of the port operations process.

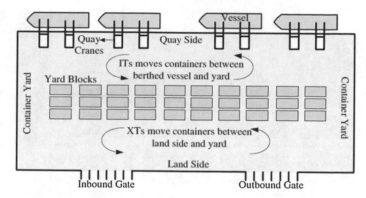

Fig. 9.1 Overview of containers terminal with operations and standard equipments

According to Stahlbock and Voss [2], the existing activities in a containers terminal can be divided in two main areas: the quay side and the yard side. We are interested to the quay side; several literature researches are dealing with the related problems to this studied area:

- Berth allocation problem (BAP);
- Schedule and stowage plan at the docked vessel;
- Quay crane allocation problem (QCAP);

Also we can found other ones related to the yard side:

- Yard crane allocation problem (YCAP): RTGC deployment;
- Storage space allocation problem;
- Yard optimization problem: location assignment;
- Internal Trucks deployment;

Based on that, Rashidi and Tsang [3] divided the operations in container terminals into 5 main operations which affect the efficiency of the operational tasks in container terminals, as mentioned in Fig. 9.2:

The majority of researches interested on the containers terminal processes can be divided into three decision levels: (1) strategic, (2) Tactical and (3) operational. Strategic decisions are long-term decisions that include the structure of the terminal, handling equipment and handling procedures. Tactical decisions are medium-term decisions that are interested in determining the number of quay cranes, yard cranes, straddles, etc. And finally operational decisions are short-term decisions and include the process to be followed by the quay cranes, the yard cranes, straddles, etc. In this paper we will focus on the operational operations, so the objective is to have storage containers planning to be flowed by gantry cranes operators in non-automated container terminals for the inbound containers by minimizing the total distance and

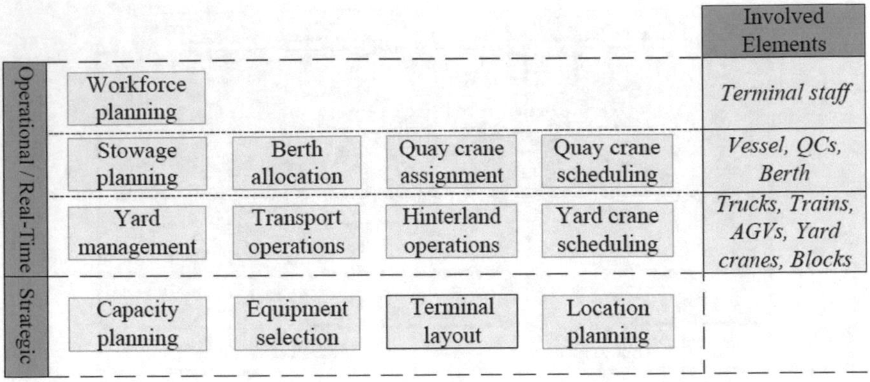

Fig. 9.2 Decision problems in container terminals

the number of reshuffles in the yard area. The rest of this paper is organized as follows: Sect. 9.2 presents a literature review, Sect. 9.3 defines mathematical formulation of our problem and Sect. 9.4 presents the solving methods to our MIP. The results of numerical simulations and a comparative study are presented in Sect. 9.5.

9.2 Literature Review

In this paper we focus on the yard bay assignment and containers transfer between yard side and sea side already presented in the previous section, and study the existing research for yard optimization problem [4]. Each yard area had a specific level to store containers; the stored containers on the last tier of the stacks must leave earlier to avoid unproductive movements (reshuffles, housekeeping). Thus the objective is to optimize the cost of the global distance between the quay side and the yard side (stacking position) while building the unloading plan, and also to reduce the number of reshuffles by reducing the housekeeping in the yard blocks (moves 0). The key factors involved to determine the stacking position are the containers and yard characteristics: stack level, weight of containers, size and EDT, stacks storage state, free positions, and stacking strategies. In our study, we will consider the below equipment: the quay crane (QC) to unload/load containers from vessels, the internal trucks (trailer) to transfer containers to the yard side, the RTGC to pick up the container from the internal vehicle to the stacking position and external trucks (XT) to transfer the container between the yard side and the gate. According to Steenken et al. and the extended study of Stahlbock and Voss [2] the main port terminal operations can be classified as follow: berth allocation, crane scheduling, internal trucks and storage yard optimization [5]. There are deterministic and dynamic cases of containers stacking problem. In the deterministic storage, the storage location of each container is known at the beginning of the planning horizon. The stacking strategy can be defined based on the list of ports to be visited, and also based on some containers characteristics: size, type and weight of containers, EDT. While in the dynamic case, containers locations are defined in the real time in parallel with the vessel berthing. The EDT and the container destination are not taken into account for the inbound containers in this study. According to Carlo et al. [4] the containers stacking problems can be divided into two types: storage of individual containers or of group of containers.

9.2.1   Individual Containers Storage

The main idea of this type of research is first to assign containers to blocks and second to specific locations within the selected block. Guldogan [6] considers an integrated policy that takes into account the work balance, number of used trucks and the travel distances, the aim of the new strategy is to define clusters which group containers according to their departure dates, weight and type of the stored containers are not considered also the number of reshuffles can grow with the big instances. Park et al. [7] first select blocks to balance the GC's workload, and second, select the specific storage location based on a weighted function of space and GC utilization so departure dates of containers are not considered which will cause additional unproductive moves. Chen and Lu [8] aim to avoid reshuffles in their approach to assign outbound containers, but workload balance of GC's are not considered. Ng et al. [5] used a new heuristic for the problem with import containers on ports with cyclical calling patterns, weight of containers are not considered.

9.2.2   Groups of Containers Storage

The main idea of the related research to this part is to propose methods to assign groups of containers to storage locations, based on the vessel destination, departure time, and type of containers. A Simulated Annealing based heuristic for this problem is proposed by Huang and Ren [9] that require enumerating all possible assignment permutations for the three container groups of import containers, result is not compared with existing storage policies. Jeong et al. [10] define a method to decide for each block how many import containers will be stored there, the consigned strategy take into account only the destination of containers the EDT is not considered. Nishimura et al. [11] propose a new MIP, and they use a new heuristic to minimize the weighted total container handling time no additional constraint for the destination type of the stored containers.

9.2.3 Design of the Bioinspired Heuristics

Our proposed approach is a new application of an existing heuristics a the YOP, for which, we used to see a lot of applications of metaheuristics and hybrid methods. The new approach aims to solve the YOP taking into account design of storage area constraint, container characteristics, and port equipments constraints. Combining all those constraints, and defining sub groups of containers gives an optimized assignment of arrived containers to the right positions, to avoid rehandling operations. This approach is discussed in detail in the next chapters.

9.3 Problem Definition: Yard Optimization Problem (YOP)

Container terminals are divided into automated and non automated ports. The automated ports (AP) use the RMGC (Rail Mounted Gantry Crane), the AGV (Automated Guided Vehicles), the ALF (Automated lifting vehicles Therefore non- automated ones (NAP), use the RTGC and the internal vehicles to transfer the containers from the berth to the storage space. So the AP needs a higher investment cost comparing with the NAP, but they have a lower operational cost, a faster crane speed and have more storage area. After the study of Lee et al. [12] 63.2% of the container terminals studied are non automated so they use the RTGC and the internal truck. So, this is the reason why we focus on the NAP in this paper. After the vessel's berth, quay cranes are assigned to each arrived docked vessel. In-bounded containers are unloaded by quay cranes (QCs) during the import operation. Then, containers are transported from berths to the storage area by trucks or other vehicles, the assignment of trucks to the storage area must verify the model constraints and also starting by the containers which will have the smaller EDT and store them as close as possible to the transfer point. Once a container arrives at its stacking yard bay in the storage area, the stacking crane lifts the container from the truck and stacks it to the storage position. Afterwards GC can re-stack containers according to their EDT and destinations. A container's position in the yard is then denoted by its block, bay, row and tier identifiers. A block is defined by its bays (length: basically [20, 30] per block) and rows (width: basically 6 per block, one is reserved to the internal truck) and the tiers (4 levels per yard bay). Stacking yards are usually divided into multiple blocks, each consisting of a number of bays. A bay is composed of several stacks (row and tiers) of a certain size, and holds containers of the same size. Figure 9.3 show a schematic overview of container terminal, the theatrical container capacity of the yard is: $15 \times 30 \times 5 \times 4 = 9000$ TEUs (twenty-feet-equivalent units), which correspond to 9000 standard 20-feet containers.

Our scope is to assign the arrived containers even for import or export operation to the storage area and to have an: optimized layout, a reduced number of the used equipment and less cost and operation time at the end of the planning horizon. This problem is known in the literature as: Yard Optimization Problem (YOP).

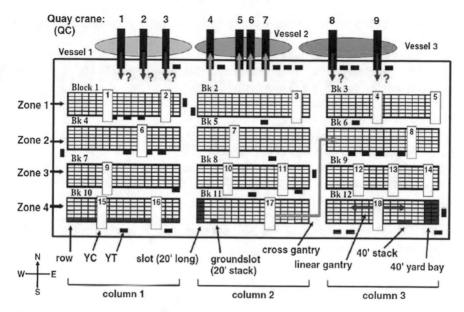

Fig. 9.3 Schematic overview of the container terminal [13]

9.3.1 Formulation

The proposed model is applied to each period of the working day in a containers terminal (shift). The planning horizon contain several shifts (3 shifts per day). In our study we will suppose that the range of shifts will be [1, 21] which means that we will prepare the planning in weekly basis. The productivity of the QC is the key factor to define which is the time needed to move containers from the vessel to the sea side. In our case of study the productivity of the QC can change between [30, 50] moves. which means for example that for a container which require 1900 moves, if we suppose that the productivity is equal to 30 moves/h we will need:

o For 1 container: $1900/30 = 64\,h$
o For 2 containers: $1900/2 * 30 = 32\,h$

The assignment of the quay cranes is following the port constraints: Number of available QCs, length of the vissel, ITs and YCs which can be assigned to a single vessel. Our objective is to reduce and optimize the related cost of the global distance for the unloading containers operation from the quay side to the yard side and then to optimize the housekeeping at the storage area. The model parameters are defined as follow:

k: Container index.
b: Yard-bay index.
r: Index of row in the yard bay b.
e: Index of the level (tier) in the bay-row.
N: Total number of inbound containers to be stored.

B: Total number of yard-bays that may contain the containers during the planning horizon.

w_k: The weight of the container k.

C_b: The free storage space in the yard-bay b (free storage positions).

r_b: Type of the yard-bay b (Three types: 20, 40, 45).

R_k: Type of container k.

D_k: Destination of container k.

D_b: Destination of yard-bay b.

t_k: Estimated departure time of container k.

C_k^b The assigned cost to choose / store a container k in the corresponding yard-bay b in the yard area.

Our decision variable is:

$X_b k^r e \in \{0, 1\}$:   equal to 1, if k is assigned to the bay b, row r and tier e, and 0 otherwise.

$$Min \quad \sum_{k \in N} \sum_{b \in B} \sum_{r \in R} \sum_{e \in E} C_k^b X_{bk}^{re} \tag{9.1}$$

$$\sum_{k \in N} \sum_{r \in R} \sum_{e \in E} X_{bk}^{re} \leq C_b, \quad b \in B \tag{9.2}$$

$$\sum_{b \in B} \sum_{r \in R} \sum_{e \in E} X_{bk}^{re} = 1, \quad k \in N \tag{9.3}$$

$$\sum_{k \in N} \sum_{r \in R} \sum_{e \in E} (R_k - r_b) X_{bk}^{re} = 0, \quad b \in B \tag{9.4}$$

$$\sum_{k \in N} \sum_{r \in R} \sum_{e \in E} (D_k - D_b) X_{bk}^{re} = 0, \quad b \in B \tag{9.5}$$

$$\sum_{k \in N} X_{bk}^{re} = 1, \quad b \in B, r \in R, e \in E \tag{9.6}$$

$$(w_k - w_k')(X_{kb}^{re'} - \sum_{e \in E} X_{kb}^{re}) \leq 0, \quad b \in B, r \in R, e \in E, k \neq k' \tag{9.7}$$

$$(t_k - t_k')(X_{kb}^{re'} - \sum_{e \in E} X_{kb}^{re}) \leq 0, \quad b \in B, r \in R, e \in E, k \neq k' \tag{9.8}$$

$$X_{bk}^{re} \in \{0, 1\} \tag{9.9}$$

In order to simplify our model, the related cost of the assignment of a specific container denoted by k to the yard bay b will be equal to the distance between the sea side and the final position in the yard bay. The reason behind this choice is to be able to compare the obtained results with the existing researches in the literature [14]. In the presented problem, only one vessel will be taken into account in each planning horizon, the planning horizon is defined before the ETA of vessel. It will be based on the vessel length and the number of the inbound containers. The aim of the objective function is to minimize the cost of the global distance for the total inbound containers during the planning horizon. This horizontal distance is calculated by adding all the distance of the stored containers between the berth location and the storage area (the assigned yard bay to each container). In the constraint (2) we define a total free positions C_b per each yard bay b, so the total number of the assigned containers must not exceed the numbers of free positions (this number is known at the beginning of the planning horizon). Constraint (3) ensures that each container is stored in a unique bay during the planning horizon. Constraint (4), (5) ensure that the size R_k and the destination D_k of the bay and the stored container are the same in each planning horizon, the aim of those constraints is to allow the proposed method to define a clustering of the available containers based on their destination and type, then we can allocate in the storage area a specific space to the those containers (consignment strategy). Each free position in the storage space can have only one container stored at the same time, this is what the constraint (6) is presenting. The constraint of the weight between the stored containers is defined in the next step (7) so the container on the top must have a weight less than the one stored in the previous level (this data is known for each inbound container). The estimated departure time is also known per each inbound container we will take it into account during the containers storage in the stacking area constraint (8). The last constraint (9) avoid empty positions between the stored containers.

9.4 Solving Methods

To proceed with solving the problem, we will suggest to test the new formulation with an exact method well known: Branch and Cut, although this method presents difficulties for large instances, it will serve to validate our new mathematical model. The adapted method for the presented formulation will be detailed afterwards.

9.4.1 Branch and Cut

The Branch and cut is a combination of two algorithms into one, namely the Branch and Bound and cutting plans, his performance is shown for the small and medium instances comparing with the basic Branch and Bound.

Algorithm01:
1- Linear relaxation is solved using the cuts method.
2- Declare the summit sterile, if the solution is feasible or it's above the BKS (best known solution).
3- The best solution found is optimal if all pending summits are sterile.
4- Otherwise, choose a non-sterile pendant summits, select a fractional variable xi, and consider two sub-problems by fixing xi at 1 and xi at 0 (Branching)
5- Solve every problem by generating new unsatisfied constraints (cutting). Go to step 3.

9.4.2 Adapted Bin Packing Algorithm to Solve the Yard Optimization Problem (ABPAYOP)

No study was interested on the application of other heuristics such as: Bin Packing to the yard optimization problem by defining a cluster with a set of containers (goods) and available positions (Bins) taking into account the defined constraints, this is what justifies the choice of this strategy. We consider several attributes of the containers while determining a stacking position. The first attribute is the expected departure time (EDT), which is defined as the time when a container will be removed from the stack. Secondly, the category of each container is considered and also the destination of each container, its weight and the type.

In this section we study the way our containers sub groups are builded, we start first by identifying the containers based on their characteristics: Container ID, Size, Type, Weight, destination, EDT. We assume that they are stored at the top of the vessel so no additional reshuffles are needed after the unloading operation case of import, and no rehandling needed case of export and transshipment. The only added constraint here is the vessel stability which we need to keep during the import, export and transshipment operations. For Import, containers are sorted first based on their destination (adjacent destinations and first assigned and stored to the adjacent sub blocks), the type of the yard bay need to checked and confirmed at first level before the assignment of adjacent sub groups.

Algorithm02:

Generate Bins

If the set CB (container Bins) does not exceed a maximum number (5000 clusters : if we assume that the storage space is empty)

1- Build all the Bins for the sub groups of the empty positions and couple of containers with same or approximate characteristics

Let's n the number of Bins, for each pair (sub group, couple of containers).

2- Generate the UB and LB for inbound containers: residence time in vessel, yard side and in the IT.

3- Define LB and UB for cranes: YC, QC(number of moves)

4- Assign the needed equipment to the arrived vessel.

5- Set the planning horizon to S shifts for the arrived vessel

a- The Bin with n container is rejected by unfeasibility.

b- If $n > 0$ we create n bins each one with $n - 1$ Bins. Otherwise $STOP$.

c- Come back to **2**

Otherwise

$C_b = 0$;

a- If the sum of containers does not exceed the available positions C_b, we add this Bin to the set CB.

b- Otherwise $C_b \leftarrow C_b + 1$, come back to **a** ;

c- $STOP$.

EndIf

EndGenerate Bins

0- Initially there are a set N of containers, a set B of empty positions and an empty set of CB.

1- For each container, sort the empty positions in ascending order in terms of assignment.

2- Assign to each container the nearest empty position taking into account containers constraints.

a- Choose containers based on container size, destination

b- Group containers in clusters. For ones go to

Repeat $|N| * (|B| - 1)$ times

3- Generate the Bins.

4- If solution is found, it can only be optimal, $STOP$.

(the algorithm stops by optimality).

5- Otherwise, add the best assignment (container, empty position) and come back to

EndRepeat

(if no solution found)

Repeat $|CB|$ times **6**- Build the Bins CB with all possible combinations.

7- If one solution is found, $STOP$. Otherwise, $CB \leftarrow CB + 1$, come back to **6**

Let's take the below example, to demonstrate how the above algorithms works. The first table illustrates the containers characteristics (Fig. 9.4).

The first field refer to the size of the used container; in this paper we are interested to the most used containers type: 20 and 40 feet. The second field refer to the final destination of the container, we have in this example 10 destinations, the adjacent ones are: 1 and 2, 3 and 4. The criteria to define the adjacent destinations is the ones which not exceed 100 km of difference. The third field is the Excepted Departure time it's defined based on the weeks of the year (52 weeks per year) and the days of the week, for example: W3-1 refer to the first day of the third week. Last and not

Container	1	2	3	4	5	6	7	8	9	10	11	12	13	14	15
Size	20'	40'	20'	20'	20'	20'	40'	40'	20'	40'	20'	20'	20'	40'	20'
Destination	1	2	5	6	9	10	3	4	4	1	2	3	3	3	4
EDT	W3-1	W3-2	W3-5	W4-4	W4-5	W2-6	W2-7	W5-3	W3-5	W2-1	W5-5	W5-4	W3-3	W4-1	W3-5
Weight	5	6.5	4.5	5.2	4.3	5.1	6.3	6.7	3.6	4.7	3.9	4.5	5.2	6.9	4.2

Fig. 9.4 Technical container characteristics with the Expected Departure Time (EDT)

least is the weight of the containers, this field has the tonne as unit of measure, the weight of the empty containers is: 2.2 tonnes for the 20 TEU and 3.7 tonnes for the 40 TEU.

9.5 Results

To demonstrate the solvability of our MIP and to evaluate the performance of the proposed strategies: ABPAYOP were evaluated using a total of 32 instances, ranging from small to large-sized problems. For benchmarking, test cases from the literature were taken to evaluate the ABPAYOP.

The used instances verify the below technical constraints which are related to our port of study. Several instances have been tested using the following data:

- We consider two types of containers (20, 40 feet).
- We consider a maximum number of 8QCs to be assigned.
- Each stack can contain a maximum of four containers.
- Expected Departure Time of containers to be stored is randomly chosen between [0, 20] days and expressed in week-day.
- The productivity of the quay crane is [30, 50] moves per hour, 50 is the upper bound.
- Planning Horizon: [0, 21sh], 3 shifts per day.
- Destination Type: 10

The computer used has the below performance: Core i7, with 2.7 CPU, Win7 and 12-GB of memory. To be able to compare our findings with other results existing in the literature, we tried first to solve the model with an exact method using ILOG Cplex, the proposed method (Branch and Cut) provide relevant result only for the small and medium instances. For the large-sized ones, the ABPAYOP is more efficient as a proposed approach. The exact method (Branch and Cut) was tested and confirmed for the instances with small and medium parameters. In fact, this exact method can be used only to verify the performance of our proposed method for this kind of instances [15]. The utilization is defined as the ratio between the total storage space occupied by all the unloaded containers and the total storage space in the storage yard [16] (Fig. 9.5).

N°	Utilizat°%	Problem size K x IT x YC x QC x V	CPLEX*	ABPAYOP	
				Make-span GAP	Time
1	4.91	40 x 5 x 2 x 1 x 1	62.04	62.04 0	560
2	6.10	50x 5 x 3 x 1 x 1	70.33	70.33 0	580
3	1.15	50x 8 x 2 x 2 x 1	77.23	77.23 0	610
4	3.2	60x 6 x 2 x 2 x 2	79.54	79.54 0	615
5	7.09	60x 6 x 3 x 2 x 2	81.54	81.54 0	644
6	6.3	60x 8 x 2 x 2 x 1	90.26	90.26 0	670
7	5.35	60x 8 x 3 x 3 x 2	91.40	91.40 0	770

Fig. 9.5 Comparison of the ABPAYOP and CPLEX

* The obtained result must be multiplied by 1000, to simplify the interpretation in the above table we put the 2 first digits. **Utilization = available positions*tiers/total positions in the yard *** GAP = (Make-span Cplex)/Cplex X 100.

In the below Table, we treat the large sized instances, the table 2 shows the results obtained after the execution of the ABPAYOP. The first and second columns show the experiment number and the utilization (already defined in the previous section). The third column defines the problem size. CPLEX runs were limited to 5h (port authority limitation to get prepared for the next shift and prepare resources) and the best solution obtained is reported in the fourth column. An N/A in the fourth columns indicates that CPLEX was not able to obtain a solution in the defined timing window. The fifth column shows the CPLEX computation time if a solution was obtained within 5h. The sixth and seventh columns show the objective function value and the computation time of the ABPAYOP. The last column shows the gap which measures the difference between the ABPAYOP solution and the CPLEX solution (GAP = (worst solution Best solution)/Best solution * 100). A negative gap means that the ABPAYOP solution was lower (better) than the CPLEX. For very large instances, CPLEX could not find the optimal solution due to either time or memory limitation. Out of the 32 instances, ABPAYOP obtained the solution much faster than CPLEX. It was observed that ABPAYOP's computation's time is not affected as CPLEX's one by increasing the number of containers and trucks. The maximum computation time for ABPAYOP is 1826 s, which explain the fact that it can used by the terminal planners. *GAP = (worst solution Best solution)/Best solution * 100.

The obtained result of the proposed method is relevant for the majority of the instances mentioned above. This means that the proposed method is more efficient to solve our MIP. For the future researches, this method can be combined with a metaheuristic method as a hybrid algorithm. As a summary of our numerical result, ABPAYOP is more efficient than ABPAYOP to solve our MIP and provide an input to terminal operators (Fig. 9.6).

Instance	Utilizat°	Problem size: K x IT x YC x QC x V	CPLEX		ABPAYOP		
			Make-span	Time	Make-span	Time	GAP%
8	4.38	70x 5 x 2 x 1 x 1	136.04	4154	133.69	45	1.76
9	4.4	70x 5 x 3 x 1 x 1	141.13	5680	137.84	47	2.39
10	7.42	80x 8 x 2 x 2 x 2	147.58	3461	143.22	49	3.04
11	6.45	80x 16 x 2 x 2 x 2	148.27	4529	144.32	51	2.74
12	5.47	80x 16 x 3 x 2 x 2	152.12	6177	146.68	53.34	3.71
13	7.5	90x 8 x 2 x 2 x 1	160.29	5594	148.4	56	8.01
14	9.52	90x 8 x 3 x 3 x 1	166.32	3529	149.84	59	11.00
15	8.55	90x 15 x 2 x 1 x 2	176.04	3521	153.69	67	14.54
16	4.56	90x 15 x 3 x 1 x 2	181.13	5780	158.84	72	14.03
17	6.57	100x 8 x 2 x 2 x 1	187.58	9800	165.69	87.12	13.21
18	4.6	100x 16 x 2 x 2x 1	188.27	3652	168.32	93.22	11.85
19	5.62	105x 16 x 3 x 2x 2	190.12	9800	169.68	107	12.05
20	3.63	110x 8 x 2 x 2x 2	190.29	5612	178.4	109.12	6.66
21	9.65	112x 8 x 3 x 3x 2	192.32	6633	179.84	89.12	6.94
22	6.67	114x 8 x 2 x 2x1	193.58	9800	185.69	65.12	4.25
23	2.7	115x 16 x 2 x 2x2	194.27	8260	188.32	67.13	3.16
24	4.75	115x 16 x 3 x 2x1	197.12	6000	189.68	69.04	3.92
25	6.8	115x 8 x 2 x 2x2	198.29	7371	190.4	84.13	4.14
26	4.81	120x 18 x 3 x 3x1	198.32	14500	191.04	87.45	3.81
27	2.83	120x 25 x 2 x 1x2	199.04	15600	191.07	92.13	4.17
28	9.85	120x 25 x 3 x 1x1	199.13	17900	192.84	95.22	3.26
29	3.87	130x 18 x 2 x 2x2	199.58	18000	192.02	107.12	3.94
30	2.9	140x 16 x 2 x 2x2	N/A	N/A	197.53	109.89	N/A
31	5.95	150x 16 x 3 x 2x1	N/A	N/A	197.57	112.23	N/A
32	1.02	160x 18 x 2 x 2x2	N/A	N/A	198.3	125.44	N/A

Fig. 9.6 ABPAYOP resolution performance against CPLEX for medium and big instances

9.6 Conclusion

In this paper we introduced the containers terminal by presenting its structure and also the related researches to each component. Then we proposed a new mathematical formulation of the yard optimization problem with new constraints and objective function. The proposed strategy defines a new method for unloading containers by using the ABPAYOP as approach of resolution. The obtained result shows a good performance for the small and medium and large sized instances. In future researches we can add additional attributes and constraints: different port terminal operations including export and transshipment, include all the components of the operations terminal in a single model. Also we may the balanced workload between adjacent bays. And finally we may also introduce the robust optimization and simulation as future works.

References

1. Steenken, D., Voss, S., & Stahlbock, R. (2004). Container terminal operation and operations research-a classification and literature review. *OR Spectrum, 26*(1), 3–49.
2. Stahlbock, R., & Voss, S. (2008). Operations research at container terminals: A literature update. *OR Spectrum, 30*(1), 1–52. Tanger Med Port Authority, Press communication, 3-6, March 2017.
3. Rashidi, H., & Tsang, E. P. K. (2013). Novel constraints satisfaction models for optimization problems in container terminals. *Applied Mathematical Modelling, 37*(6), 3601–3634.
4. Carlo, et al. (2014). Storage yard operations in container terminals: Literature overview, trends, and research directions. *European Journal of Operational Research, 235*(2014), 412–430.
5. Ng, W. C., Mak, K. L., & Li, M. K. (2010). Yard planning for vessel services with a cyclical calling pattern. *Engineering Optimization, 42*(11), 1039–1054.
6. Guldogan, E. U. (2010). Simulation-based analysis for hierarchical storage assignment policies in a container terminal. *Simulation, 87*(6), 523–537.
7. Park, T., Choe, R., Kim, Y. H., & Ryu, K. R. (2011). Dynamic adjustment of container stacking policy in an automated container terminal. *International Journal of Production Economics, 133*, 385–392.
8. Chen, L., & Lu, Z. (2012). The storage location assignment problem for outbound containers in a maritime terminal. *International Journal of Production Economics, 135*(1), 73–80.
9. Huang, J. J., & Ren, Z. Z. (2011). Research on SA-based addressing model of slot in container terminal. *Applied Mechanics and Materials, 9798*, 985–989.
10. Jeong, Y. H., Kim, K. H., Woo, Y. J., & Seo, B. H. (2012). A simulation study on a workload-based operation planning. *Industrial Engineering and Management Systems, 11*(1), 103–113.
11. Nishimura, E., Imai, A., Janssens, G. K., & Papadimitriou, S. (2009). Container storage and transshipment marine terminals. *Transportation Research Part E, 45*, 771–786.
12. Lee, L. H., Chew, E. P., Tan, K. C., & Han, Y. (2006). An optimization model for storage yard management in transshipment hubs. *OR Spectrum, 28*, 539–561.
13. Houa, D.-L., & Chen, F.-R. (2012). Constraint satisfaction technology for stacking problem with ordered constraints. *Procedia Engineering*, 3317–3321.
14. Moussi, R., et al. (2015). A hybrid ant colony and simulated annealing algorithm to solve the container stacking problem at seaport terminal, 14–19.
15. Laik, N., & Hadjiconstantinou, E. (2008). Container assignment and gantry crane deployment in a container terminal: A case study. *Maritime Economics and Logistics, 10*, 90–107.

16. Moussi Ryadh, A., & Yassine, T. Galinho. (2011). Modern methods of combinatorial optimization for solving optimization problems in a containers terminal. *Thesis published in, 2012,* 73–79.
17. Murty, K. G. (2007). Gantry crane pools and optimum layouts for storage yards of container terminals. *Journal of Industrial and Systems Engineering, 1*(3), 190–199.
18. Lee, B. K., & Kim, K. H. (2013). Optimizing the yard layout in container terminals. *OR Spectrum, 35,* 363–398.
19. Gven, C., & Eliiyi D. E. (2014). Trip allocation and stacking policies at a container terminal. *Transportation Research Procedia3*(2014), 565–573.

Chapter 10
Hidden Markov Model Classifier for the Adaptive ACS-TSP Pheromone Parameters

Safae Bouzbita, Abdellatif El Afia and Rdouan Faizi

Abstract The Hidden Markov Models (HMM) are a powerful statistical techniques for modeling complex sequences of data. In this paper a Hidden Markov Model classifier is a special kind of these models that aims to find the posterior probability of each state given a sequence of observations and predicts the state with the highest probability. The purpose of this work is to enhance the performance of Ant Colony System algorithm applied to the Travelling Salesman Problem (ACS-TSP) by varying dynamically both local and global pheromone decay parameters based on the Hidden Markov Model algorithm, using two indicators: Diversity and Iteration that reflect the state of research space in a given moment. The proposed method was tested on several TSP benchmark instances, which compared with the basic ACS, the combination of Fuzzy Logic Controller (FLC) and ACS to prove the efficiency of its performance.

10.1 Introduction

The Ant Colony System (ACS) is one of the powerful variants of the original Ant System introduced by Dorigo and Gambardella (1997) [5–7], that characterized by the addition of local pheromone updating rule to shuffle the ants' tours by changing dynamically the amount of pheromone $\tau(r, s)$ on the visited edges (r, s). In this way, edges become slightly less desirable and the search becomes more diversified [4, 8, 9, 11]. In fact, edges are initialized with an initial amount of pheromone τ_0,

S. Bouzbita (✉) · A. El Afia · R. Faizi
ENSIAS - Mohammed V University, Rabat, Morocco
e-mail: safae.bouzbita@gmail.com

A. El Afia
e-mail: a.elafia@um5s.net.ma

R. Faizi
e-mail: rdfaizi@gmail.com

© Springer International Publishing AG, part of Springer Nature 2019
E.-G. Talbi and A. Nakib (eds.), *Bioinspired Heuristics for Optimization*,
Studies in Computational Intelligence 774,
https://doi.org/10.1007/978-3-319-95104-1_10

then every time an ant k chooses an edge (r_k, s_k), it applies the local pheromone updating rule using the local pheromone decay parameter $\xi \in [0, 1]$ according to this equation:

$$\tau(r_k, s_k) := (1 - \xi)\tau(r_k, s_k) + \xi\tau_0 \tag{10.1}$$

To make the search more directed, ACS algorithm perform the global pheromone updating rule to the globally best ant's tour L_{best}, that is to say only the best ant k is allowed to modify the amount of pheromone. Global pheromone updating is performed using the global pheromone parameter $\rho \in [0, 1]$ after all ants have terminated their tours according to the following equation:

$$\tau(r_k, s_k) := (1 - \rho)\tau(r_k, s_k) + \frac{\rho}{L_{best}} \tag{10.2}$$

As it is obvious, the performance of ACS depends strongly to the pheromone information. In most applications of ACS algorithm, the ξ and ρ parameters which are corresponding of the pheromone updating rules are set constant throughout the run of the algorithm. However, adapting ACS parameters can improve the performance of the algorithm at computational time. In this work, instead of using constant values of ξ and ρ parameters, we performed a dynamic adaptation using the Hidden Markov Model (HMM) algorithm.

The HMM is a generative probabilistic model that tries to model the process generating the training sequences, or more precisely, the distribution over the sequences of observations, unlike the other proposed deterministic methods reviewed in this paper: The fuzzy logic system which is characterized by a gradually stable behaviour, the genetic algorithm that depends strongly on the initial population where a wrong initialization of population can lead to very bad solutions, and the linear regression which looks only at linear relation between variables which is not always correct, also, the linear regression assumes that the variables are independent which is not the case of ACS algorithm.

The remainder of this paper is organized as follows: First, we summarize the related work, then we present the proposed method in Sects. 10.3, 10.4 we discuss the results and experiments on the Travelling Salesman Problem (TSP). Finally, in Sect. 10.5 conclusions and future work are presented.

10.2 Related Work

In the last years, adapting the ACS parameters become a hot topic. Many studies have been conducted to propose improved ACS algorithm by adapting one or several parameters. In this section we review the adaptation of pheromone decay parameters ξ and ρ chronologically.

Pilat and White [21] used a genetic algorithm to evolve the values of ρ, β and q_0. The main idea consist on initializing the algorithm with 4 ants, where each ant is characterized by its own parameters' values, then the two ants with the best fitness are chosen to produce two children by applying the genetic algorithm, after that the two worst ants' parameters replaced by the two produced.

Randall [22] proposed a self-adaptation method using the following equation $p_i = l_i + \dfrac{w_i}{P}(u_i - l_i)$ to identify suitable values for ρ, β, q_0 and ξ for the TSP and the Quadratic Assignment Problems (QAP). Where, P is a constant that defines the granularity of parameter values, p_i is the value of the ith parameter, l_i is the lower bound value of the ith parameter, w_i is the discretized division chosen for the ith parameter and u_i is the upper bound value of the ith parameter.

Gaertner and Clark in [10] proposed a genetic algorithm for automatically determining the optimal combination of β, ρ and q_0 parameters for a given TSP problem. In this approach the ants communicate with the environment using the genetic process to replace the old population of ants with the new one.

Hao et al. [14] used a particle swarm optimization (PSO) approach for dynamically adapting ρ, β and q_0 parameters, in which each ant is characterized by its own parameter setting and the particle position X_k presents the parameters ρ_k, β_k and q_{0k} of the ant k. If the particle's best solution of the current iteration is better than its previous best solution replace the best previous by the current one.

Ling and Luo [17] proposed the use of an Artificial Fish Swarm Algorithm (AFSA) for adapting ρ, α and Q. As in the proposed method of Hao et al. [14]. the state X_k of an Artificial Fish presents the parameters ρ_k, α_k and Q_k of an ant k of the colony, then the adaptation of parameters is done according to AFSA's strategies and formulas by replacing the previous global solution with the current best found one. The main difference between this work and the Hao's [14] is that Ling and Luo use the same parameter setting for all ants.

Gomez et al. [12] proposed a PSO method to perform the search over a wide range of available sets of parameters ξ, ρ, α, β, q_0 and the number of ants m. The PSO was used to compute the best fitness of ACS algorithm and return the set of parameters according to this fitness.

Hao et al. [13] presented a tuning method for the ρ parameter depending on the quality of solution built by an artificial ant, according to this equation: $\rho_m = \dfrac{L_m^{-1}}{(L_m^{-1} + P_m^{-1})}$. Where, L_m is the length of the tour found by the ant m, L_p is the length of the tour constructed based on the pheromone matrix.

Cai and Huang in [3] proposed an automatic parameters configuration to adapt the parameter β and ρ, by using two adaptive strategies. First, a new transition rule with adaptive weight parameter is set, to adjust the relative weight of pheromone trail and heuristic value. Second, a tuning rule for ρ parameter is ran based on the quality of the solution generated by artificial ants.

Melo et al. [19] proposed a multi-colony ACS algorithm, where various autonomous colonies of ants try to solve the same problem simultaneously. Each colony has its parameter settings for α, β, ρ and q_0. The exchanging of information among the colonies is ensured by a migration mechanism. The proposed algorithm also includes a mutation operator that modifies the parameter settings of the worst colony by the value of the same parameter of the best colony.

Cai et al. in [26] proposed an adaptive pheromone decay parameter ρ based on the quality of the solutions constructed by ants, using the same equation proposed by Hao [13], the only difference between the two approaches is the demonstration of the convergence of the proposed approach.

Kumar et al. In [15] proposed an adaptive pheromone values for the 1-ANT variant according to some suggested theorems that depend on the edges contained in the path of the best solution. Also, they discussed how to select the pheromone parameter ρ using some proposed formulas.

Liu-ai and Wen-qing [18] used a genetic algorithm to adapt a combination of four parameters which are: ρ, α, β, Q by balancing exploration and exploitation abilities of the search space. In other words, the genetic algorithm's fitness was built according to the objective function of the ACS algorithm.

Bouzbita et al. [2] proposed an improved Ant Colony System (ACS) algorithm based on a Hidden Markov Model (HMM) so as dynamically adapt the local pheromone decay parameter ξ. The proposed algorithm uses two indicators as hidden states and five values for ξ parameter as observation.

Aoun et al. [1] integrated Hidden Markov Model Particle swarm optimization (HMM) in APSO to have a stochastic state classification at each iteration using Expectation-Maximization as an online learning for HMM parameters.

Olivas et al. in [20] introduced an approach for adapting the ρ parameter in a dynamical manner in rank-based Ant colony variant, using a Fuzzy Logic system which controls the abilities for the diversification and intensification of the search space.

Unlike the Fuzzy Logic, the genetic algorithm and the PSO, HMM algorithm is a learnable stochastic automate that has the ability to learn from the data produced by the process described by the HMM itself. The HMM is widely used for dealing with ranging and sequences from speech recognition, process monitoring, pattern recognition, and computational biology due to its stochastic modeling.

Thereby, the use of the HMM algorithm in our approach consist on learning from the exploitation and exploration in the search space by two chosen indicators to find good solution in an appropriate time.

10.3   Proposed Method

In this section we describe the dynamical adaptation of the pheromone decay parameters ξ and ρ using the Hidden Markov Model (HMM). To perform this adaptation two HMM alternatives were proposed. The first alternative consists in the Baum–

Welch algorithm which is considered as a learning method for the best presentation of HMM parameters. The second alternative is the Viterbi algorithm which is used here as a classification method. The main idea of the proposed method is to find at each iteration of the ACS algorithm the most suitable state that best explains the solution found by ants in which, we increase or decrease the values of ξ and ρ according to the found state. Thus, to determine the values of ξ and ρ several tests were done to find the appropriate values.

10.3.1   Hidden Markov Model

The Hidden Markov Model (HMM) is a powerful statistical tool for modeling sequences of data that can be characterized by an underlying stochastic process that is hidden producing a sequence of observations.The HMM is defined by five elements: (S, V, A, B, π), where

- $S = \{S_1, \ldots, S_T\}$ is the set of the hidden states
- $V = \{V_1, \ldots, V_M\}$ is the set of the observation symbols per state
- $A = [a_{ij}]$ is a matrix of transition probabilities from state S_i at time t to state S_j at time t+1,
- $B = [b_{jk}]$ is the emission matrix of observing a symbol V_k from a state S_i
- $\Pi = [\pi_{i \in \{1,T\}}]$ is the initial probability where π_i is the probability of being in the state S_i.

In this work, we define two Hidden Markov Models corresponding respectively to the pheromone decay parameters ξ and ρ, which are characterized by five hidden states corresponding to the values of evaporation parameters ξ and ρ respectively: High (H), Medium High (MH), Medium (M), Medium Low (ML), Low (L), then

$$S = \{H, MH, M, ML, L\}.$$

And the observation sequence is a combination of two parameters named Iteration and Diversity.

$$Iteration = \frac{Current\ iteration}{Total\ of\ iterations}; \tag{10.3}$$

$$Diversity = \frac{1}{m} \sum_{i=1}^{m} \sqrt{\sum_{j=1}^{n} (x_{ij}(t) - \bar{x}_j(t))^2} \tag{10.4}$$

Where, Current iteration is the number of fulfilled iterations, total of iteration is the total number of iterations for testing the algorithm, m is the population size, i is the number of the ant, n is the total number of dimensions, j is the number of the dimension, x_{ij} is the j dimension of the ant i, $\bar{x}_j$ is the j dimension of the current best ant of the colony.

The diversity function used to measure the degree of dispersion of the ants with respect to the current best ant of the colony. This measure reflects the degree of exploration among the search space which considers the dimensions of the ants to compute the value of variety. Thus, when the diversity of ants is low, we need to use more exploration to find another solutions by increasing the value of ρ and decreasing the value of ξ and vice versa.

In this work, we divided each of the two indicators into three equal intervals. These are Low (L), Medium (M) and High (H). The combination of these two indicators gives nine possible observations then we have

$$V = \{LL, LM, LH, ML, MM, MH, HL, HM, HH\}.$$

The initial model for ξ parameter is as follows:

The array of the initial state probabilities is defined by equiprobable values:

$$\Pi = [\pi_1, \pi_2, \pi_3, \pi_4, \pi_5] = \begin{bmatrix} \dfrac{1}{5} & \dfrac{1}{5} & \dfrac{1}{5} & \dfrac{1}{5} & \dfrac{1}{5} \end{bmatrix}$$

We define the state transition matrix and the emission matrix respectively as following:

$$A = (a_{ij}) = \begin{bmatrix} 0.5 & 0 & 0 & 0 & 0.5 \\ 0.5 & 0.5 & 0 & 0 & 0 \\ 0 & 0.5 & 0.5 & 0 & 0 \\ 0 & 0 & 0.5 & 0.5 & 0 \\ 0 & 0 & 0 & 0.5 & 0.5 \end{bmatrix} \quad B = (b_{jk}) = \begin{bmatrix} 0 & 0 & 0 & 0 & 0 & 0 & 0 & 0 & 1 \\ 0 & 0 & 0 & 0 & 0 & 0.5 & 0 & 0.5 & 0 \\ 0 & 0 & \dfrac{1}{3} & 0 & \dfrac{1}{3} & 0 & \dfrac{1}{3} & 0 & 0 \\ 0 & 0.5 & 0 & 0.5 & 0 & 0 & 0 & 0 & 0 \\ 1 & 0 & 0 & 0 & 0 & 0 & 0 & 0 & \end{bmatrix}$$

The initial model for ρ parameter is as follows:

The array of the initial state probabilities is defined as the initial state probabilities for ξ parameter:

$$\Pi = [\pi_1, \pi_2, \pi_3, \pi_4, \pi_5] = [\dfrac{1}{5} \ \dfrac{1}{5} \ \dfrac{1}{5} \ \dfrac{1}{5} \ \dfrac{1}{5}]$$

The state transition matrix and the emission matrix are defined respectively as following:

$$
A = (a_{ij}) = \begin{bmatrix} 0.5 & 0.5 & 0 & 0 & 0 \\ 0 & 0.5 & 0.5 & 0 & 0 \\ 0 & 0 & 0.5 & 0.5 & 0 \\ 0 & 0 & 0 & 0.5 & 0.5 \\ 0.5 & 0 & 0 & 0 & 0.5 \end{bmatrix} \quad B = (b_{jk}) = \begin{bmatrix} 1 & 0 & 0 & 0 & 0 & 0 & 0 & 0 & 0 \\ 0 & 0.5 & 0 & 0.5 & 0 & 0 & 0 & 0 & 0 \\ 0 & 0 & \frac{1}{3} & 0 & \frac{1}{3} & 0 & \frac{1}{3} & 0 & 0 \\ 0 & 0 & 0 & 0 & 0 & 0.5 & 0 & 0.5 & 0 \\ 0 & 0 & 0 & 0 & 0 & 0 & 0 & 0 & 1 \end{bmatrix}
$$

This model was developed according to the concept, that is, a low value of ξ accelerate the convergence speed, while high value of ξ weak the attraction of short edges. In contrast, a low value of ρ allows a long persistence of pheromone. Thus ants can exploit the best solutions while high values of ρ encourage the exploration of search space by forgetting previously attractive solutions, and focusing on new information. After building the model and setting the parameters, we use the Baum–Welch algorithm to re-estimate the HMM parameters $\lambda = (A, B, \pi)$ in an appropriate way.

10.3.1.1   Learning HMM Parameters

At the end of each iteration we perform an Online Learning for HMM parameters by the Baum–Welch algorithm. The Baum–Welch algorithm uses the expectation maximization algorithm to find the maximum likelihood estimate of the HMM parameters given a sequence of observed data.

First, we initialize the model $\lambda = (A, B, \pi)$ with best guess values, in our case the values were concluded from the rule base of FLC, then we execute an iterative process to re-estimate the initial, transition and emission probabilities matrices which mostly fit the observed data set.

At first, Baum–Welch algorithm uses the Forward algorithm to find $\alpha_{t,i}$: the probability of the subsequence $(o_1, o_2, ..., o_t)$ ending at time t, in state S_i and $P(O/\lambda)$: the probability of the occurrence of the observation sequence O given the model λ. The second algorithm used by Baum–Welch is the Backward algorithm which returns $\beta_{t,i}$: the probability of observing $(o_{t+1}, ..., o_T)$ that starts at time (t+1) and ends at time t at state S_i.

To estimate the new probabilities, Baum–Welch algorithm uses two new matrices: ξ and γ where the coefficients $\xi_t(i, j)$ represent the probability to be at state S_i at time t and moving to state S_j at time (t+1) given the model λ and the set of observation O. The coefficients $\gamma_{t,i}$ represent the probability to be at state S_i at time t given the model λ and the set of observation O.

Algorithm 7: Baum–Welch

Input: O=o_1,o_2,...,o_T, S={H,MH,M,ML,L}, $\lambda = (A, B, \pi)$
Output: Re-estimated λ
repeat
1   **Forward:**
       for *i=1 to 5* **do** $\alpha_{1,i} = \pi_i * b_{i,o_1}$;
       for *t=1 to T-1* **do**
              for *j=1 to 5* **do** $\alpha_{t+1,j} = \sum_{i=1}^{N} \alpha_{t,i} * a_{ij} * b_{j,o_{t+1}}$;
       end
       P(O/λ)=$\sum_{i=1}^{3} \alpha_{T,i}$
2   **Backward:**
       for *i=1 to 5* **do** $\beta_{T,i}$=1;
       for *t=T to 1* **do**
              for *i=1 to 5* **do** $\beta_{t,i} = \sum_{j=1}^{5} a_{ij} * \beta_{t+1,j} * b_{j,o_{t+1}}$;
       end
3   **Update:**
       for *t=1 to T* **do**
              for *i=1 to 5* **do**
                     for *j=1 to 5* **do** $\xi_{t,i,j} = \dfrac{\alpha_{t,i} a_{ij} * b_{j,o_{t+1}} * \beta_{t+1,j}}{P(O/\lambda)}$;
              end
              $\gamma_{t,i} = \dfrac{\alpha_{t,i} * \beta_{t,i}}{P(O/\lambda)}$
       end
       $\pi_i = \gamma_{1,i}$, $a_{ij} = \dfrac{\sum_{t=1}^{T-1} \xi_t(i,j)}{\sum_{t=1}^{T-1} \gamma_{t,i}}$, $b_{i,k} = \dfrac{\sum_{t=1 \cap o_t=q_k}^{T} \gamma_{t,i}}{\sum_{t=1}^{T} \gamma_{t,i}}$
       until *no increase of P(O/λ) or no more iterations are possible to do*;

10.3.2 HMM Classifier

The Viterbi algorithm is used to find the most likely explanation or the most likely state that generated a particular sequence of observations by finding a maximum over all possible state sequences. This algorithm has the ability to deal with the entire sequences of hidden states from the beginning of the algorithm till the current iteration,and then make a decision base on the whole history, which makes it advantageous compared to other algorithms that depend only on the information of the current iteration. The main idea of the proposed method is to find at each iteration of the ACS algorithm the most suitable hidden state that best explains the solution found by ants in which, we increase or decrease the values of ξ and ρ according to the found state.

Algorithm 8: Viterbi

Input: $O=\{o_1,o_2,...,o_T\}$, S, A, B, π
Output: The classified state z_t
1 [A, B, π]=Baum–Welch($O=\{o_1,o_2,..., o_T\}$,S, A, B, π)

2 **Initialization:**

  for $i:=1$ to 5 **do** $\alpha_{1,i} = \pi_i * b_{i,o_1}$, $\phi_{1,i} = 0$;
3 **Recursion:**

  for $t:=2$ to T, $j:=1$ to 5 **do** $\alpha_{t,j} = max_{i\in S}[\alpha_{t-1,i} * a_{ij} * b_{j,o_t}]$
  $\phi_{t,j} = argmax_{i\in S}[\alpha_{t-1,i} * a_{ij}]$;
4 **Termination:**

  $p = max_{i\in S}[\alpha_{T,i}]$, $z_T = argmax_{i\in S}[\alpha_{T,i}]$
5 **Reconstruction:**

  for $t = T - 1, T - 2, ..., 1$, **do** $z_t = \phi_{t+1,z_{t+1}}$;

  Result: The classified state z_t

10.3.3 Proposed Algorithm

This section gives more details about the hybridization of the ACS and the HMM algorithms. It is divided into three pseudo-codes: For the first algorithm, after creating an HMM model, we proceed with the combination of Iteration and diversity parameters calculated at the end of each iteration to the Baum–Welch algorithm as observation sequence to best fit the observed data of Hidden Markov Model. The combination of the two indicators that belongs to $V = \{LL, LM, LH, ML, MM, MH, HL, HM, HH\}$ is done according to the following classification:

Algorithm 9: Classification of indicators

Input: {Iteration, Diversity, L_{best} , MaxIter}

Output: {Observation: O}

Initialization:
if $0 < Iteration \leq \dfrac{MaxIter}{3}$ **then** Iteration = L;
else if $\dfrac{MaxIter}{3} < Iteration \leq \dfrac{2 * MaxIter}{3}$ **then** Iteration = M;
else Iteration= H;
if $0 < Diversity \leq \dfrac{1}{3}$ **then** Diversity= L;
else if $\dfrac{1}{3} < Diversity \leq \dfrac{2}{3}$ **then** Diversity= M;
else Diversity= H;
Return O=Iteration Diversity;

According to the observation sequence returned by the Algorithm 2, the following algorithm tries to find the suitable state's path that most explain the obtained observation sequence. In our case we take the last state of the found path as the desired state:

Algorithm 10: Classification of state

Input: {state, S}

Output: $\{\xi, \rho\}$

if *state=L* then $\xi = \dfrac{5}{6}$ and $\rho = \dfrac{1}{6}$;

else if *state=ML* then $\xi = \dfrac{4}{6}$ and $\rho = \dfrac{2}{6}$;

else if *state=M* then $\xi = \dfrac{3}{6}$ and $\rho = \dfrac{3}{6}$;

else if *state=MH* then $\xi = \dfrac{2}{6}$ and $\rho = \dfrac{4}{6}$;

else $\xi = \dfrac{1}{6}$ and $\rho = \dfrac{5}{6}$;

To better understand the hybridization of the ACS and the HMMM algorithms we made, we propose the next algorithm which resumes everything we have come to say above. First we run ACS algorithm, where ants build their tours by choosing the next node using the pseudo random-proportional action choice rule. At the end of each iteration we build an HMM using the data obtained from calculating the Iteration and the Diversity indicators. Then, we use the Baum–Welch algorithm to re-estimate the parameters of the proposed HMM. After that, the Viterbi algorithm was used as a classifier method to find the appropriate sequence of state that generated a particular sequence of observation, the concerned state is the last one in the path.

10.4   Experimental Results and Comparison

To test the performance of the proposed adaptive ACS algorithm, we compared it with the standard ACS algorithm on several TSP instances.

10.4.1   Experiment Setup

According to [24] the best known values of ACS algorithm parameters are $\beta = 2$, $\rho = 0.1$, *and* $q_0 = 0.9$. We have also compared it with the Fuzzy Logic Controller with the same initial parameters values to give logical comparisons with the proposed method. The initial position of ants is set randomly on all experiments.

Algorithm 11: The pseudocode of our proposed algorithm

---

1 **Initialization:**

Initialize the ant number m,set parameters α,β, ρ, ξ and q_0,
Initialize pheromone trails with $\tau_0 = 1/L_{NN}$

2 **Construction Solution:**

while *termination condition not met* **do**
    O:={};

    Place ants in starting node randomly

    for *t:=1 to Maxiter* **do**
        foreach *ant k in the population* **do**
            if $q < q_0$ **then**
                $s_k = argmax_{s_k \in J_k(r_k)} [\tau(r_k, s_k)]^\alpha [\eta(r_k, s_k)]^\beta$

            end

            else $p_{rs}^k = \dfrac{[\tau(r_k, s_k)]^\alpha . [\eta(r_k, s_k)]^\beta}{\sum_{u_k \in J_k(r_k)} [\tau(r_k, u_k)]^\alpha [\eta(r_k, u_k)]^\beta}, \quad if \quad s_k \in J_k(r_k)$

            determinate state s_k from wheel roulette algorithm;

            $\tau(r_k, s_k) := (1 - \xi)\tau(r_k, s_k) + \xi\tau_0$
        end
        for *k:=1 to m* **do** Compute L_k;

        Compute L_{best}

        Compute Iteration and Diversity according to (3)

        o_t:= algorithm 3 (Iteration, Diversity, L_{best} , MaxIter);

        O= {O, o_t};

        state=Viterbi(O, S, A, B, π)

        Update ξ and ρ values according to Algorithm 4 (state, S)

    end
    Update the global best solution;

3   **Global pheromone update:**

    $\tau(r_k, s_k) := (1 - \rho)\tau(r_k, s_k) + \dfrac{\rho}{L_{best}}$
end
Return L_{best}

Table 10.1 shows the TSP benchmark instances used in this study which were chosen from the TSPLIB [23] according to the most common used instances in the literature. The algorithm is developed on MATLAB.

Each result in the Table 10.2 above is a best found length over 30 runs with 1000 iterations and different population sizes (m = 10, 20, 30, n/4, n), where n is the size of the problem.

Table 10.1 Characteristics of TSP benchmark instances

TSP	Number of cities	Best known solutions
att48	48	10628
berlin52	52	7542
ch130	130	6110
D198	198	15780
eil51	51	426
eil76	76	538
eil101	101	629
kroA100	100	21282
lin105	105	14379
pr226	226	80369

10.4.2 Comparison on the Solution Accuracy

Before moving on, we should mention in passing that we have implemented the HMM into the ACS algorithm in three ways: First, we have modified only the local pheromone decay parameter ξ which is named in the table by "LC". Second, we have modified only the global pheromone decay parameter ρ which is named in the table by "GB". Third, we have modified both parameters in the same computation time which is mentioned in the table by "LC+GB". The same procedure was applied to the integration of fuzzy logic (FL) into the ACS algorithm.

For the solution accuracy, we can observe from the Table 10.2 that instances of small sizes have found the same minimum length. On the other hand, the proposed algorithm attains a better solution accuracy comparing to the basic ACS with a

Table 10.2 Summary of results using ACSHMM algorithm for TSP instances

TSP	ACS	ACSFL			ACSHMM		
		LC	GB	LC+GB	LC	GB	LC+GB
att48	33523.70	33523.70	33523.70	33523.70	33523.70	33523.70	33523.70
berlin52	7544.36	7544.36	7544.36	7544.36	7544.36	7544.36	7544.36
ch130	6176	6200.34	6224.31	6246.39	6235.33	6187.08	6172.50
D198	16206.75	16096.55	16233.42	16228.65	16118.06	16041.18	16099.27
eil51	431.67	428.98	432.72	431.11	430.60	432.73	428.98
eil76	547.39	547.73	553.61	550.60	549.79	547.54	544.36
eil101	653.92	647.11	651.44	657.83	661.21	643.29	647.83
kroA100	21445.83	21294.39	21519.75	21362.97	21285.44	21285.44	21285.44
lin105	14382.99	14410.19	14416.64	14382.99	14382.99	14382.99	14382.99
pr226	81257.96	80632.94	80691	81355.92	81311.20	80410.82	80398.69

constant set of parameters and the combination of FL and ACS for the big size problems.

For the ACSFL algorithm, we can see from the results that varying the ξ parameter gives better results than varying the ρ parameter or varying both parameters. Also, the adaptation of varying the two parameters in the same time gives better results than varying the global parameter ρ in most of the instances.

However, For the ACSHMM algorithm there is a contrast in the results, where we obtain better results when varying only the local parameter ξ or varying both parameters ξ and ρ.

In general, it can be noted that the adaptation of ξ and ρ in the same time for the ACSHMM shows better solutions in most of the instances. Also, the adaptation of ξ and ρ parameters separately outperform the original ACS and the FL system in all its variants.

10.4.3 Comparison on the Convergence Speed

In this section, we display some figures obtained from running the basic ACS algorithm and the proposed one with the same parameters, that draws the best found solution in each iteration for each algorithm.

From the figures above, we can observe that the proposed algorithm ACSHMM in the case of varying ξ and ρ simultaneously found better solutions than the basic ACS with faster convergence.

From Fig. 10.1 we can observe a clear difference between the obtained solutions in both convergence speed and solution accuracy. For Fig. 10.2 the basic ACS converged earlier than the proposed algorithm but to a bad solution.

As for Fig. 10.3 the ACSHMM could achieve better results compared to the standard ACS and Fuzzy logic.

From Fig. 10.4 we can observe that ACSHMM achieve better results in both convergence speed and solution accuracy.

10.4.4 Statistical Test

In Table 10.3 we report the p-value for the Wilcoxon Rank Test in a pair-wise comparison statistical procedure under significance level $\alpha = 0.05$. This is the recommended statistical test method used in many other researches [16, 25]. The preference of using the Wilcoxon procedure came from its consideration to the quantitative differences in the algorithms performance. From the results in Table 10.3 we can assume that

Fig. 10.1 Sample run on
D198.tsp

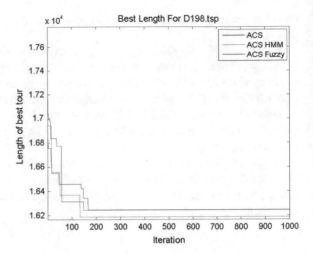

Fig. 10.2 Sample run on
pr226.tsp

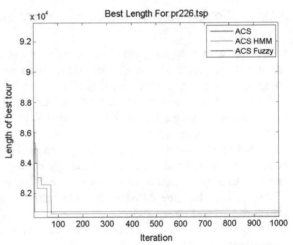

our proposed approach can achieve better results with level of significance of 5 percent compared with other methods. Thus, the calculated p-value for those algorithms is below the significance level in most benchmark instances comparisons which is mean the control algorithm ACSHMM (LC+GB) outperforms the other algorithms, on the other hand just few times the p-value fail to reveal significant differences which means that the proposed algorithm could not achieve better solutions.

Fig. 10.3 Sample run on lin105.tsp

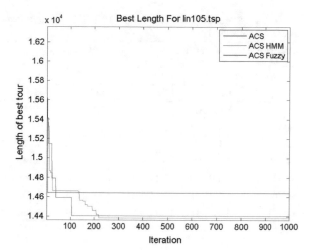

Fig. 10.4 Sample run on eil101.tsp

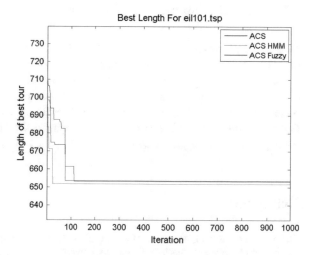

Table 10.3 Statistical validation for the 10 TSP benchmark instances with ACSHMM (LG+LC) as control algorithm

ACSHMM LC+GB versus	ACSHMM		ACSFL			ACS
	LC	GB	LC	GB	LC+GB	
att48	3.05E-01	7.70E-01	1.71E-01	5.34E-02	9.52E-01	3.90E-01
berlin52	7.00E-03	3.67E-02	1.91E-02	9.20E-01	1.84E-01	1.90E-01
ch130	1.15E-01	9.17E-01	3.00E-04	6.79E-02	2.34E-01	3.18E-01
D198	5.99E-01	9.77E-04	5.26E-05	2.14E-01	1.69E-02	8.11E-04
eil51	2.38E-04	2.11E-02	6.35E-01	5.08E-06	4.20E-04	6.19E-04
eil76	4.55E-01	6.15E-02	9.35E-01	1.12E-02	8.50E-02	2.97E-01
eil101	3.01E-11	2.71E-11	1.22E-01	4.42E-06	2.13E-05	3.99E-02
kroA100	8.30E-01	2.42E-01	6.66E-02	4.41E-06	9.02E-04	2.56E-02
lin105	7.70E-02	2.80E-01	9.85E-02	1.40E-04	4.77E-01	3.91E01
pr226	3.25E-01	3.60E-03	9.29E-01	1.38E-06	6.91E-04	6.70E-03

10.5   Conclusion and Future Work

This paper proposed a new method to control the pheromone parameters ξ and ρ that have a crucial impact on the performance of ACS algorithm using the HMM algorithm. This approach outperforms the basic ACS and FL algorithm in terms of both solution quality and convergence speed. In our future works we specify other indicators that best represent the exploration and exploitation abilities of Ant Colony System search space in order to enhance the control of HMM into ACS. Also, problems of big sizes will be treated.

References

1. Aoun, O., Sarhani, M., & El Afia, A. (2016). Investigation of hidden markov model for the tuning of metaheuristics in airline scheduling problems. *IFAC-PapersOnLine, 49*(3), 347–352.
2. Bouzbita, S., El Afia, A., Faizi, R., & Zbakh, M. (2016). Dynamic adaptation of the acs-tsp local pheromone decay parameter based on the hidden markov model. In *2016 2nd international conference on cloud computing technologies and applications (CloudTech)* (pp. 344–349). New York: IEEE.
3. Cai, Z., & Huang, H. (2008). Ant colony optimization algorithm based on adaptive weight and volatility parameters. In *Second international symposium on intelligent information technology application, 2008. IITA'08* (Vol. 2, pp. 75–79). New York: IEEE.
4. Dorigo, M., Birattari, M., & Stutzle, T. (2006). Ant colony optimization. *IEEE Computational Intelligence Magazine, 1*(4), 28–39.
5. Dorigo, M., & Blum, C. (2005). Ant colony optimization theory: A survey. *Theoretical Computer Science, 344*(2–3), 243–278.

6. Dorigo, M., & Gambardella, L. M. (1997). Ant colonies for the travelling salesman problem. *biosystems*, *43*(2), 73–81.
7. Dorigo, M., & Gambardella, L. M. (1997). Ant colony system: A cooperative learning approach to the traveling salesman problem. *IEEE Transactions on Evolutionary Computation*, *1*(1), 53–66.
8. Dorigo, M., & Stützle, T. (2010). Ant colony optimization: Overview and recent advances. In *Handbook of metaheuristics* (pp. 227–263). Berlin: Springer.
9. Erol, A. H., Er, M., & Bulkan, S. (2012). Optimizing the ant colony optimization algorithm using neural network for the traveling salesman problem. In *Actas de la Conferencia Internacional de* (2012)
10. Gaertner, D., & Clark, K. L. (2005). On optimal parameters for ant colony optimization algorithms. In *IC-AI* (pp. 83–89).
11. Gilmour, S., & Dras, M. (2005). Understanding the pheromone system within ant colony optimization. In *AI 2005: Advances in Artificial Intelligence* (pp. 786–789).
12. Gomez-Cabrero, D., Armero, C., & Ranasinghe, D. N. (2007). The travelling salesmans problem: A self-adapting pso-acs algorithm. In *International conference on industrial and information systems, 2007. ICIIS 2007* (pp. 479–484). New York: IEEE.
13. Hao, Z., Huang, H., Qin, Y., & Cai, R. (2007). An aco algorithm with adaptive volatility rate of pheromone trail. *Computational Science-ICCS, 2007*, 1167–1170.
14. Hao, Z. F., Cai, R. C., & Huang, H. (2006). An adaptive parameter control strategy for aco. In *2006 International Conference on Machine Learning and Cybernetics* (pp. 203–206). New York: IEEE.
15. Kumar, P., & Raghavendra, G. (2011). A note on the parameter of evaporation in the ant colony optimization algorithm. *International Mathematical Forum*, *6*, 1655–1659.
16. LaTorre, A., Muelas, S., & Peña, J. M. (2015). A comprehensive comparison of large scale global optimizers. *Information Sciences*, *316*, 517–549.
17. Ling, W., & Luo, H. (2007). An adaptive parameter control strategy for ant colony optimization. In *2007 international conference on computational intelligence and security* (pp. 142–146). New York: IEEE.
18. Liu-ai, W., & Wen-Qing, F. (2012). A parameter model of genetic algorithm regulating ant colony algorithm. In *2012 IEEE ninth international conference on e-business engineering (ICEBE)* (pp. 50–54). New York: IEEE.
19. Melo, L., Pereira, F., & Costa, E. (2009). Mc-ant: A multi-colony ant algorithm. In *International conference on artificial evolution (evolution artificielle)* (pp. 25–36). Berlin: Springer.
20. Olivas, F., Valdez, F., & Castillo, O. (2015). Ant colony optimization with parameter adaptation using fuzzy logic for tsp problems. In *Design of intelligent systems based on fuzzy logic, neural networks and nature-inspired optimization* (pp. 593–603). Berlin: Springer.
21. Pilat, M. L., & White, T. (2002). Using genetic algorithms to optimize acs-tsp. In *International workshop on ant algorithms* (pp. 282–287). Berlin: Springer.
22. Randall, M. (2004). Near parameter free ant colony optimisation. In *International workshop on ant colony optimization and swarm intelligence* (pp. 374–381). Berlin: Springer.
23. Reinelt, G. (1995). *Tsplib discrete and combinatorial optimization*.
24. Stützle, T., López-Ibánez, M., Pellegrini, P., Maur, M., De Oca, M. M., Birattari, M., & Dorigo, M. (2011). Parameter adaptation in ant colony optimization. In *Autonomous search* (pp. 191–215). Berlin: Springer.
25. Veček, N., Črepinšek, M., & Mernik, M. (2017). On the influence of the number of algorithms, problems, and independent runs in the comparison of evolutionary algorithms. *Applied Soft Computing*, *54*, 23–45.
26. Zhaoquan, C., Huang, H., Yong, Q., & Xianheng, M. (2009). Ant colony optimization based on adaptive volatility rate of pheromone trail. *International Journal of Communications, Network and System Sciences*, *2*(08), 792.

Chapter 11
CHN and Min-Conflict Heuristic to Solve Scheduling Meeting Problems

Adil Bouhouch, Chakir Loqman and Abderrahim El Qadi

Abstract Meetings are important for teem works, However, scheduling a meeting that involves persons with different preferences and engagements remains a difficult task. This paper proposes a new hybrid approach to solve meeting scheduling problem (MSP). The proposed network combines the characteristics of neural networks and minimizing conflicts approach. Her the MSP is considerate as Constraint Satisfaction Problem, then we apply Continuous Hopfield Neural Netwok (CHN) and Conflicts Minimization Heuristics to solve a quadratic reformulation of the CSP-MSP model in other words the Min-Conflict heuristic will improve the given CHN solution. So, the performance of the network is compared with the existing scheduling algorithms under various experimental conditions.

11.1 Introduction

The Meeting Scheduling problem (MSP) is a very relevant problem for large organizations meetings. Each meeting involves group of agents and includes at least two participants. Agents represent the peoples participating in meeting, and the solution aims to found starting time for each meeting with the respect of some constraints. Each constraint limits for two meetings do not overlap each other if they share a last agent, they can't be started at the same times. The MSP can be viewed as a set of temporal and disjunctive constraints, so it's easier to formulate it as a Constraint

A. Bouhouch (✉)
Team TIM, High School of Technology, CEDoc-SFA Faculty of Sciences,
Moulay Ismail University, Meknes, Morocco
e-mail: bouhouch.adil@gmail.com

C. Loqman
Department of Informatics, Sciences Faculty, Dhar Mehraz,
Sidi Mohammed Ben Abdellah University, Fez, Morocco

A. El Qadi
LASTIMI, High School of Technology, Mohammed V University of Rabat Morocco,
Rabat, Morocco

© Springer International Publishing AG, part of Springer Nature 2019   171
E.-G. Talbi and A. Nakib (eds.), *Bioinspired Heuristics for Optimization*,
Studies in Computational Intelligence 774,
https://doi.org/10.1007/978-3-319-95104-1_11

Satisfaction Problem. An instance of the CSP involves a set of variables, a domain for each variable and a set of constraints. The goal is to assign values to variables so that all constraints are simultaneously satisfied. There has been a long history of using neural networks for combinatorial optimization and constraint satisfaction problems. [5, 17]. Recently, a new approaches based on continuous Hopfield neural network [12] have proved an efficient to solve optimization problem [6, 7]. However, symmetric Hopfield networks and similar approaches use steepest descent dynamics that converge to the closest local minimum of the energy, which result of its deterministic input-output interaction of the units in the network. Consequently, the network is not able to escape from local minima closed to starting point. Also, for Hopfield neural network with continuous dynamics case, each output unit can take any value between 0 and 1. So, the network can be stranded at a local minima which contains some units that remains in real values. If the last problem appears, we get incomplete solution especially for the affectation problem kind such as MSP. To overcome those problems, we introduce a new hybrid approach to solve MSP called CHN-MNC, the main idea of CHN-MNC is to improve the solution given by CHN, this improvement is enhanced by the known Min-conflict algorithm [21].

In this paper, our main objective is to improve our proposed approach in [3] (CHN-BMNC). The main idea of our previous work was to solve the MSP problem by operating on it CSP reformulation, then we involved the continuous Hopfield network many times, next we improve just the best solution founded by all CHN runs. This paper is organized as follows: In Sect. 11.2, we provide a formulation of MSP problem as a CSP and the CHN model to solve it. In Sect. 11.3, we compare two hybrid approaches to solve MSP (CHN-MNC and CHN-BMNC). The experimental results are presented in the last section.

11.2 CSP and Hopfield Neural Network to Solve Meeting Scheduling Problems

11.2.1 Meeting Scheduling Problem

MSP is a one of recurrent and real problems. It has been widely studied in several works for a long time. It is also easier to reformulate it as a constraints satisfaction problem (CSP). But many of work resolve MSP as Distribute Constraint Satisfaction Problem as most of them take into consideration only time and not location/geometry, [1, 25, 26]. On the other hand, there are many work like [2] which propose an efficient meeting-location algorithm that considers the time between two consecutive meetings. However, all private information about users is public. The systems in [4, 15] apply distributed meeting scheduling algorithms. Rasmussen and Trick [22] define the timetable constrained distance minimization problem, which is a sports scheduling problem applicable for tournaments where the total travel distance must be minimized. They use integer programming and constraint programming formulation for the problem. Integer programming is also applied by Wang et al. [30] for meeting scheduling among students and teachers in universities. For another scheduling approach to school meetings (between teachers and parents), Rinaldi et al. [23] set weights and build directed graphs based on the time slots, then they find shortest paths in the resulting graphs. Another kind of resolving approaches use meta-heuristic methods, Tsuchiya [29] uses a parallel algorithm for solving meeting schedule problems; the proposed system is composed of two maximum neural networks that interacts with each other. [18] The general definition of the MSP is as follows:

- A group S of m agents
- A set T of n meetings
- The duration of each meeting m_i is $duration_i$
- Each meeting m_i is associated with a set s_i of agents in S, that attend it.

So, each meeting attends a set of agent, and each meeting is associated with a location. An agent must participle at all meetings who attend them, since the scheduled time-slots for meetings in T must enable the participating agents to travel among their meetings.

Example 1 The table below presents an example of a MSP: And travelling time in time-units between different meeting locations can be modelled as:

Meeting	Location	Attending agents
m1	L1	A1, A3
m2	L2	A2, A3, A4
m3	L3	A1, A4
m4	L4	A1, A2

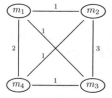

11.2.2 Quadratic Model of Constraint Satisfaction Problem

A large number of real problems such as artificial intelligence, scheduling, an assignment problem can be formulated as a Constraint Satisfaction Problem. A CSP consists in finding an assignment of all variable problems under constraints restriction. The CSP can be formulated as three sets:

- Set of N variables: $X = \{X_i, 1 \leq i \leq N\}$.
- Set of N variables domains: $D = \{D_i, 1 \leq i \leq N\}$ where each D_i contains the set of d_i range values for X_i.
- Set of M constraints: $C = \{C_i, 1 \leq i \leq M\}$.

Each constraint C_i associates an ordered subset variables which called the scope of C_i. The cardinality of this subset is noted the arity. The meeting scheduling problem as described above can be naturally represented as a constraint satisfaction problem (CSP) in the following way:

- A set of variables $T = m_1, m_2 \ldots, m_n$ the meetings to be scheduled
- Domains of values D - all weekly time-slots
- A set of constraints $C = C_{ij}, 1 < i < j < n$ - for every pair of meetings m_i, m_j that attend at last the same agent. For each constraint C_{ij} is given as : let t_i, t_j the selected time-slots for m_i and m_j respectively, there is a conflict if $|t_i - t_j| - duration_i < \tilde{D}(L_i, L_j)$

With $\tilde{D}$ is traveling time on unite slot-times between two meeting location. Then, we can easily reformulate CSP as a Quadratic Problem, by introducing for each CSP variable $m_i \in T$ a binary variable x_{ik}, where k varies over the range of m_i, given as follows:

$$x_{ik} = \begin{cases} 1, & \text{if the time-slot } t_k \text{ is assined to } m_i; \\ 0, & \text{otherwise.} \end{cases} \tag{11.1}$$

We deduce the objective function of its equivalent QP:

$$f(x) = \sum_{i=1}^{N} \sum_{j=1}^{N} \sum_{r=1}^{d_i} \sum_{s=1}^{d_j} x_{ir} x_{js} Q_{ijrs} \tag{11.2}$$

With the quadratic term:

$$Q_{ijrs} = \begin{cases} 1, & \text{if}(|r - s| - duration_i < \tilde{D}(L_i, L_j)) \\ 0, & \text{otherwise.} \end{cases} \tag{11.3}$$

Where r is time slot a Valid solutions must satisfy some strict constraints that can be written as linear equations:

$$\sum_{r=1}^{d_i} x_{ir} = 1 \quad , \quad for\ i = 1..N$$

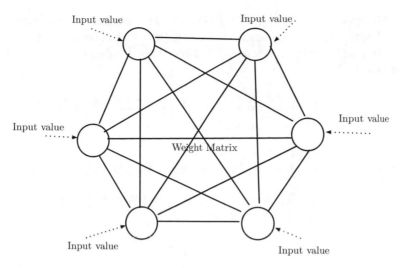

Fig. 11.1 Hopfield neural network architecture

11.2.3   Hopfield Neural Network

Hopfield neural network was proposed by Hopfield and Tank [12–14] (see Fig. 11.1). It was first applied to solve the combinatorial optimization problems.

This neural network model consists completely with n interconnected neurons and its dynamic is governed by the following differential equation:

$$\frac{dy}{dt} = -\frac{x}{\tau} + T x + i^b \qquad (11.4)$$

where

x : vector of neurons input
y : vector of output
T : the Matrix of weight between each neurones pairs

For each neurone, the input is governed by an activation function $x = g(y)$ which varies between 0 and 1. g(y) is given by:

$$g(y) = \frac{1}{2}\left(1 + \tanh\left(\frac{y}{u_0}\right)\right)$$

We define the energy function of CHN by:

$$E(x) = x^t T x + (i^b)^t x \qquad (11.5)$$

In this paper we use the capacity of CHN to resolve Quadratic model optimization. So we choose a CHN energy function similar to [8], which include the objectif function and relaxed by linear constraints. It is adapted to solve the quadratic formulation of the binary CSPs:

$$E(x) = \alpha \sum_{i=1}^{N} \sum_{r=1}^{d_i} \sum_{j=i}^{N} \sum_{s=1}^{d_j} x_{ir} x_{js} \, Q_{irjs} + \phi \sum_{i=1}^{N} \sum_{r=1}^{d_i} \sum_{s=r}^{d_i} x_{ir} x_{is} + \beta \sum_{i=1}^{N} \sum_{r=1}^{d_i} x_{ir} + \gamma \sum_{i=1}^{N} \sum_{r=1}^{d_i} x_{ir} (1 - x_{ir})$$

(11.6)

With $\alpha > 0$, β, ϕ and γ are real constants.

By corresponding (11.5) and (11.6), we deduce the link weight and biases of the network:

$$\begin{cases} T_{irjs} = -\alpha(1 - \delta_{ij})q_{irjs} - \delta_{ij}\phi + 2\delta_{ij}\delta_{rs}\gamma \\ i_{ir}^b = -\beta - \gamma \end{cases}$$

with $\delta_{ij} = \begin{cases} 1 \text{ si } i = j \\ 0 \text{ si } i \neq j \end{cases}$ is the Kronecker delta.

To ensure the convergence we make impose two conditions :

- The first one we must escape the case when two values are assigned to the same variable, $\exists i, j \in D(X_k)$ with $i \neq j$ and there corresponding neurons $x_k i = x_k j = 1$, so E(x) must verify:

$$E_{ir}(x) \geq 2\phi + \beta - \gamma \geq \varepsilon$$

- All variable must be assigned and escape the case where : $x_{ir} = 0 \quad \forall r \in \{1, \ldots, d_i\}$ so E(x) must verify also:

$$E_{ir}(x) \leq \alpha d + \beta + \gamma \leq -\varepsilon$$

with $d = Max \left\{ \sum_{j=1}^{N} \sum_{s=1}^{d_j} q_{irjs} \; / \, i \in \{1, \ldots, N\} \; et \; r \in \{1, \ldots, d_i\} \right\}$

The best value of ε is 10^{-5} and the parameter settings are given by solving [9]:

$$\begin{cases} \phi \geq 0, \quad \alpha > 0 \\ -\phi + 2\gamma \geq 0 \\ 2\phi + \beta - \gamma = \varepsilon \\ \alpha d + \beta + \gamma = -\varepsilon \end{cases}$$

We notice that this function relaxed by the aggregation of all linear constraints cited above. Empirically, we note that applying CHN to solve this kind of problem network have 70% of success rate and low quality of the given solution. To overcome this unsatisfied result, we propose to use a new approach, known as Min-conflict algorithm, to improve and repair the solution.

11.3 CHN and Min Conflict Heuristic to Solve CSPs

The MNC [20] algorithm is very simple and fast local search to resolve CSPs. The principle of MNC is based firstly in assigning all variables randomly, and selecting iteratively one variable from the set of variables with conflicts as well, the fact that violates one or more constraints of the CSP, then we assign to this variable the value that minimizes the number of conflicts. MNC have demonstrated to be able to solve the million queens problem in minutes [21]. Thus, MNC is widely used to construct hybrid algorithms with other optimization [10, 11, 19, 31]. In this way, the basic idea of our proposed approach is using MNC to improve the solution reached by CHN (Fig. 11.2). This will be done in two steps. First, MNC visits all assigned $X_i \in V_a$ variables as shown in Fig. 11.3, For each one the decision will be taken by computing penalty $p(k)$ of its associated neurones. Second, we propagate this assignment to the other set of not assigned variables yet. We study variant executions of our methods. The first one is CHN-MNC, which execute CHN and MNC consecutively for each instance of a problem. The second is CHN-BMNC [3], in which we run CHN many times to solve the given instance. Then, we execute min-conflict on the best solution founded by all run (Fig. 11.4).

Remark 1 We run CHN as described in [8] with the same values setting and the starting point.

Function CHNMNC (CSP : **Problem**) :

$$V_a = CHN(CSPs)$$
For (each cluster $x_j \in V_a$) **do**

$\quad V_a = V_a \setminus (x_j, a)$ {a is the current affected value to x_j }
$\quad a = Min\text{-}Conf(x_j, V_a)$ { new value assigned to x_j }
$\quad V_a = V_a \bigcup (x_j, a)$

end For

{propagate current sub assignment to others
variables not yet get decision if they exist }
For (each cluster $x_j \notin V_a$) **do**

$\quad a' = Min - Conf(x_j, V_a)$
$\quad V_a = V_a \bigcup (x_j, a')$
end For
return V_a

End

Fig. 11.2 Main function witch improve solution by Min-conflict algorithm

Function Min-Conf (x_i : **Current variable** , V_a : **Assigned variables set**) :

   let v_{ir}^* current position of x_i variable affectation
   For (each value $v_k \in Dom(x_i)$) **do**
      $p(k) = \sum_{j \in V_a} v_{js}^* T_{ikjs}$
      { $v_{js}^* = 1$ the position of value which assigned to variable j}
   end For
   let P_{max} set of maximum output penalization
   If ($v_{ir}^* \in P_{max}$) **then**
      return r
   else
      return random value from P_{max}
   end If
End

Fig. 11.3 Selected the most coherent neuron of current cluster with others variables clusters already affected

Fig. 11.4 Diagram of the proposed algorithm CHN-MNC

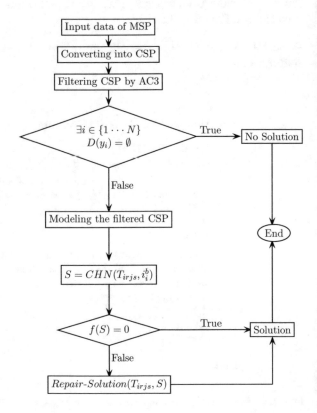

11.4 Results and Discussion

For showing the practical interest of our approach, we also study its performance over meeting instances provided in [27]. Table 11.1 show the comparison between our approach CHN-MNC and the original CHN; with [9] parameters settings and for stabilisation point, we have used the Variable Updating Step (VUS) technique proposed by Talavánn and Yánnez in [28]. The description of table columns is as follow:

- M: the number of instance meetings
- C: the number of arrival constraints
- Ratio mean: the better results obtained by each solver
- Ratio mean: the average of the optimal value in a number of run.
- Ratio mode: the most repetitive (mode) optimal value obtained by CHN in number of run

Remark 2 For each instance we run solvers 200 times.

Figures 11.5, 11.6 and 11.7 plot the performance of CHN-BMNC against CHN-MNC and CHN alone, the study concerns two classes of instance, the classification is done by considering the number of instance variables, so the first class of instances has $N = 20$, and $N = 40$ for the second class. Results show the effectiveness of CHN-MNC over CHN-BMNC. This means that it is not necessary the good minimum is closed to the best solution founded by a CHN in the first phase.

In Fig. 11.5, we observe that according to a minimum values over 200 run our approach variant CHN-MNC out perform significatively CHN alone and the second variant CHN-BMNC.

According to many values founded by two variants, Fig. 11.6 reports that CHN-MNC improves the solution meanly with 81%. Also for class $N = 20$ and $N = 40$ variant CHN-MNC is better than CHN-BMNC by 50%.

To evaluate our approach with other evolutionary algorithm, we copmare it perfor-mance with the Genetic Algorithm (GA) [16] and the Particular Swarm Optimisation (PSO) [24]. In practice, rather than authors settings we have empirically searching the Genetic Algorithm and PSO goods ones, adopted to our CSP formulation of MSP problem. So, for GA the population was 200 mutation rate equal to 5% and crossing rate equal to 72%, as for PSO [24] we choose $\varphi_1 = \varphi_2 = 1$ and population size fixed at 100. We run also 200 times each one.

From Fig. 11.8 we learn that GA and CHN-MNC are closed and both better than PSO, but regarding the mean time taken by all compared algorithms CHN-MNC is the short one, as we can constant in Table 11.2. We note also that the istance 18 and instances from 25 to 27 don't have any solution, over this kind of unsatisfied CHN-MNC give good results, therefore, we can say that approach based CHN are more oriented Max-CSP solver.

Table 11.1 CHN performance of CHN-MNC and CHN alone [27]

Instances names	M	C	CHN-MNC			CHN			CPU(s)	Success rate	Number of iteration
			Ratio min	Ratio means	Ratio mod	Ratio min	Ratio means	Ratio mod			
Instance #1	20	78	0	2.6	3	10	18.36	20	0.42496	0.68	128.0
Instance #2	20	72	1	1.9	2	14	19.2	15	0.39195	0.62	112.0
Instance #3	20	75	0	1.5	1	12	15.54	15	0.41117	0.78	116.0
Instance #4	20	70	0	0.6	1	8	15.72	14	0.43679	0.83	155.0
Instance #5	20	66	0	0	0	8	13.48	13	0.42822	0.78	122.0
Instance #6	20	97	2	3.8	2	16	21.98	23	0.37535	0.70	126.0
Instance #7	20	93	2	4.8	4	15	21.96	21	0.43264	0.65	124.0
Instance #8	20	71	1	1.7	1	12	16.08	16	0.44782	0.67	127.0
Instance #9	20	80	0	1.9	1	12	17.91	17	0.38811	0.88	115.0
Instance #10	20	70	0	1.5	2	10	16.42	13	0.4034	0.79	117.0
Instance #11	20	72	0	2.2	3	11	17.15	19	0.42603	0.79	120.0
Instance #12	20	73	1	3.5	2	14	20.17	19	0.34246	0.72	96.0
Instance #13	20	64	1	1.9	1	11	17.35	16	0.37739	0.87	112.0
Instance #14	20	66	4	4.8	4	14	18.92	20	0.40289	0.81	125.0
Instance #15	20	70	0	2	2	9	14.88	15	0.42734	0.88	121.0
Instance #16	20	73	0	2.5	2	11	16.82	19	0.39188	0.67	121.0
Instance #17	20	74	1	2.7	1	12	16.81	17	0.41901	0.90	117.0
Instance #18	20	104	5	6.3	8	19	25.5	26	0.34146	0.60	102.0
Instance #19	40	125	3	6.3	6	20	29.47	30	0.48332	0.79	576.0
Instance #20	40	125	4	5.6	6	20	28.55	28	0.47398	0.74	553.0
Instance #21	40	121	4	6	8	23	30.09	33	0.48786	0.78	573.0
Instance #22	40	123	3	5.3	4	19	27.48	26	0.4796	0.84	572.0
Instance #23	40	96	0	3.2	4	14	23.5	22	0.47992	0.72	558.0
Instance #24	40	96	2	4.3	2	16	26.78	26	0.47745	0.77	577.0
Instance #25	40	96	2	4.6	7	10	26.65	25	0.47181	0.71	544.0
Instance #26	40	157	7	9.7	11	27	36.87	30	0.45777	0.81	532.0
Instance #27	40	75	4	6.8	8	18	26.86	23	0.50308	0.72	598.0

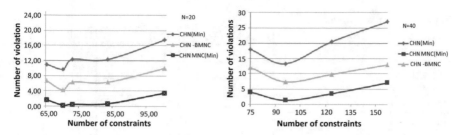

Fig. 11.5 Comparison between minimums founded by CHN-MNC and CHN against CHN-BMN

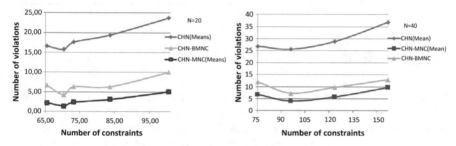

Fig. 11.6 Comparison between means values founded by CHN-MNC and CHN against CHN-BMN

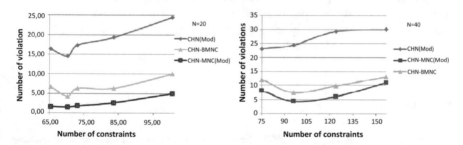

Fig. 11.7 Comparison between the most values founded by CHN-MNC and CHN against CHN-BMN

Fig. 11.8 Comparison between the means values founded by CHN-MNC and CHN against CHN-BMN

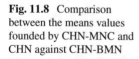

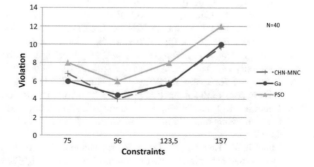

Table 11.2 Times performance

Instances	M	CHN-MNC(s)	PSO(s)	GA(s)
Instance #16	20	0.39	1.21	0.85
Instance #17	20	0.42	1.38	0.95
Instance #18	20	0.34	1.40	1.02
Instance #19	40	0.48	1.35	0.97
Instance #20	40	0.47	1.65	1.23
Instance #21	40	0.49	1.65	1.33
Instance #22	40	0.48	1.40	1.24

11.5 Conclusion

In this paper, we have considered the meeting-scheduling problem. Hence, we have solved it as a constraint satisfaction problem. After reformulation, we apply on AC3 an algorithm to reduce the search space. Then, we use the hybrid meta-heuristic search to solve it, which is based on Continuous Hopfield network and Mon-conflict method. Some numerical examples assess the effectiveness of the theoretical results that are shown in this paper; more than that the advantages of this new approach improve considerably the quality solution and avoid network crush as well. Other studies are in progress to apply this approach to many problems such as time-tabling problem and resource allocation.

References

1. BenHassine, A., & Ho, T. B. (2007). An agent-based approach to solve dynamic meeting scheduling problems with preferences. *Engineering Applications of Artificial Intelligence*, *20*(6), 857–873.
2. Berger, F., Klein, R., Nussbaum, D., Sack, J. R., & Yi, J. (2009). A meeting scheduling problem respecting time and space. *GeoInformatica*, *13*(4), 453–481.
3. Bouhouch, A., Chakir, L., & El Qadi, A. (2016). Scheduling meeting solved by neural network and min-conflict heuristic. In *2016 4th IEEE International Colloquium on Information Science and Technology (CiSt)* (pp. 773–778). New York: IEEE.
4. Chun, A., Wai, H., & Wong, R. Y. (2003). Optimizing agent-based meeting scheduling through preference estimation. *Engineering Applications of Artificial Intelligence*, *16*(7), 727–743.
5. Davenport, A., Tsang, E., Wang, C. J., Zhu, K. (1994). Genet: A connectionist architecture for solving constraint satisfaction problems by iterative improvement. In *AAAI* (pp. 325–330).
6. Ettaouil, M., Loqman, C., & Haddouch, K. (2010). Job shop scheduling problem solved by the continuous hopfield networks. *Journal of Advanced Research in Computer Science*, *2*(1), 31–47.
7. Ettaouil, M., Loqman, C., Hami, Y., & Haddouch, K. (2012). Task assignment problem solved by continuous hopfield network. *International Journal of Computer Science Issues*, *9*(1), 206–2012.

8. Haddouch, K., Ettaouil Mohamed Loqman, C., (2013). Continuous hopfield network and quadratic programming for solving the binary constraint satisfaction problems. *Journal of Theoretical and Applied Information Technology, 56*(3), 362–372.

9. Haddouch, K., Ettaouil, M., & Loqman, C. (2013). Continuous hopfield network and quadratic programming for solving the binary constraint satisfaction problems. *Journal of Theoretical and Applied Information Technology, 56*(3), 362–372.

10. Handa, H. (2003). Hybridization of estimation of distribution algorithms with a repair method for solving constraint satisfaction problems. In *Genetic and Evolutionary ComputationGECCO 2003* (pp. 991–1002). Berlin: Springer.

11. Hayakawa, D., Mizuno, K., Sasaki, H., & Nishihara, S. (2011). Improving search efficiency adopting hill-climbing to ant colony optimization for constraint satisfaction problems. In *2011 Third International Conference on Knowledge and Systems Engineering (KSE)* (pp. 200–204). New York: IEEE.

12. Hopfield, J. J. (1982). Neural networks and physical systems with emergent collective computational abilities. *Proceedings of the National Academy of Sciences, 79*(8), 2554–2558.

13. Hopfield, J. J. (1984). Neurons with graded response have collective computational properties like those of two-state neurons. *Proceedings of the National Academy of Sciences, 81*(10), 3088–3092.

14. Hopfield, J. J., & Tank, D. W. (1985). neural computation of decisions in optimization problems. *Biological Cybernetics, 52*(3), 141–152.

15. Jennings, N. R., & Jackson, A. (1995). Agent-based meeting scheduling: A design and implementation. *IEE Electronics Letters, 31*(5), 350–352.

16. Kanoh, H., Matsumoto, M., Nishihara, S. (1995). Genetic algorithms for constraint satisfaction problems. In *IEEE International Conference on Systems, Man and Cybernetics, 1995. Intelligent Systems for the 21st Century* (Vol. 1, pp. 626–631). New York: IEEE.

17. Lee, J. H., Leung, H., Won, H. (1995). Extending genet for non-binary csp's. In *Seventh International Conference on Tools with Artificial Intelligence, 1995. Proceedings* (pp. 338–343). New York: IEEE.

18. Li, X., Zhang, X., Qian, Q., Wang, Z., Shang, H., Huang, M., Lu, Z. (2016). Cloud tasks scheduling meeting with qos. In *Proceedings of the 2015 International Conference on Electrical and Information Technologies for Rail Transportation* (pp. 289–297). Berlin: Springer.

19. Liu, T., Liu, M., Zhang, Y.B., Zhang, L. (2005). Hybrid genetic algorithm based on synthetical level of resource conflict for complex construction project scheduling problem. In *Proceedings of 2005 International Conference on Machine Learning and Cybernetics, 2005* (Vol. 9, pp. 5699–5703). New York: IEEE.

20. Minton, S., Johnston, M. D., Philips, A. B., & Laird, P. (1992). Minimizing conflicts: a heuristic repair method for constraint satisfaction and scheduling problems. *Artificial Intelligence, 58*(1), 161–205.

21. Minton, S., Philips, A., Johnston, M. D., & Laird, P. (1993). Minimizing conflicts: A heuristic repair method for constraint-satisfaction and scheduling problems. *Journal of Artificial Intelligence Research, 1*, 1–15.

22. Rasmussen, R. V., & Trick, M. A. (2006). The timetable constrained distance minimization problem. In *International Conference on Integration of Artificial Intelligence (AI) and Operations Research (OR) Techniques in Constraint Programming* (pp. 167–181). Berlin: Springer.

23. Rinaldi, F., & Serafini, P. (2006). Scheduling school meetings. In *International Conference on the Practice and Theory of Automated Timetabling* (pp. 280–293). Berlin: Springer.

24. Schoofs, L., & Naudts, B. (2002). Swarm intelligence on the binary constraint satisfaction problem. In *Proceedings of the 2002 Congress on Evolutionary Computation, 2002. CEC'02* (Vol. 2, pp. 1444–1449). New York: IEEE.

25. Sen, S. (1997). Developing an automated distributed meeting scheduler. *IEEE Expert, 12*(4), 41–45.

26. Shakshuki, E., Koo, H. H., Benoit, D., & Silver, D. (2008). A distributed multi-agent meeting scheduler. *Journal of Computer and System Sciences, 74*(2), 279–296.

27. Shapen, U., Zivan, R., & Meisels, A. CSPLib problem 046: Meeting scheduling. http://www.csplib.org/Problems/prob046
28. Talaván, P. M., & Yáñez, J. (2005). A continuous hopfield network equilibrium points algorithm. *Computers and Operations Research, 32*(8), 2179–2196.
29. Tsuchiya, K., & Takefuji, Y. (1997). A neural network parallel algorithm for meeting schedule problems. *Applied Intelligence, 7*(3), 205–213.
30. Wang, J., Niu, C., & Shen, R. (2007). Scheduling meetings in distance learning. In *International Workshop on Advanced Parallel Processing Technologies* (pp. 580–589). Berlin: Springer.
31. Zou, H., & Choueiry, B. Y. (2003). Characterizing the behavior of a multi-agent search by using it to solve a tight, real-world resource allocation problem. In *Workshop on Applications of Constraint Programming, Kinsale, County Cork, Ireland* (pp. 81–101).

Chapter 12
A Probabilistic Finite State Machine Design of Particle Swarm Optimization

Abdellatif El Afia, Malek Sarhani and Oussama Aoun

Abstract Nowadays, control is the main concern with emergent behaviours of multi-agent systems and state machine reasoning. This paper focuses on the restriction of this general issue to swarm intelligence approaches designed for solving complex optimization problems. Indeed, we propose a probabilistic finite state machine for controlling particles behaviour of the particle swarm optimization algorithm. That is, our multi-agent approach consists of assigning different roles to each particle based on its probabilistic finite state machine control which is used to address this issue. We performed evaluations on ten benchmark functions to test our control scheme for particles. Experimental results show that our proposed scheme gives a distinguishable out-performance on a number of state of the art of PSO variants.

12.1 Introduction

Decentralized approaches have become promising to solve complex problems, especially in artificial intelligence. Among such approaches fall into the area of multi-agent systems in which a number of agents have to solve together cooperatively problems. In particular, various efforts have been made to integrate multi-agent approaches for modeling the collective search behavior of optimization algorithms mainly in swarm intelligence approaches such as particle swarm optimization [22] and ant colony optimization [15]. The main purpose of these essays is to enable swarms to collaborate and share information with each other and learn from each other. Furthermore, it may help agents to react rapidly to unexpected variations and

A. El Afia · M. Sarhani (✉) · O. Aoun
Departement Informatics and Decision support,
Mohammed V University, Rabat, Morocco
e-mail: malek.sarhani@um5s.net.ma

A. El Afia
e-mail: a.elafia@um5s.net.ma

O. Aoun
e-mail: oussama.aoun@um5s.net.ma

© Springer International Publishing AG, part of Springer Nature 2019
E.-G. Talbi and A. Nakib (eds.), *Bioinspired Heuristics for Optimization*,
Studies in Computational Intelligence 774,
https://doi.org/10.1007/978-3-319-95104-1_12

185

control the variables of their subsystems when they are communicating with other agents. Concerning particle swarm optimization (PSO), attempts have been made to formalize the design of such cooperative multi-agent behavior of particles as a mean to enhance the diversity of the algorithm or to achieve a trade-off between exploration and exploitation. A commonly used cooperation form of PSO is based on the idea of considering multi-swarms (multi-populations), it consists of dividing the whole search space into local subspaces, each of which might cover one or a small number of local optima, and then separately searches within these subspaces. Another way to define multi-agent in PSO is to assign different roles to particles. Thus, different particles can play different roles, and each one of these particles can play different roles during the search processes. A challenging task within this PSO variant is how each particle has to decide which role it will assume. In this paper, Finite state machines (FSM) is applied to model the decision making of an agent with the aim of guiding particles to move toward different promising sub-regions. To do that, the swarm behavior can be represented as a finite state machine based on very simple agents and simple interaction rules. That is, a behavior specification defines a set of finite state machines, called options and a set of predefined behavior routines, called basic behaviors. Several approaches have been used to implement probabilistic FSM. In this paper, we integrate Hidden Markov Model (HMM) -a commonly used probabilistic FSM [37]- to learn and predict the most probable states of the probabilistic FSM in order to control particles behaviour of PSO. This process is performed through the Viterbi algorithm that gives the most likely path of states for each particle in each PSO iteration. The rest of the paper is organized as follows: in the next section, we outline the related works. In Sect. 12.3, we present our approach. Section 12.4 presents the obtained results for the experiments. Finally, we conclude and present perspectives to our work.

12.2 Literature Review

In recent years, there has been increased interest in the use of automation approaches inside PSO to control its behaviour and then to improve its performance. That is, various methods have been proposed to control PSO behaviour and to improve learning capability of particles. We can differentiate between two kinds of control approaches which have been used inside PSO in the literature. In the first one, the control depends on the iteration and then the whole swarm follow the same strategy. In the second one, the control depends on the particle itself. That is, at each iteration, particles are grouped into sub-swarms, and the particles of each swarm have a specific role in the swarm (as in multi-agent systems). This type corresponds to the adaptivity control. On the one hand, an example of the first case can be found in our previous work [3] in which hidden markov model (HMM) has been used inside PSO to have a stochastic control on the state classification at each iteration. The mentioned paper has extended the adaptive PSO proposed by [41] which has proposed to update the values of the acceleration factors c_1 and c_2 according to four defined states which are: exploration, exploitation, convergence and jumping-out. On the other hand, concerning the second type, [26] proposed four operators which play similar roles as the four of states the adaptive PSO defined in [41]. Their approach is based on the idea of assigning to each particle one among different operators based on their rewards. Furthermore, this type of control is related to the concept of cooperative swarms which has been introduced by [35]. This principle has been achieved in their paper by using multiple swarms to optimize different components of the solution vector cooperatively. This issue can also be treated by clustering approaches as proposed by [40]. Their approach consists of assigning particles to different promising sub-regions basing on a hierarchical clustering method. We can see that this variant of PSO (multi-swarm PSO) has been presented in the literature as a specific and separate algorithm known by multi-swarm optimization [8]. in the proposed variant, the authors has been inspired by the quantum model of atoms to define the quantum swarm. Also, another grouping approach has been suggested by [29]. More generally, four main categories have been proposed to improve PSO performance, which are: configuration of the parameters (adaptivity control), the study of neighbourhood topology of particles in the swarm of PSO, hybridization with other optimization algorithms and integration of learning strategies (diversity control). Concerning the two types mentioned at the beginning of this section. The former correspond to the first type, while the latter is related to the second one. Furthermore, the control the PSO parameters has been proposed in a number of papers with the purpose of achieving a trade-off between the diversity and the convergence speed. It has generally been done using learning strategies such as the comprehensive learning [27] approach in where each particle learns from another particle which is chosen according to a learning probability. Concerning hybridization, it is a long standing of PSO and example of improvement can be found in [32]. The issue of the interaction between swarm intelligence and multi-agent systems has been given much attention in the last few years in particular by the popularization of the swarm robotic field. In particular, [7] affirmed the concept of swarm appears nowadays closely associated with intelligent systems in order to carry out useful tasks. The author also misanalysed qualitatively the impact of automation concepts to define the intelligent swarms. Moreover, [11] have outlined the main characteristics of swarm robotics and analyzed the collective behavior of individuals in some fields. They affirmed that finite state machines are one of the most used adequate approaches to model this behaviour. Another commonly used approach for this purpose is reinforcement learning. A practical example of using FSM in swarm robotics is described in [23]. More generally, [31] presented a detailed description about using Hierarchical FSM for behavior Control, The use of FSM for this purpose can be justified by the fact that there are a finite number of priori pre-defined organizational forms of multi-agent systems. In particular, probabilistic FSM has been used [34] for aggregation purpose in swarm robotics. Concerning HMM (which is a type of probabilistic FSM used especially for learning [16]), it has been successfully applied for controlling purpose in a number of engineering problems [19]. An example of using the multi-agent concept in PSO can be found in [13]. That is, incremental social learning which is often used to improve the scalability of systems composed of multiple learning agents has been used to improve the performance of PSO. Furthermore, [4] proposed a multi-agent approach which combines simulated annealing (SA) and PSO, we can remark that their idea is related to the generic notion of hyper-heuristics which consists of finding the most suitable configuration of heuristic algorithms. Reference [14] has cited the may features obtained by using agents in configuring metaheuristics which are distributed execution, remote execution, cooperation and autonomy. The using of multi-agent concepts can be useful to self-organize particles in PSO using simple rules as defined by [6].Their main idea was to define six states which are cohesion, alignment, separation, seeking, clearance, and avoidance. Furthermore, the finite state machine has been used for movement control. That is, the states of the FSM has been defined by a collection of velocity components and their behavior specific parameters. Furthermore, the population has been divided into two swarms in order to introduce the divide and conquer concept using genetic operators.Another automation approach which can be used inside PSO is cellular automata (CA). CA can be considered as an arrangement of FSM. It can be used for instance split the population of particles into different groups across cells of cellular automata. Reference [33] has integrated it in the velocity update to modify the trajectories of particles. Regarding Hybridization between PSO and FSM, it has also be done in most cases with the aim of improving FSM performance by PSO as in [18]. More general, concerning the integration of HMM inside PSO and other metaheuristics, it has been done for instance in [1, 2, 9, 10, 17, 24].

12.3   Probabilistic FSM Design of PSO

12.3.1   Approach Background

Firstly, the native PSO has been introduced by [22]. The PSO concept consists of changing the velocity of each particle at each iteration toward its pBest and gBest locations. Then the velocity and the position of each particle is updated according to Eqs. (12.1) and (12.2)

$$v_i = w\, v_i + c_1 r_1 (pBest - x_i) + c_2 r_2 (gBest - x_i) \tag{12.1}$$

$$x_i = x_i + v_i \tag{12.2}$$

The detail of these parameters can be found for instance in [3].

In our approach, we interest especially on introducing probabilistic FSM to control particles state in PSO based on the defined states in [26]: each particle is viewed as automata having four finite states which are exploration, exploitation, convergence and jumping-out.More detail on these states is described in the following sections.

Among the existing probabilistic FSM [37], we have chosen HMM to address thisissue as the most common type of probabilistic FSM. Indeed, HMM has the ability to learn states of our automata from hidden observation based on the maximum likelihood estimation [16], this learning feature of HMM is used to control the particles individually cross PSO iterations.

12.3.2   Probabilistic FSMs Definition

We define each particle as a machine having different finite states. A particle of PSO can have many alternative traces for a given input because of the random behavior of the PSO algorithm. FSM associated with a particle is a little machine that feeds with inputs and provides many outcomes across iterations. We can say that the particle is associated with a random process.

During iterations, a particle is a probabilistic FSM related to a state $\{x_i\}_{i \in N}$ that generates outcome or also called observation $\{y_i\}_{i \in N}$.

This definition yield to have at each iteration several groups of particles, each one plays a defined role according to its identified state machine. So, we can have for instance 40 particles, in which 20 particles explore throughout the search space, 10 others are exploiting, 7 are converging, and 3 are jumping out. Thus, particles are divided to sub-swarms with different states. The change of state or role of particles during iterations is governed by their associated probabilistic FSMs which is defined by the following formalism for each individual particle:

1. Outcomes $\{y_i\}_{i \in N}$
2. State $\{x_i\}_{i \in N}$
3. $A = (a_{ij})$ The state transition matrix: $P(x_t = i \mid x_{t-1} = j) i, j \in N$, t : iteration number.
4. $B = (b_{jk})$ The emission probabilities of outcomes: $P(Y_t = k \mid X_t = j) \ k, j \in N$, t : iteration number.

Our approach consists of finding the most suitable current state by finding the most probable trace for the given input of states across iterations. This problem constitutes a classical decoding problem in HMM theory and resolved by Viterbi algorithm.

12.3.3   Probabilistic FSM Parameters

As mentioned earlier, four states of the Probabilistic FSM have been defined which are: exploration, exploitation, convergence and jumping-out. These states generate an outcome to define below.

Particles are moving around the search space and change position and velocity at each iteration step according to Eqs. 12.1 and 12.2. Consider the mean distance of each particle i to all the other particles as d_i. And calculate the measure :

$l = \left| \frac{d_{Pbest} - d_i}{d_{max} - d_{min}} \right|$, $Pbest$ is the best particle of current iteration.

l is considered as outcome for as particle viewed as an FSM. l belong to subintervals of [0, 1] ([0, 0.2], [0.2, 0.3], [0.3, 0.4], [0.4, 0.6], [0.6, 0.7], [0.7, 0.8], [0.8, 1]). We divide [0,1] to seven subintervals, so the outcomes $\{y_i\}_{i \in [1,7]}$ will be number of subintervals which belong l. Emission probabilities are deduced from defuzzification process of an evolutionary factor in[41] as follow:

$$P = \begin{bmatrix} 0 & 0 & 0 & 0.5 & 0.25 & 0.25 & 0 \\ 0 & 0.25 & 0.25 & 0.5 & 0 & 0 & 0 \\ 2/3 & 1/3 & 0 & 0 & 0 & 0 & 0 \\ 0 & 0 & 0 & 0 & 0 & 1/3 & 2/3 \end{bmatrix} \tag{12.3}$$

We take state transitions probabilities as equiprobable for all possible transitions, the same as in [3].

12.3.3.1   Our Proposed Algorithm

An online EM learning is first performed at each iteration to calculate and update HMM parameters that are re-estimated upon observing each new sub-sequence.

Particles positions and velocities vary over iterations, impacting also the evolutionary factor. Then, the classification environment for HMM changes during operations. Online learning of new data sequences allows adapting HMM parameters as new data become available as shown if Fig. 12.2, where $(o_i)_{i \in [0,n]}$ are data observations and $(\lambda_i)_{i \in [0,n]}$ parameters values update.

At each iteration t, a new classifier r_t is performed with new updated parameters. We choose online learning EM algorithm [12] instead of Bach learning (classical Baum-Welch algorithm [5], because this last one needs to run on all observation sequence which this is not our case.

The online Expectation-Maximization algorithm used for HMM parameters learning can be summarized as follows:

12.3.3.2   HMM Classification

The Viterbi Algorithm is used with online parameters setting to find the most probable sequence of hidden states with a given sequence of observed states. The Viterbi algorithm does not simply accept the most likely state at a particular time instant, but also takes a decision based on the whole observation sequence. The algorithm will find the max Q(state sequence $Q = q_1 q_2 \ldots q_T$) for a given observation sequence by the means of induction. An array $\psi_t(j)$ is used to store the highest probability paths [30].

Algorithm 12: Expectations-Maximization algorithm

Data: observation sequence $O = (o_1 o_2 \ldots o_N)$
Initialization: initial parameters set λ_0;
for $i = 1$ to N_{step} **do**

   E-step : find conditionally optimal hidden trace S^i :
   $S^i = \arg\max_s P(O|S, \lambda_{i-1})$;
   Compute likelihood :
   $L^{i-1}(\lambda) = P(O|S^i, \lambda_{i-1})$;
   if ($i \leq N_{step}$ & *likelihood not yet converged*) **then**
      M-step - find conditionally optimal parameter set λ :
      $\lambda_i = \arg\max_s P(O|S, \lambda)$
   end

end
Result: Estimated parameters $\lambda_{N_{step}}$

HMM parameters, HMM classification is done by the Viterbi algorithm as defined in the next algorithm:

Algorithm 13: Viterbi algorithm

Data: Observations of length T, state-graph of length N
Initialization: Observations of length T, state-graph of length N ;
create a path probability matrix viterbi [N + 2,T] ;
create a path backpointer matrix backpointeri [N + 2, T] ;
for *state s from 1 to N* **do**
   $forward[s, 1] \longrightarrow a_{o,s} \times b_s(o_1)$ $backpointer[s, 1] \longrightarrow 0$;
end
for *time step t from 2 to T* **do**
   for *state s from 1 to N* **do**
      $viterbi[s, t] \longrightarrow \max_{s=1}^N viterbi[s', t-1] \times a_{s',s} \times b_s(o_t)$;
      $backpointer[s, 1] \longrightarrow \arg\max_{s=1}^N viterbi[s', t-1] \times a_{s',s}$;
   end
end
$t \longrightarrow t + 1$;
Result: Best-path of states: the classified current state

12.3.3.3   Our Algorithm

Therefore, we delegate choosing states of APSO iterations to online HMM classification, transitions between states is represented by online probabilities transitions. At each iteration transition and observation probabilities are updated according to the online EM algorithm. The Viterbi Algorithm is then used for state classification of APSO iteration. Then, we update positions and velocities according to the classified state. The complete hybrid APSO with HMM is depicted below (Algorithm 3).

Algorithm 14: HMM-APSO algorithm

Data: The objective function
Initialization: positions, velocities of particles, accelerations factors and HMM parameters;
Set t value to 0 ;
while *(number of iterations $t \leq t_{max}$ not met)* **do**

    Update HMM parameters by online EM process (Algorithm 1) ;
    Classification of PSO state by HMM classifier (Algorithm 2) ;
    Update c_1 , c_2 and w values according to the corresponding state ;
    for *i = 1 to number of particles* **do**

        compute f ;
        Update velocities and positions according to Eqs. (2) and (3) ;
        if ($f \leq f_{best}$) **then**
            $f_{best} \longrightarrow f$;
            $p_{best} \longrightarrow X$;
        end
        if ($f(p_{best}) \leq f(g_{best})$) **then**
            $f(g_{best}) \longrightarrow f_{best}$;
            $g_{best} \longrightarrow X_{best}$;
        end
        if *state = convergence* **then**
            Elistic learning
        end

    end
    $t \longrightarrow t + 1$

end
Result: The solution based on the best particle in the population and corresponding fitness
        value

12.4 Experiment

In this part, tests and validations of the proposed hybrid approach HMM-APSO are performed. Experimentations are done with several benchmark functions and compared with other PSO's variants of literature.

12.4.1 Parameters Setting

For each of benchmark functions shown in Table 12.1, we perform ten executions, and compare for each function the best and the average value.

Table 12.2 shows ten chosen PSO variant from literature. The same initial values of acceleration and inertia weight coefficients ($c_1 = c_2 = 2$, $\omega = 0.9$) are used. Swarm size is 30 with dimension of 30. Each run contains 1000 generation of optimization process.

Table 12.1 Description of Benchmark functions

Test functions	Name	Type
f_1	Rotated Ackley	Multimodal
f_2	Ackley	Multimodal
f_3	Dropwave	Multimodal
f_4	Rotated Elliptic	Unimodal
f_5	Elliptic	Unimodal
f_6	Griewank	Multimodal
f_7	Rotated Griewank	Multimodal
f_8	Quadric	Multimodal
f_9	Rotated Rastrigin	Multimodal
f_{10}	ShiftedRastrigin	Multimodal
f_{11}	Rastrigrin	Multimodal
f_{12}	Shifted Rosenbrock	Unimodal
f_{13}	Rosenbrock	Unimodal
f_{14}	Schewefel	Multimodal
f_{15}	ShiftedSchwefel	Multimodal
f_{16}	ShiftedSphere	Multimodal
f_{17}	Sphere	Multimodal
f_{18}	Step	Unimodal
f_{19}	Tablet	Unimodal
f_{20}	Rotated Weierstrass	Multimodal

Table 12.2 Compared variants of PSO

Algorithm	Name	Reference
YSPSO	PSO with compressibility factor	[28]
SELPSO	Natural selection based PSO	[20]
SecVibratPSO	Order oscillating PSO	[21]
SecPSO	Swarm-core evolutionary PSO	[38]
SAPSO	Self-adaptive PSO	[20]
RandWPSO	Random inertia weight PSO	[39]
LinWPSO	Linear decreasing weights PSO	[39]
CLSPSO	Cooperative line search PSO	[38]
AsyLnCPSO	Asynchrous PSO	[20]
SimuAPSO	PSO with Simulated annealing	[38]

Table 12.3 Results comparisons with other variants of PSO

Functions		APSO	PSO	SimuA-PSO	Sec-PSO	RandWPSO	YSPSO	SelPSO	SecVibratPSO	SAPSO	LinWPSO	AsyLnCPSO	PFSM-PSO
f_1	Best	20.83	20.83	20.96	20.84	20.89	20.71	20.83	20.81	20.79	20.83	20.52	20.52
	Mean	20.98	20.98	21.03	21.01	21.02	20.82	21.01	21.03	20.96	21.02	20.78	**20.68**
f_2	Best	1.456	1.456	7.098	4.011	4.857	3.659	5.174	1.835	4.254	4.463	4.596	**0.001**
	Mean	2.254	2.254	9.071	5.022	5.888	4.466	6.267	5.186	5.511	5.467	5.951	**0.004**
f_3	Best	−1	−1	−0.93624	−1	−0.99297	−1	−0.99981	−0.9943	−1	−1	−1	−1
	Mean	−0.986	−0.986	−0.793	−0.958	−0.943	−0.990	−0.967	−0.823	−0.961	−0.963	−0.994	**−0.996**
f_4	Best	1.59.E+5	1.59.E+5	6.75.E+6	1.31.E+6	3.22.E+6	2.42.E+5	2.54.E+6	5.45.E+6	1.54.E+6	2.23.E+6	3.77.E+5	**1.24.E+4**
	Mean	6.42.E+5	6.42.E+5	2.24.E+7	5.28.E+6	8.07.E+6	9.20.E+5	4.69.E+6	2.14.E+7	5.48.E+6	5.74.E+6	1.07.E+6	**3.77.E+4**
f_5	Best	2.96E+3	2.96E+3	1.13E+6	1.69E+5	4.45E+5	3.67E+4	3.75E+5	9.12E+4	8.65E+4	3.73E+4	2.22E+4	**0.004**
	Mean	1.34E+4	1.34E+4	4.88E+6	4.45E+5	1.26E+6	1.07E+5	9.09E+5	2.23E+6	3.74E+5	6.24E+5	1.00E+5	**6.706**
f_6	Best	0.006	0.006	0.503	0.107	0.157	0.071	0.182	0.039	0.118	0.130	0.068	**0.000**
	Mean	0.026	0.026	0.919	0.293	0.535	0.167	0.408	0.465	0.341	0.323	0.195	**0.012**
f_7	Best	0.657	0.657	1.335	1.141	1.148	1.052	1.143	1.311	1.104	1.131	1.050	**0.002**
	Mean	0.852	0.852	1.521	1.227	1.278	1.080	1.236	1.477	1.202	1.241	1.100	**0.036**
f_8	Best	8.71E+4	8.71E+4	4.51E+8	2.13E+7	6.59E+7	1.71E+6	3.80E+7	1.33E+7	2.81E+7	3.48E+6	3.70E+6	**59.22**
	Mean	8.93E+5	8.93E+5	1.46E+9	1.52E+8	3.11E+8	6.95E+6	1.49E+8	6.23E+8	9.41E+7	9.93E+7	1.86E+7	**932.33**
f_9	Best	445.55	445.55	1134.15	718.55	703.37	544.54	727.49	1250.14	713.11	838.73	612.94	**77.74**
	Mean	580.54	580.54	1642.53	1005.86	1077.63	688.26	983.55	1603.80	972.70	1068.78	878.03	**278.16**
f_{10}	Best	20.21	20.21	637.66	370.84	514.18	306.51	456.01	688.28	427.16	496.76	377.54	**0.00**
	Mean	54.62	54.62	847.51	596.74	691.50	448.77	604.19	856.43	591.19	620.23	506.11	**80.93**
f_{11}	Best	37.39	37.39	328.92	166.76	204.31	109.96	232.42	209.51	184.92	170.47	197.77	**0.00**
	Mean	57.36	57.36	473.35	253.05	313.34	182.18	308.80	328.07	246.16	258.39	290.52	54.66

(continued)

Table 12.3 (continued)

Functions		APSO	PSO	SimuA-PSO	Sec-PSO	RandWPSO	YSPSO	SeIPSO	SecVibratPSO	SAPSO	LinWPSO	AsyLnCPSO	PFSM-PSO
f_{12}	Best	299.73	299.73	1.13E+6	2.68E+5	5.53E+5	4.13E+4	3.04E+5	1.19E+6	2.08E+5	2.85E+5	3.41E+4	0.54
	Mean	2151.73	2151.73	2.31E+6	7.65E+5	9.51E+5	1.46E+5	7.68E+5	2.34E+6	5.40E+5	6.49E+5	1.24E+5	27.98
f_{13}	Best	377.54	377.54	76023.74	3199.49	6897.23	646.98	7122.80	581.75	3899.11	4034.21	1005.38	0.47
	Mean	1038	1038	295411	11206	41622	2676	30657.91	34762.82	15669	22428	8823	49.63
f_{14}	Best	33.24	33.24	162.89	29.81	68.25	15.51	82.98	10.87	38.24	32.56	18.61	0.59
	Mean	56.33	56.33	456.22	84.95	176.98	24.71	140.74	233.83	66.77	66.34	58.79	1.20
f_{15}	Best	111.30	111.30	681.07	387.95	642.76	260.79	492.64	822.37	384.93	428.14	343.36	0.51
	Mean	243.92	243.92	1306.04	668.86	1003.57	374.87	795.69	1766.13	622.24	697.59	469.77	1.46
f_{16}	Best	1.35	1.35	387.18	130.82	206.81	87.26	188.43	442.44	187.64	223.45	79.21	5.12E-07
	Mean	3.75	3.75	615.97	304.52	365.06	128.92	296.61	558.07	266.69	311.65	113.44	4.60E-05
f_{17}	Best	0.49	0.49	108.73	18.32	27.28	7.31	25.68	4.61	13.16	15.91	16.53	1.64E-06
	Mean	2.50	2.50	186.93	34.14	60.72	14.45	53.07	64.75	29.65	31.62	40.88	2.56E-05
f_{18}	Best	0	0	8.79E-05	4.93E-32	1.17E-09	0.00E+0	1.18E-11	1.69E-04	0	0	0	0
	Mean	0	0	5.26E-02	2.41E-09	2.38E-04	0.00E+0	6.73E-05	9.28E-02	0	0	1.42E-29	0
f_{19}	Best	1.14	1.14	283.61	40.73	118.63	22.07	58.79	3.89	19.06	34.46	24.10	5.35E-07
	Mean	4.23	4.23	494.37	96.05	211.17	41.88	93.82	195.12	84.60	105.43	43.91	8.12E-04
f_{20}	Best	30.21	30.21	36.55	36.11	39.54	33.21	39.78	37.75	35.35	36.15	33.04	28.05
	Mean	37.59	37.59	40.87	41.06	41.71	38.74	42.18	41.23	40.21	40.04	37.99	34.57

4.2 Performance evaluation

For each of benchmark functions shown in Table 12.1, and each PSO variant of Table 12.2, ten executions are done with 1000 generations. The best and the average value resulted from experimentations are given in the table below:

The proposed approach has very distinguishable in Table 12.3, better results than the majority of the state of the art. It exceeds in some cases the order of 10^4 improvements compared to other PSO variants. Very impressive PSO performances in term of solution accuracy are clearly observable.

In term of convergence speed, plots from Figs. 12.1, 12.2, 12.3, 12.4 and 12.5 gives good convergence in favor of PFSM-PSO. In most cases, the solution is found after less than 80 iterations except for Rastrigrin function.

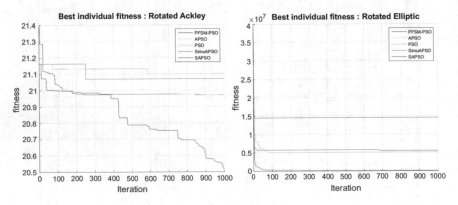

Fig. 12.1 Comparison on Rosenbrock and step functions

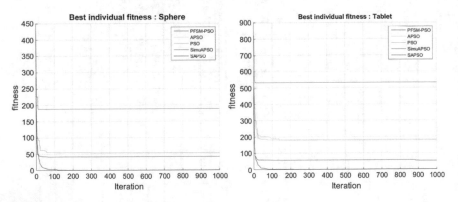

Fig. 12.2 Comparison on sphere and tablet functions

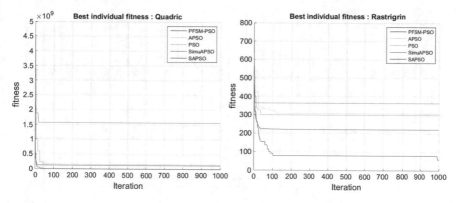

Fig. 12.3 Comparison on quadric and Rastrigrin functions

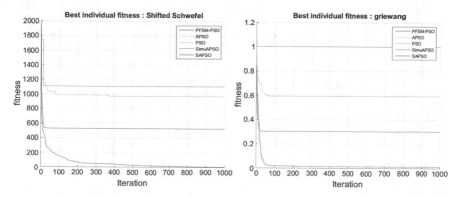

Fig. 12.4 Comparison on Ackley and Griewang functions

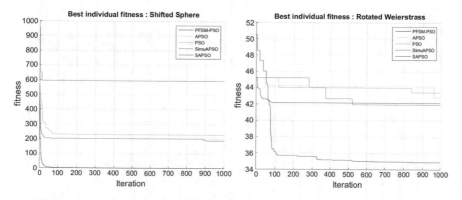

Fig. 12.5 Comparison on Schwefel and drop wave functions

Table 12.4 Statistical test

Function	APSO	PSO	SimuA-PSO	Sec-PSO	RandWPSO	YSPSO	SelPSO	SecVibratPSO	SAPSO	LinWPSO	AsyLnCPSO
f_1	0	0	0	0	0	0.015	0	0	0	0	0
f_2	0	0	0	0	0	0	0	0	0	0	0
f_3	0	0.1797	0.0124	0	0	0	0.0014	0.0076	0	0.5488	0
f_4	0	0	0	0	0	0	0	0	0	0	0
f_5	0	0	0	0	0	0	0	0	0	0	0
f_6	0.07	0	0	0	0	0	0	0	0	0	0
f_7	0	0	0	0	0	0	0	0	0	0	0
f_8	0	0	0	0	0	0	0	0	0	0	0
f_9	0.12	0	0	0	0	0	0	0	0	0	0
f_{10}	0	0	0	0	0	0	0	0	0	0	0
f_{11}	0	0	0	0	0	0	0	0	0	0	0
f_{12}	0	0	0	0	0	0	0	0	0	0	0
f_{13}	0	0	0	0	0	0	0	0	0	0	0
f_{14}	0	0	0	0	0	0	0	0	0	0	0
f_{15}	0	0	0	0	0	0	0	0	0	0	0
f_{16}	0	0	0	0	0	0	0	0	0	0	0
f_{17}	0	0	0	0	0	0	0	0	0	0	0
f_{18}	1	1	1	1	1	1	1	1	1	1	1
f_{19}	0	0	0	0	0	0	0	0	0	0	0
f_{20}	0.005	0	0	0	0	0	0	0	0	0	0
+1	17	20	19	20	20	20	20	19	20	19	20
0	3	0	1	0	0	0	0	1	0	1	0
−1	0	0	0	0	0	0	0	0	0	0	0

12.4.2   Statistical Test

For further comparison of the algorithms, we have considered the non-parametric two-sided test Wilcoxon ([25, 36]) with a significance level of 0.05 between the BMCPSO and the native PSO. As in [1], we present the P-values on every function of this two-tailed test with a significance level of 0.05. The test is performed using the statistical toolbox of Matlab (Table 12.4).

it can be seen that our approach outperforms the other algorithms in terms of Wiloxocon test results.

Even if our approach has given more accurate results, it is much more computationally expensive because the control mechanism must be done for each particle at each iteration. Therefore, our approach may be useful especially in swarm robotics which may complex optimization problems.

12.5   Conclusion

In this paper, we have shown how a probabilistic FSM, which has been often designed for the control of multi-agent systems, can be used for the control of particles behavior in PSO. To do that, individual state control by PFSM is associated to each particle during the search process. Experimental results have shown that has given competitive computational performance than some PSO variants. Future research should attempt to define collaborative rules between sub-swarms of particles and illustrate its applicability to real problems.

References

1. Aoun, O., Sarhani, M., & El Afia, A. (2017). Particle swarm optimisation with population size and acceleration coefficients adaptation using hidden markov model state classification. In *International Journal of Metaheuristics*
2. Aoun, O., Sarhani, M., & El Afia, A. (2016). Investigation of hidden markov model for the tuning of metaheuristics in airline scheduling problems. *IFAC-PapersOnLine, 49*(3), 347–352.
3. Aoun, O., Sarhani, M., & El Afia, A. (2017). Hidden markov model classifier for the adaptive particle swarm optimization. In Talbi, E.-G., Yalaoui F., Amodeo, L. (Eds.), *Recent developments of metaheuristics*, Chap. 1. Berlin: Springer.
4. Aydin, M. E. (2012). Coordinating metaheuristic agents with swarm intelligence. *Journal of Intelligent Manufacturing, 23*(4), 991–999.
5. Baum, L. E., Petrie, T., Soules, G., & Weiss, N. (1970). A maximization technique occurring in the statistical analysis of probabilistic functions of markov chains. *The Annals of Mathematical Statistics*, 164–171.
6. Bengfort, B., Kim, P. Y., Harrison, K., & Reggia, J. A. (2014). Evolutionary design of self-organizing particle systems for collective problem solving. *2014 IEEE symposium on swarm intelligence (SIS)* (pp. 1–8). New York: IEEE.
7. Beni, G. (2004). From swarm intelligence to swarm robotics. *Swarm robotics* (pp. 1–9). Berlin: Springer.

8. Blackwell, T., Branke, J. et al. (2004). Multi-swarm optimization in dynamic environments. In *EvoWorkshops* (Vol. 3005, pp. 489–500). Berlin: Springer. 2004.

9. Bouzbita, S., El Afia, A., & Faizi, R. (2016). A novel based hidden markov model approach for controlling the acs-tsp evaporation parameter. *2016 5th international conference on multimedia computing and systems (ICMCS)* (pp. 633–638). New York: IEEE.

10. Bouzbita, S., El Afia, A., Faizi, R., & Zbakh, M. (2016). Dynamic adaptation of the acs-tsp local pheromone decay parameter based on the hidden markov model. *2016 2nd international conference on cloud computing technologies and applications (CloudTech)* (pp. 344–349). New York: IEEE.

11. Brambilla, M., Ferrante, E., Birattari, M., & Dorigo, M. (2013). Swarm robotics: a review from the swarm engineering perspective. *Swarm Intelligence, 7*(1), 1–41.

12. Cappé, O. (2011). Online em algorithm for hidden markov models. *Journal of Computational and Graphical Statistics, 20*(3),

13. De Oca, M. A. M., Stützle, T., Van den Enden, K., & Dorigo, M. (2011). Incremental social learning in particle swarms. *IEEE Transactions on Systems, Man, and Cybernetics, Part B: Cybernetics, 41*(2), 368–384.

14. Monett Diaz, D. (2005). *Agent-Based Configuration of (Metaheuristic) Algorithms*. PhD thesis, Humboldt University of Berlin.

15. Dorigo, M., & Gambardella, L. M. (1997). Ant colony system: A cooperative learning approach to the traveling salesman problem. *IEEE Transactions on Evolutionary Computation, 1*(1), 53–66.

16. Dupont, P., Denis, F., & Esposito, Y. (2005). Links between probabilistic automata and hidden markov models: Probability distributions, learning models and induction algorithms. *Pattern Recognition, 38*(9), 1349–1371.

17. El Afia, A., Sarhani, M., & Aoun, O. (2017). Hidden markov model control of inertia weight adaptation for particle swarm optimization. In *IFAC2017 World Congress*, Toulouse, France.

18. El-Maleh., A. H., Sheikh, A. T., & Sait, S. M. (2013). Binary particle swarm optimization (BPSO) based state assignment for area minimization of sequential circuits. *Applied Soft Computing, 13*(12), 4832–4840.

19. Elliott, R. J., Aggoun, L., & Moore, J. B. (2008). *Hidden Markov models: estimation and control* (Vol. 29). Berlin: Springer Science & Business Media.

20. Jiang, W., Zhang, Y., & Wang, R. (2014). Comparative study on several pso algorithms. In *Control and Decision Conference (2014 CCDC), The 26th Chinese* (pp. 1117–1119).

21. Jianxiu, H., & Jianchao, Z. (2007). A two-order particle swarm optimization model [j]. *Journal of Computer Research and Development, 11*, 004.

22. Kennedy, J., & Eberhart, R. C. (1995). Particle swarm optimization. In *Proceedings of IEEE international conference neural networks* (pp. 1942–1948). New York: IEEE.

23. Knig, L., Mostaghim, S., & Schmeck, H. (2009). Decentralized evolution of robotic behavior using finite state machines. *International Journal of Intelligent Computing and Cybernetics, 2*(4), 695–723.

24. Lalaoui, M., El Afia, A., & Chiheb, R. (2016). Hidden markov model for a self-learning of simulated annealing cooling law. In *2016 5th international conference on multimedia computing and systems (ICMCS)* (pp. 558–563). New York: IEEE.

25. LaTorre, A., Muelas, S., & Peña, J.-M. (2015). A comprehensive comparison of large scale global optimizers. *Information Sciences, 316*, 517–549.

26. Li, C., Yang, S., & Nguyen, T. T. (2012). A self-learning particle swarm optimizer for global optimization problems. *IEEE Transactions on Systems, Man, and Cybernetics, Part B: Cybernetics, 42*(3), 627–646.

27. Liang, J. J., Qin, A. K., Suganthan, P. N., & Baskar, S. (2006). Comprehensive learning particle swarm optimizer for global optimization of multimodal functions. *IEEE Transactions on Evolutionary Computation, 10*(3), 281–295.

28. Li-li, L. I. U., & Xing-bao, G. A. O. (2012). An adaptive simulation of bacterial foraging algorithm. *Basic Sciences Journal of Textile Universities, 4*, 022.

29. Mirjalili, S., Lewis, A., & Sadiq, A. S. (2014). Autonomous particles groups for particle swarm optimization. *Arabian Journal for Science and Engineering, 39*(6), 4683–4697.
30. Rabiner, L. (1989). A tutorial on hidden markov models and selected applications in speech recognition. *Proceedings of the IEEE, 77*(2), 257–286.
31. Risler, M. (2010). *Behavior control for single and multiple autonomous agents based on hierarchical finite state machines.* PhD thesis, tuprints.
32. Sarhani, M., & El Afia, A. (2015). Facing the feature selection problem with a binary pso-gsa approach. In *The XI metaheuristics international conference (MIC)*, Agadir, Morocco.
33. Shi, Y., Liu, H., Gao, L., & Zhang, G. (2011). Cellular particle swarm optimization. *Inf. Sci., 181*(20), 4460–4493.
34. Soysal, O., & Şahin, E. (2006). A macroscopic model for self-organized aggregation in swarm robotic systems. In *Swarm robotics* (pp. 27–42. Berlin: Springer.
35. van den Bergh, F., & Engelbrecht, A. (2004). A cooperative approach to particle swarm optimization. *IEEE Transactions on Evolutionary Computation, 8*(3), 225–239.
36. Veček, N., Črepinšek, M., & Mernik, M. (2017). On the influence of the number of algorithms, problems, and independent runs in the comparison of evolutionary algorithms. *Applied Soft Computing, 54*, 23–45.
37. Vidal, E., Thollard, F., De La Higuera, C., Casacuberta, F., & Carrasco, R. C. (2005). Probabilistic finite-state machines-part ii. *IEEE Transactions on Pattern Analysis and Machine Intelligence, 27*(7), 1026–1039.
38. Wang, S., Chen, M., Huang, D., Guo, X., & Wang, C. (2014). Dream effected particle swarm optimization algorithm.
39. Wu, Z. (2014). Optimization of distribution route selection based on particle swarm algorithm. *International Journal of Simulation Modelling (IJSIMM), 13*(2),
40. Yang, S., & Li, C. (2010). A clustering particle swarm optimizer for locating and tracking multiple optima in dynamic environments. *IEEE Transactions on Evolutionary Computation.*
41. Zhan, Z. H., Zhang, J., Li, Y., & Chung, H. S.-H. (2009). Adaptive particle swarm optimization. *IEEE Transactions on Systems, Man, and Cybernetics, Part B: Cybernetics, 39*(6), 1362–1381.

Chapter 13
Algorithm Selector and Prescheduler in the ICON Challenge

François Gonard, Marc Schoenauer and Michèle Sebag

Abstract Algorithm portfolios are known to offer robust performances, efficiently overcoming the weakness of every single algorithm on some particular problem instances. Two complementary approaches to get the best out of an algorithm portfolio are to achieve algorithm selection (AS), and to define a scheduler, sequentially launching a few algorithms on a limited computational budget each. The presented system relies on the joint optimization of a pre-scheduler and a per-instance AS, selecting an algorithm well-suited to the problem instance at hand. ASAP has been thoroughly evaluated against the state-of-the-art during the ICON challenge for algorithm selection, receiving an honorable mention. Its evaluation on several combinatorial optimization benchmarks exposes surprisingly good results of the simple heuristics used; some extensions thereof are presented and discussed in the paper.

Keywords Algorithm selection · Algorithm portfolios · Combinatorial optimization

F. Gonard (✉)
IRT SystemX, 8 avenue de la Vauve, 91127 Palaiseau Cedex, France
e-mail: Francois.gonard@inria.fr

F. Gonard · M. Schoenauer · M. Sebag
LRI, CNRS, INRIA, University of Paris-Sud, University of Paris-Saclay, Bat. 660 Claude Shannon, 91405 Orsay Cedex, France
e-mail: Marc.schoenauer@inria.fr

M. Sebag
e-mail: Michele.sebag@inria.fr

© Springer International Publishing AG, part of Springer Nature 2019
E.-G. Talbi and A. Nakib (eds.), *Bioinspired Heuristics for Optimization*,
Studies in Computational Intelligence 774,
https://doi.org/10.1007/978-3-319-95104-1_13

13.1 Introduction

In quite a few domains related to combinatorial optimization, such as satisfiability, constraint solving or operations research, it has been acknowledged for some decades that there exists no universal algorithm, dominating all other algorithms on all problem instances [21]. This result has prompted the scientific community to design algorithm portfolios addressing the various types of difficulties involved in the problem instances, i.e., such that at least one algorithm in the portfolio can efficiently handle any problem instance [6, 8]. Algorithm portfolios thus raise a new issue, that of selecting a priori an algorithm well suited to the application domain [12]. This issue, referred to as *Algorithm Selection* (AS) [19], is key to the successful transfer of algorithms outside of research labs. It has been tackled by a number of authors in the last years [9, 15–17, 24] (more in Sect. 13.2).

Algorithm selection comes in different flavors, depending on whether the goal is to yield an optimal performance in expectation with respect to a given distribution of problem instances (global AS), or an optimal performance on a particular problem instance (*per-instance* AS). Note that the joint problems of selecting an algorithm and the optimal hyper-parameters thereof, referred to as *Algorithm Configuration* (AC), are often considered together in the literature, as the choice of the hyper-parameter values governs the algorithm performance. Only Algorithm Selection will be considered in the following; the focus is on the *per-instance* setting, aimed at achieving peak performance on every problem instance.

Noticing that some problem instances can be solved in no time by some algorithms, it makes sense to allocate a fraction of the computational budget to a *pre-scheduler*, sequentially launching a few algorithms with a small computational budget each. The pre-scheduler is expected to solve "easy" instances in a first stage; in a second stage, AS is only launched on problem instances which have not been solved in the pre-scheduler phase. Note that the pre-scheduler yields some additional information characterizing the problem at hand, which can be used together with the initial information about the problem instance, to support the AS phase.

This paper presents the *Algorithm Selector And Prescheduler* system (ASAP), aimed at algorithm selection in the domain of combinatorial optimization (Sect. 13.3). The main contribution lies in the joint optimization of both a pre-scheduler and a per-instance algorithm selector. The extensive empirical validation of ASAP is conducted on the ICON challenge on algorithm selection [10]. This challenge leverages the Algorithm Selection library [1], aimed at the fair, comprehensive and reproducible benchmarking of AS approaches on 13 domains ranging from satisfiability to operations research (Sect. 13.4). The comparative empirical validation of ASAP demonstrates its good performances comparatively to state-of-art pre-schedulers and AS approaches (Sect. 13.5), and its complementarity with respect to the prominent zilla algorithm (based on SATzilla [23]). The paper concludes with a discussion of the limitations of the ASAP approach, and some perspectives for further research.

13.2   Related Work

13.2.1   Algorithm Selectors

The algorithm selection issue, aimed at selecting the algorithm best suited to the problem at hand, was first formalized by Rice [19]. Given a problem space mapping each problem instance onto a description $\mathbf{x}$ thereof (usually $\mathbf{x}$ in $\mathbf{R}^d$) and the set $\mathscr{A}$ of algorithms in the portfolio, let $\mathscr{G}(\mathbf{x}, a)$ be a performance model estimating the performance of algorithm a onto problem instance $\mathbf{x}$ for each $(\mathbf{x}, a)$ pair. Such a performance model yields an AS strategy, by selecting for problem instance $\mathbf{x}$ the algorithm a with optimal $\mathscr{G}(\mathbf{x}, a)$.

$$AS(\mathbf{x}) = \arg\max_{a \in \mathscr{A}}\{\mathscr{G}(\mathbf{x}, a)\} \qquad (13.1)$$

The performance model is usually built by applying machine learning approaches onto a dataset reporting the algorithm performances on a comprehensive set of benchmark problem instances (with the exception of [5], using a multi-armed bandit approach). Such machine learning approaches range from k-nearest neighbors [16] to ridge regression [23], random forests [24], collaborative filtering [15, 20], or learning to rank approaches [17]. The interested reader is referred to [11] for a more comprehensive review of algorithm selectors.

As expected, the efficiency of the machine learning approaches critically depends on the quality of the training data: i.e. the representativity of the problem instances used to train the performance model and the description of the problem instances. Considerable care has been devoted to the definition of descriptive features in the SAT and Constraint domains [22].

13.2.2   Schedulers

An algorithm portfolio can also take advantage of parallel computer architectures by launching several algorithms working independently or in cooperation on the considered problem instance (see, e.g., [9, 25]). Schedulers embed parallel solving strategies within a sequential setting, by defining a sequence of κ (algorithm a_i, timeout τ_i) pairs, such that each problem instance is successively tackled by algorithm a_i with a computational budget τ_i, until being solved. Note that the famed restart strategy — launching a same algorithm with different random seeds or different initial conditions — can be viewed as a particular case of scheduling strategy [6]. Likewise, AS can be viewed as a particular case of scheduler with $\kappa = 1$ and τ_1 set to the overall computational budget.

A multi-stage process, where a scheduler solves easy instances in a first stage, and remaining instances are handled by the AS and tackled by the selected algorithm in the next stage, is described by [24]. In [9], hybrid *per-instance* schedules are proposed, with AS as one of the components.

13.3   Overview of ASAP

After discussing the rationale for the presented approach, this section presents the pre-scheduler and AS components forming the ASAP system, Versions 1 and 2.

13.3.1   Analysis

It is notorious that the hardness of a problem instance often depends on the considered algorithm. As shown on Fig. 13.1 in the case of the SAT11-HAND dataset (Sect. 13.4), while several algorithms might solve 20% of the problem instances within seconds, the oracle (selecting the best one out of these algorithms for each problem instance) solves about 40% of the problem instances within seconds. Along this line, the pre-scheduler problem thus consists of selecting a few algorithms, such that running each of these algorithms for a few seconds would solve a significant fraction of the problem instances.

Definition 1   (*Pre-scheduler*) Let $\mathscr{A}$ be a set of algorithms. A κ-pre-scheduler component, defined as a sequence of κ (algorithm a_i, time-out τ_i) pairs,

$$\left((a_i, \tau_i)_{i=1}^{\kappa}\right) \text{ with } (a_i, \tau_i) \in \mathscr{A} \times \mathbb{R}^+, \ \forall i \in 1, \dots, \kappa$$

sequentially launches algorithm a_j on any problem instance $\mathbf{x}$ until either a_j solves $\mathbf{x}$, or time τ_j is reached, or a_j stops without solving $\mathbf{x}$. If $\mathbf{x}$ has been solved, the execution stops. Otherwise, j is incremented while $j \leq \kappa$.

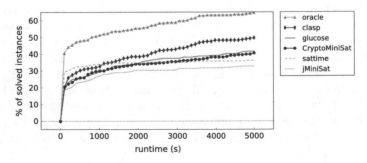

Fig. 13.1 Percentage of solved instances versus runtime on the SAT11-HAND dataset, for 5 algorithms and the oracle (selecting the best algorithm out of 5 for each problem instance)

A pre-scheduler can contribute to better peak performances [13]. It can also increase the overall robustness of the resolution process and mitigate the impact of AS failures (where the selected algorithm requires much computational resources to solve a problem instance or fails to solve it), as it increases the chance for each problem instance to be solved in no time, everything else being equal.

Accordingly, the ASAP system involves: (i) a pre-scheduler aimed at solving as many problem instances as possible in a first stage; and (ii) an AS taking care of the remaining instances. A first decision regards the division of labor between both components: how to split the available runtime between the two, and how many algorithms are involved in the pre-scheduler (parameter κ). For simplicity and tractability, the maximal runtime allocated to the pre-scheduler is fixed to T_{ps}^{max} (10% of the overall computational budget in the experiments, Sect. 13.5), and the number κ of algorithms in the pre-scheduler is set to 3. A similar setup is used in [9, 14], with the difference that the pre-scheduler uses a prescribed fraction of the computational budget.

Given T_{ps}^{max} and κ, ASAP tackles the optimization of the pre-scheduler and the AS components. Both optimization problems are interdependent: the AS must focus on the problem instances which are not solved by the pre-scheduler, while the pre-scheduler must symmetrically focus on the problem instances which are most uncertain or badly addressed by the AS. Formally, this interdependence is handled as follows:

- A performance model $\mathscr{G}(\mathbf{x}, a)$ is built for each algorithm over all training problem instances, defining AS_{init} (Eq. 13.1);
- A pre-scheduler is built to optimize the joint performance (pre-scheduler, AS_{init}) over all training problem instances;
- Another performance model $\mathscr{G}_2(\mathbf{x}, a)$ is built over all training problem instances, using an additional Boolean feature that indicates for each problem instance whether it was solved by the above pre-scheduler; let AS_{post} denote the AS based on performance model $\mathscr{G}_2(\mathbf{x}, a)$.

ASAP finally is composed of the pre-scheduler followed by AS_{post}.

13.3.2 ASAP.V1 pre-scheduler

Let $(a_i, \tau_i)_{i=1}^{\kappa}$ denote a pre-scheduler, with overall computational budget $T_{ps} = \sum_{i=1}^{\kappa} \tau_i$, and let $\mathscr{F}\left((a_i, \tau_i)_{i=1}^{\kappa}\right)$ denote the associated domain-dependent performance (e.g., number of solved instances or time-to-solution). ASAP.V1 considers for simplicity equal time-outs ($a_i = \frac{1}{\kappa} T_{ps}, i = 1 \ldots \kappa$). The pre-scheduler is thus obtained by solving the following optimization problem:

$$\max_{T_{ps} \leq T_{ps}^{max}, a_1, \ldots a_\kappa} \left\{ \mathscr{F}\left((a_i, \frac{T_{ps}}{\kappa})_{i=1}^{\kappa}\right) \right\} \tag{13.2}$$

This mixed optimization problem is tackled in a hierarchical way, determining for each value of T_{ps} the optimal κ-uple of algorithms $a_1 \ldots a_\kappa$. Thanks to both small κ values ($\kappa = 3$ in the experiments) and small number of algorithms (≤ 31 in the ICON challenge, Sect. 13.4), the optimal κ-uple is determined by exhaustive search conditionally to the T_{ps} value.

The ASAP.V1 pre-scheduler finally relies on the 1-dimensional optimization of the overall computational budget T_{ps} allocated to the pre-scheduler. In all generality, the optimization of T_{ps} is a multi-objective optimization problem, e.g., balancing the overall number of problems solved and the overall computational budget. Multi-objective optimization commonly proceeds by determining the so-called Pareto front, made of non-dominated solutions. In our case, the Pareto front depicts how the performance varies with the overall computational budget, as illustrated on Fig. 13.2, where the performance is set to the number of solved instances.

In multi-objective decision making [2], the choice of a solution on the Pareto front is tackled using post-optimal techniques [4], including: (i) compromise programming, where one wants to find the point the closest to an ideal target in the objective space; (ii) aggregation of the objectives into a single one, e.g., using linear combination; or (iii) marginal rate of return. The last heuristics consists of identifying the so-called "knees", that is, the points where any small improvement on a given criterion is obtained at the expense of a large decrease on another criterion, defining the so-called marginal rate of return. The vanilla marginal rate of return is however sensitive to strong local discontinuities; for instance, it would select point A in Fig. 13.2. Therefore, a variant taking into account the global shape of the curve, and measuring the marginal rate of improvement w.r.t. the extreme solutions on the Pareto front is used (e.g., selecting point K instead of point A in Fig. 13.2).

13.3.3 ASAP.V1 Algorithm Selector

As detailed in Sect. 13.3.1, the AS relies on the performance model learned from the training problem instances. Two machine learning (ML) algorithms are considered in

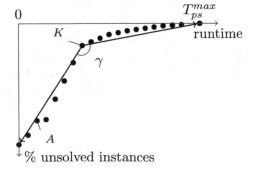

Fig. 13.2 Among a set of Pareto-optimal solutions, solution A has the best marginal rate of return; solution K, which maximizes the average rate of return w.r.t. the extreme solutions of the Pareto front (minimizing angle γ), is the knee selected in ASAP

this paper: random forests and k-nearest neighbors (where the considered distance is the Euclidean distance in the description space of the problem instances). One hyper-parameter was adapted for each ML approach (all other hyper-parameters being set to their default value, using the Python scikit-learn library [18]), based on a few preliminary experiments: 35 trees are used for the RandomForest algorithm and the number of neighbors is set to $k = 3$ for the k-nearest neighbors. In the latter case, the predicted value associated to problem instance $\mathbf{x}$ is set to the weighted sum of the performance of its nearest neighbors, weighted by their relative distance to $\mathbf{x}$:

$$\widehat{\mathscr{G}}(\mathbf{x}, a) = \frac{\sum_i ||\mathbf{x} - \mathbf{x}_i|| \mathscr{G}(a, \mathbf{x}_i)}{\sum_i ||\mathbf{x} - \mathbf{x}_i||}$$

where $\mathbf{x}_i$ ranges over the 3 nearest neighbors of $\mathbf{x}$. The description of the instances is normalized (each coordinate having zero mean and unit variance).

A main difficulty comes from the descriptive features forming the representation of problem instances. Typically, the feature values are missing for some groups of features, for quite a few problem instances, due to diverse causes (computation exceeded time limit, exceeded memory, presolved the instance, crashed, other, unknown). Missing feature values are handled by (i) replacing the missing value by the feature average value; (ii) adding to the set of descriptive features 7 additional Boolean features per group of initial features, indicating whether the feature group values are available or the reason why they are missing otherwise.[1]

13.3.4 ASAP.V2

Several extensions of ASAP.V1 have been considered after the closing of the ICON challenge, aimed at exploring a richer pre-scheduler-AS search space while preventing the risk of overfitting induced by a larger search space.

We investigated the use of different time-outs for each algorithm in the pre-scheduler, while keeping the set of algorithms $(a_1, \ldots, a_\kappa)$ and the overall computational budget T_{ps}. The sequential optimization strategy (Sect. 13.3.2), deterministically selecting T_{ps} as the solution with maximal average return rate, exhaustively determining the κ-uple of algorithms conditionally to T_{ps}, is thus extended to optimize the $(\tau_1, \ldots \tau_{\kappa-1})$ vector conditionally to $\sum_{i=1}^{\kappa-1} \tau_i \leq T_{ps}$, using a prominent continuous black-box optimizer, specifically the Covariance-Matrix Adaptation-Evolution Strategy (CMA-ES) [7].

---

[1] The increase in the overall number of features is handled by an embedded feature selection mechanism, removing all features with negligible importance criterion ($<10^{-5}$ in the experiments) in a independently learned 10-trees random forest regression model.

This extended search space is first investigated by considering the **raw optimization criterion** $\mathscr{F}_{raw}(\tau_1, \ldots \tau_\kappa)$ measuring the cumulative performance of ASAP over all training problem instances, defined as follows:

$$\mathscr{F}_{raw}(\tau_1, \ldots \tau_\kappa) = \sum_j Solv(\tau_1, \ldots, \tau_\kappa, \mathbf{x}_j) \tag{13.3}$$

where $Solv(\tau_1, \ldots, \tau_\kappa, \mathbf{x}_j)$ is the time required by the pre-scheduler $(a_i, \tau_i)_{i=1}^\kappa$ followed by AS_{init} to solve the jth instance, or 10 times the dataset time-out.

However a richer search space entails some risk of overfitting, where the higher performance on data used to optimize ASAP (training data) is obtained at the expense of a lower performance on test data. Generally speaking, the datasets used to train an AS are small ones (a few hundred to a few thousand instances).

A **penalized optimization criterion** is thus considered:

$$\mathscr{F}_{L_2}\left((a_i, \tau_i)_{i=1}^\kappa\right) = \mathscr{F}\left((a_i, \tau_i)_{i=1}^\kappa\right) + w \sqrt{\sum_{i=1}^\kappa \left(\tau_i / T_{ps}\right)^2}$$

which penalizes uneven time sharing within the pre-scheduler (the regularization term is minimized when $\tau_i = \frac{1}{\kappa} T_{ps}$). The rationale for this penalization is to prevent brittle improvements on the training set due to opportunistic adjustments of the τ_is, at the expense of stable performances on further instances. The penalization weight w is adjusted using a preliminary cross-validation process.

A **randomized optimization criterion** is also considered. By construction, the ideal fitness function to be minimized is the expected performance over the problem domain. Only the empirical average performance over the problem instances is available, defining a noisy optimization problem. Sophisticated approaches have been proposed to address the noisy optimization issue (see e.g. [3]). Another approach is proposed here, based on the bootstrap principle: in each CMA-ES generation, the set of n problem instances used to compute the performance is uniformly drawn with replacement from the n-size training set. In this manner, each optimization generation considers a slightly different optimization objective noted $\mathscr{F}_{rand}$, thereby discouraging hazardous improvements and contributing to a more robust search.

Finally, a **probabilistic optimization criterion** is considered, handling the ASAP performance on a single problem instance as a random variable with a triangle-shape distribution (Fig. 13.3) centered on the actual performance $p(\mathbf{x})$, with support in $[p(\mathbf{x}) - \theta, p(\mathbf{x}) + \theta]$, and taking the expectation thereof. The merit of this triangular probability distribution function is to allow for an analytical computation of the overall fitness expectation, noted $\mathscr{F}_{dfp}$.

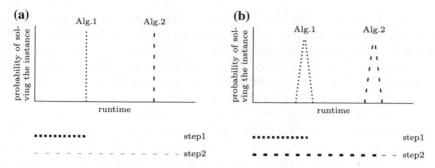

Fig. 13.3 Impact of a probabilistic optimization criterion: Difference between deterministic and probabilistic execution time. **a** the schedule deterministically stops as Algorithm 1 solves the instance. **b** with some probability, Algorithm 1 does not solve the instance and the execution proceeds

13.4   Experimental Setting: The ICON Challenge

13.4.1   ASlib Data Format

Due to the difficulty of comparing the many algorithm selection systems and the high entry ticket to the AS field, a joint effort was undertaken to build the Algorithm Selection Library (ASlib), providing comprehensive resources to facilitate the design, sharing and comparison of AS systems [1]. ASlib (version 1.0.1) involves 13 datasets, also called scenarios (Table 13.1), gathered from recent challenges and surveys in the operations research, artificial intelligence and optimization fields. The interested reader is referred to [1] for a more comprehensive presentation.

Each dataset includes (i) the performance and computation status of each algorithm on each problem instance; (ii) the description of each problem instance, as a vector of the expert-designed feature values (as said, this description considerably facilitates the comparison of the AS systems); (iii) the computational status of each such feature (e.g., indicating whether the feature could be computed, or if it failed due to insufficient computational or memory resources). Last but not least, each dataset is equi-partitioned into 10 subsets, to enforce the reproducibility of the 10 fold cross-validation assessment of every AS algorithm.

13.4.2   The ICON Challenge on Algorithm Selection

The ICON Challenge on Algorithm Selection, within the ASlib framework, was carried on between February and July 2015 to evaluate AS systems in a fair, comprehensive and reproducible manner.[2] Each submitted system was assessed on the 13 ASlib datasets [1] with respect to three measures: (i) number of problem instances solved; (ii) extra runtime compared with the virtual best solver (VBS, also called oracle); and (iii) Penalized Average Time-10 (PAR10) which is the cumulative runtime needed to solve all problem instances (set to ten times the overall computational budget whenever the problem instance is unsolved).

As the whole datasets were available to the community from the start, the evaluation was based on hidden splits between training and test set. Each submitted system provides a dataset-dependent, instance-dependent schedule of algorithms, optionally preceded by a dataset-dependent presolver (single algorithm running on all instances during a given runtime before the per-instance schedule runs). Each system can also, in a dataset-dependent manner, specify the groups of features to be used (in order to save the time needed to compute useless features).

Two baselines are considered: the oracle, selecting the best algorithm for each problem instance; and the single best (SB) algorithm, with best average performance over all problem instances in the dataset. The baselines are used to normalize every system performance over all datasets, associating performance 0 to the oracle (respectively performance 1 to the single best), supporting the aggregation of the system results over all datasets.

Table 13.1 ASlib datasets (V1.0.1)

Dataset	# Instances	# Algorithms	# Features
ASP-POTASSCO	1294	11	138
CSP-2010	2024	2	86
MAXSAT12-PMS	876	6	37
PREMARSHALLING-ASTAR-2013	527	4	16
PROTEUS-2014	4021	22	198
QBF-2011	1368	5	46
SAT11-HAND	296	15	115
SAT11-INDU	300	18	115
SAT11-RAND	600	9	115
SAT12-ALL	1614	31	115
SAT12-HAND	767	31	115
SAT12-INDU	1167	31	115
SAT12-RAND	1362	31	115

---

[2]The codes of all submitted systems and the results are publicly available, http://challenge.icon-fet.eu/challengeas.

Table 13.2 Normalized performances of submitted systems, aggregated across all folds and all measures (the lower, the better). Ranks of zilla (challenge winner) and ASAP_RF.V1 (honourable mention) are given in parenthesis. Numbers were computed from the challenge outputs. Note: "PREMARSHALLING" stands for "PREMARSHALLING-ASTAR-2013"

	ASAP_RF.V1	ASAP_kNN.V1	Autofolio	Flexfolio	Sunny	Sunny-presolv	Zilla	Zillafolio
ASP-POTASSCO	0.294 (2)	0.359	0.299	0.314	0.37	0.336	0.319 (5)	**0.283**
CSP-2010	**0.146** (1)	0.247	0.288	0.223	0.263	0.406	0.2 (3)	0.157
MAXSAT12-PMS	0.168 (4)	0.159	0.45	**0.149**	0.166	0.224	0.201 (5)	0.233
PREMARSHALLING	0.349 (4)	0.369	0.359	0.307	0.325	**0.296**	0.374 (7)	0.385
PROTEUS-2014	0.16 (4)	0.177	0.222	**0.056**	0.134	0.103	0.245 (8)	0.223
QBF-2011	0.097 (2)	**0.091**	0.169	0.096	0.142	0.162	0.191 (7)	0.194
SAT11-HAND	0.341 (4)	0.318	0.342	0.342	0.466	0.464	0.328 (3)	**0.302**
SAT11-INDU	1.036 (5)	0.957	**0.875**	1.144	1.13	1.236	0.905 (2)	0.966
SAT11-RAND	0.104 (6)	0.09	**0.046**	0.226	0.116	0.088	0.053 (2)	0.067
SAT12-ALL	0.392 (5)	0.383	0.306	0.502	0.509	0.532	**0.273** (1)	0.322
SAT12-HAND	0.334 (5)	0.31	**0.256**	0.434	0.45	0.467	0.272 (2)	0.296
SAT12-INDU	0.955 (6)	0.919	0.604	0.884	1.074	1.018	0.618 (3)	**0.594**
SAT12-RAND	1.032 (5)	1.122	0.862	1.073	1.126	0.97	**0.779** (1)	0.79

13.5 Experimental Validation

13.5.1 Comparative Results

Table 13.2 reports the results of all submitted systems on all datasets (the statistical significance tests are reported in Fig. 13.4). The general trend is that zilla algorithms dominate all other algorithms on the SAT datasets, as expected since they have consistently dominated the SAT contests in the last decade. On non-SAT problems however, zilla algorithms are dominated by ASAP_RF.V1.

The robustness of the ASAP approach is demonstrated as they never rank last; they however perform slightly worse than the single best on some datasets. The rescaled performances of ASAP_RF.V1 is compared to zilla and autofolio (Fig. 13.5, on the

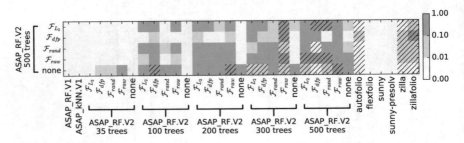

Fig. 13.4 Significance analysis after Wilcoxon signed-rank test p-value: ASAP_RF.V2 (fitness variants, 500 trees) against all other systems. Color indicates the significance; hatched indicates that the line algorithm is outperformed by the column algorithm

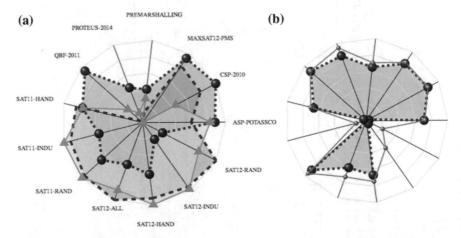

Fig. 13.5 Comparative performances. **a** Per-dataset performances of ASAP RF.V1 (balls, dotted line), zilla (no marker, dashed line) and autofolio (triangles, solid line); the scale is such that 0 corresponds to the worst submitted AS and 1 to the best submitted AS. **b** Comparison of ASAP RF.V1 and the best submitted AS per dataset in normalized performance (small balls, solid line). On all scenarios except 3, ASAP_RF reaches similar performances as the best submitted AS on this scenario

left), demonstrating that ASAP_RF.V1 offers a balanced performance, significantly lower than for zilla and autofolio on the SAT problems, but significantly higher on the other datasets; in this respect it defines an overall robust AS approach.

13.5.2   Sensitivity Analysis

The sensitivity analysis conducted after the closing of the challenge compares ASAP.V2 (with different time-outs in the pre-scheduler) and ASAP.V1, and examines the impact of the different optimization criteria, aimed at avoiding overfitting: the raw fitness, the L_2-penalized fitness, the randomized fitness and the probabilistic fitness (Sect. 13.3.4).

The impact of the hyper-parameters used in the AS (number of trees set to 35, 100, 200, 300 and 500 trees in the Random Forest) is also investigated.

Table 13.3 summarizes the experimental results that each ASAP.V2 configuration would have obtained in the ICON challenge framework,[3] together with the actual submissions results, including systems that were not competing in the challenge: llama-regr and llama-regrPairs from the organizers, and autofolio-48 which is identical to autofolio but with 48h time for training (12h was the time limit authorized in the challenge) [10].

The significance analysis, using a Wilcoxon signed-rank test, is reported in Fig. 13.4. A first result is that all ASAP.V2 variants improve on ASAP.V1 with significance level 1%. A second result is that ASAP.V2 with the probabilistic optimization criterion is not statistically significantly different from zilla, autofolio and zillafolio.

A third and most surprising result is that the difference between the challenge-winner zilla and most of ASAP.V2 variants is not statistically significant.

Figure 13.6 details per dataset the performance improvement between ASAP.V2 (500 trees, $\mathscr{F}_{L_2}$ version) and ASAP.V2 (500 trees, $\mathscr{F}_{dfp}$ version) and on the other hand ASAP.V1_RF (35 trees). Note that ASAP.V2 outperforms the per-dataset best submission to the challenge for 3 datasets.

13.6   ASAP.V2 optimized pre-scheduler behavior analysis

ASAP.V1 was designed after the observation that its pre-scheduler and AS components should be complementary. ASAP.V2 introduces a specific tuning of the pre-scheduler to strengthen the division of labor between them. Specifically, the optimized pre-scheduler should i) focus on trying to solve instances ill-handled by the selector and ii) be more efficient than its non-optimized counterpart to solve "easy"

---

[3] For CSP-2010 dataset, only two algorithms are available: the pre-scheduler thus consists of a single algorithm, and all ASAP_RF.V2 variants with the same selector hyperparameter are identical.

Table 13.3 Optimized pre-scheduler performances aggregated across all datasets, all splits and all measures (the lower, the better). The hyperparameters for $\mathscr{F}_{L_2}$ and $\mathscr{F}_{dfp}$ were chosen after preliminary experiments using the cross validation provided with ASlib. For each configuration of the selector, the best-evaluated fitness function appears in bold

fitness function (if relevant)	$\mathscr{F}_{L_2}$	$\mathscr{F}_{dfp}$	$\mathscr{F}_{rand}$	$\mathscr{F}_{raw}$	none
ASAP_RF.V2 35 trees	0.416	0.414	0.412	**0.410**	0.414
ASAP_RF.V2 100 trees	0.404	**0.398**	0.405	0.402	0.414
ASAP_RF.V2 200 trees	0.404	0.402	0.402	**0.399**	0.405
ASAP_RF.V2 300 trees	0.399	0.399	0.402	**0.393**	0.405
ASAP_RF.V2 500 trees	0.398	**0.394**	0.398	0.398	0.401
ASAP_RF.V1	0.416				
ASAP_kNN.V1	0.423				
autofolio	0.391			equivalent to the means	
flexfolio	0.442			over the columns of	
sunny	0.482			Table 13.2	
sunny-presolv	0.485				
zilla	0.366				
zillafolio	0.37				
autofolio-48			0.375		
llama-regrPairs			0.395		
llama-regr			0.425		

(a)   **(b)**

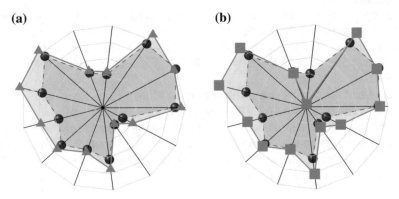

Fig. 13.6 Left: Comparison of ASAP RF.V2 ($\mathscr{F}_{L_2}$, 500 trees, triangles) with ASAP RF.V1. Right: Comparison of ASAP RF.V2 ($\mathscr{F}_{dfp}$, 500 trees, squares) with ASAP RF.V1

instances. The comparison of ASAP.V2 variants and the non-optimized prescheduler (rightmost column in Table 13.3, with equal time-outs) shows that these goals are met to some extent.

On the one hand, the pre-scheduler fine-tuning does improve the pre-scheduler performance; the overlap between instances that each component standalone can solve[4] (within T_{ps} for the pre-schedule, within the remaining time for the AS) tends to diminish when optimizing the pre-scheduler for most datasets, as depicted on

---

[4]Remind that these instances are not actually passed to the AS in the challenge evaluation setup

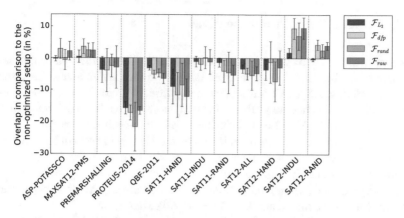

Fig. 13.7 Per-dataset number of instances that can be solved by both the pre-scheduler and the selector components in comparison to the non-optimized pre-scheduler variant

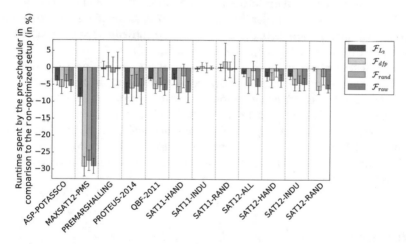

Fig. 13.8 Per-dataset runtime spent in the pre-scheduler phase in comparison to the non-optimized pre-scheduler variant

Fig. 13.7. On the other hand, the full ASAP.V2 systems (optimized pre-scheduler + AS) solve roughly as many instances as the non-optimized setup (difference <1%). It follows that the pre-scheduler fine-tuning leads to a better specialization of both components, though it does not translate into a global improvement.

The inclusion of T_{ps} in the set of optimized variables is expected to further strengthen this specialization.

The time spent in the pre-scheduling phase is reduced (up to 29%) by the prescheduler fine-tuning, as illustrated in Fig. 13.8. As one could have expected, the use of the L_2 regularization mitigates this effect (remind this setup prevents the optimized τ_is to be far apart from the equal time-outs of the the non-optimized pre-scheduler): it is low-risk, low-reward. No clear winner emerges from the other 3 variants.

13.7 Conclusion and Perspectives

This paper presents two contributions. The first one is a new hybrid algorithm selection approach, the ASAP system, combining a pre-scheduler and a per-instance algorithm selector. ASAP.V1 introduces an algorithm selector learned conditionally to a predetermined schedule so that it focuses on instances that were not solved by the pre-scheduler. ASAP.V2 completes the loop as it re-optimizes the pre-scheduler conditionally to the new AS. A main lesson learned is that the scheduler and the AS must be optimized jointly to achieve an effective division of labor.

ASAP.V1, thoroughly evaluated in the ICON challenge on algorithm selection (ranked 4th) received an honourable mention, due to its novelty and good performance comparatively to the famed and long-known *zilla algorithms.

The second contribution is the ASAP.V2 extension, that achieves significantly better results along the same challenge setting. A main lesson learned is the importance of considering regularized optimization objectives: the amount of available data does not permit to consider richer AS search spaces without incurring a high risk of overfitting. In particular, the probabilistic performance criterion successfully contributed to a more stable optimization problem.

Further research will first comfort these good results by additional experiments on fresh data. A mid-term research is concerned with extending the optimization search space, and specifically adjusting the pre-scheduler parameters (number κ of algorithms and budget) depending on the scenario. Another perspective is concerned with learning the margin between any two algorithms depending on the problem instance, in order to yield a better AS, taking inspiration from [24]. The long term research will be devoted to move from alternate optimization of ASAP.V2 components, to a global optimization problem.

References

1. Bischl, B., Kerschke, P., Kotthoff, L., Lindauer, M., Malitsky, Y., Fréchette, A., et al. (2016). Aslib: a benchmark library for algorithm selection. *Artificial Intelligence, 237*, 41–58.
2. Branke, J., Deb, K., Dierolf, H., & Osswald, M. (2004). Finding knees in multi-objective optimization. *Parallel problem solving from nature-PPSN VIII* (pp. 722–731). Berlin: Springer.
3. Cauwet, M., Liu, J., Rozière, B., & Teytaud, O. (2016). Algorithm portfolios for noisy optimization. *Annals of Mathematics and Artificial Intelligence, 76*(1–2), 143–172.
4. Deb, K. (2003). Multi-objective evolutionary algorithms: introducing bias among pareto-optimal solutions. *Advances in evolutionary computing* (pp. 263–292). Berlin: Springer.
5. Gagliolo, M., & Schmidhuber, J. (2011). Algorithm portfolio selection as a bandit problem with unbounded losses. *Annals of Mathematics and Artificial Intelligence, 61*(2), 49–86.
6. Gomes, C. P., & Selman, B. (2001). Algorithm portfolios. *Artificial Intelligence, 126*(1), 43–62.
7. Hansen, N., Müller, S. D., & Koumoutsakos, P. (2003). Reducing the time complexity of the derandomized evolution strategy with covariance matrix adaptation. *Evolutionary Computation, 11*(1), 1–18.
8. Huberman, B. A., Lukose, R. M., & Hogg, T. (1997). An economics approach to hard computational problems. *Science, 275*(5296), 51–54.

9. Kadioglu, S., Malitsky, Y., Sabharwal, A., Samulowitz, H., & Sellmann, M. (2011). Algorithm selection and scheduling. In *Proceedings of the 17th CP* (pp. 454–469). LNCS. Berlin: Springer.
10. Kotthoff, L. (2015). ICON challenge on algorithm selection. CoRR arXiv:abs/1511.04326.
11. Kotthoff, L. (2016). Algorithm selection for combinatorial search problems: a survey (pp. 149–190). Berlin: Springer International Publishing.
12. Leyton-Brown, K., Nudelman, E., Andrew, G., McFadden, J., & Shoham, Y. (2003). A portfolio approach to algorithm selection. In *Proceedings of the IJCAI* (pp. 1542–1543).
13. Lindauer, M., Bergdoll, R. D., & Hutter, F. (2016). An empirical study of per-instance algorithm scheduling. In *Proceedings of the LION10* (pp. 253–259). Berlin: Springer.
14. Malitsky, Y., Sabharwal, A., & Samulowitz, H. (2013). Algorithm portfolios based on cost-sensitive hierarchical clustering. In *Proceedings of the 23rd IJCAI* (pp. 608–614). California: AAAI Press.
15. Mısır, M., & Sebag, M. (2017). Alors: an algorithm recommender system. *Artificial Intelligence*, *244*, 291–314. Published on-line Dec. 2016.
16. O'Mahony, E., Hebrard, E., Holland, A., Nugent, C., & O'Sullivan, B. (2008). Using case-based reasoning in an algorithm portfolio for constraint solving. In *Proceedings of the ICAICS* (pp. 210–216).
17. Oentaryo, R. J., Handoko, S. D., & Lau, H. C. (2015). Algorithm selection via ranking. In *Proceedings of the 29th AAAI Conference on Artificial Intelligence (AAAI)*.
18. Pedregosa, F., et al. (2011). Scikit-learn: machine learning in python. *Journal of Machine Learning Research*, *12*, 2825–2830.
19. Rice, J. R. (1976). The algorithm selection problem. *Advances in Computers*, *15*, 65–118.
20. Stern, D., Herbrich, R., Graepel, T., Samulowitz, H., Pulina, L., & Tacchella, A. (2010). Collaborative expert portfolio management. In *Proceedings of the 24th AAAI* (pp. 179–184).
21. Wolpert, D., & Macready, W. (1997). No free lunch theorems for optimization. *IEEE Transactions on Evolutionary Computation*, *1*(1), 67–82.
22. Xu, L., Hutter, F., Hoos, H., & Leyton-Brown, K. (2012). Features for SAT. University of British Columbia.
23. Xu, L., Hutter, F., & Hoos, H. H. (2008). Satzilla: portfolio-based algorithm selection for sat. *Journal of Artificial Intelligence Research*, pp. 565–606.
24. Xu, L., Hutter, F., Shen, J., Hoos, H. H., & Leyton-Brown, K. (2012). Satzilla2012: improved algorithm selection based on cost-sensitive classification models. In *Proceedings of the SAT Challenge* (pp. 57–58).
25. Yun, X., & Epstein, S. L. (2012). Learning algorithm portfolios for parallel execution. In Y. Hamadi & M. Schoenauer (Eds.), In *Proceedings of the LION 6* (Vol. 7219, pp. 323–338). LNCS. Berlin: Springer.

Chapter 14
A Metaheuristic Approach to Compute Pure Nash Equilibria

Niels Elgers, Nguyen Dang and Patrick De Causmaecker

Abstract A pure Nash equilibrium is a famous concept in the field of game theory and has a wide range of applications. In the last decade, a lot of progress has been made in determining the computational complexity of finding equilibria in games. Deciding if a pure Nash equilibrium exists in n-player normal form games and several subclasses has been shown to be NP-complete. Current exact approaches to compute them are impractical and only able to solve small instances. In this paper, we apply three local search-based metaheuristics for solving the problem. Results on 280 randomly generated instances with different sizes show the clear outperformance of the metaheuristic approaches over the exact method based on Mixed Integer Linear Programming.

Keywords Local search-based metaheuristics · Approximate pure Nash equilibria

14.1 Introduction

Game theory is the study of mathematical models (*games*) that represent situations where multiple rational agents (*players*) receive payoffs which depend on the choices (*actions*) they make. In most contexts, each player strives to achieve an as high as possible payoff. Problems are modeled as games in an effort to predict

N. Elgers (✉)
KU Leuven KULAK, Kortrijk, Belgium
e-mail: niels.elgers@student.kuleuven.be

N. Dang · P. De Causmaecker
Department of Computer Science, KU Leuven, CODeS & imec-ITEC, Kortrijk, Belgium
e-mail: nguyenthithanh.dang@kuleuven.be

P. De Causmaecker
e-mail: patrick.decausmaecker@kuleuven.be

© Springer International Publishing AG, part of Springer Nature 2019
E.-G. Talbi and A. Nakib (eds.), *Bioinspired Heuristics for Optimization*,
Studies in Computational Intelligence 774,
https://doi.org/10.1007/978-3-319-95104-1_14

how the players will behave. There are multiple solution concepts (*equilibrium*) that attempt to predict this behavior. Applications of game theory include, but are not limited to, the modeling of psychological theories; resource allocation, networking and artificial intelligence in computer science; competition and coexistence between species in biology and the modeling of many economic interactions. We consider *non-cooperative games* in which players make decisions independent of each other as opposed to *cooperative games* in which players may form coalitions or reach consensus as to achieve the best possible outcome. Players have full information of the game, i.e., they know what actions are available to the other players and what payoff values result from them. Computer scientists study game theory to describe the computational complexity of computing equilibria in games.

It is possible to define a probability distribution over the actions available to a player. A set containing one such probability distribution for each player is called a *mixed strategy*. The expected payoff for a player is the payoff value a player can expect from playing the game when the other players sample actions from their distributions. A mixed strategy is called a *mixed Nash equilibrium (MNE)* if no player can alter his/her distribution as to increase his/her expected payoff, assuming that the probability distributions of the other players are held fixed. A *pure strategy* is a special case of a mixed strategy when all players limit themselves to playing only one action, i.e., that action has a probability of one of being sampled. A pure strategy is called a *pure Nash equilibrium (PNE)* if no player can alter his action as to increase his payoff, assuming that the actions of the other players are held fixed.

John Forbes Nash, Jr. showed that in each finite game, there exists at least one MNE [10, 17]. Later, it has been shown that computing a MNE is *PPAD-complete* [4], i.e., no efficient solution method has been found, although the evidence of intractability is weaker than the one for NP-completeness. There are a number of exact methods [12] and local search metaheuristics [3] for computing MNEs in two-player games, and a collection of polynomial-time approximation algorithms [8, 19] for specific types of multi-player games.

Games do not always have a PNE, and it has been shown to be NP-complete to decide if there exists at least one in a finite game [9]. PNE is appealing, nevertheless, because it is simple to describe and execute and thus is a more plausible explanation of how agents may behave in a game. Similarly to MNE, there exists a collection of polynomial-time approximation algorithms [1, 2] to decide if a PNE exists for games with specific settings. However, for general multi-player games, as far as we are aware, there is only one exact method (other than brute force), which is based on Mixed Integer Linear Programming (MILP) [20], for solving the problem. In this paper, we propose using three local search-based metaheuristic approaches, namely Random Restart Hill Climbing, Simulated Annealing [11] and Chained Local Optimization [15], for computing a PNE in general games. Due to the nature of the approaches, if a PNE is not found in a game, we won't be able to prove that PNEs do not exist, instead, the algorithms return an approximate PNE. We demonstrate the outperformance of the metaheuristics over the MILP approach on a number of games generated by GAMUT [18] with up to 9 players and 5 possible actions per player. We also report results of the metaheuristics on larger games, where the MILP approach becomes infeasible.

The paper is organized as follows. We present the formal formulation of PNE and the definition of approximate PNEs used in our metaheuristics in Sect. 14.2. The game representations used in our experiments are described in Sect. 14.3. The three local search metaheuristics for computing PNE are shown in Sect. 14.4. We report experimental results in Sect. 14.5 and conclude in Sect. 14.6.

14.2 Pure Nash Equilibria and a Definition of Approximation

A game $G = (N, S, U)$ is a tuple in which N represents the set of players, S represents the possible outcomes of the game and U a set of functions for each player that maps an outcome to a payoff value for the player. The players are represented as unique integers; e.g. $N = \{1, 2, \ldots, n\}$ in which n is the number of players that participate in the game. Each player $i \in N$ has an action set $S_i = \{s_i^1, s_i^2, \ldots, s_i^{m_i}\}$. It represents the choices available for player i in the game. m_i is the number of available actions to player i. We denote $m = \max_{i \in N} m_i$. An outcome is represented by $s = \{s_1, s_2, \ldots, s_n\}$, in which $s_i \in S_i$. The set of all possible outcomes s is denoted by S. In other words, $S = \Pi_{i \in N} S_i$. $U = \{u_1(s), u_2(s), \ldots, u_n(s)\}$ is the set of payoff functions. That is, functions $u_i(s)$ determining the payoff value for player i when all players choose the action proposed for them in the outcome s. For ease of notation, $u_i(s)$ is sometimes written as $u_i(s_i, s_{-i})$ in which s_{-i} represents the set of actions played by the players that are not i. Each player strives to obtain an as high as possible payoff value from the game.

An outcome s is called a pure Nash equilibrium (PNE) if and only if no player i can improve his payoff $u_i(s)$ by solely altering his action. Formally, s is PNE if and only if:

$$\forall i \in N, \forall s' \in S_i : u_i(s_i, s_{-i}) \geq u_i(s', s_{-i}) \tag{14.1}$$

The number of constraints in the PNE definition is polynomial in the number of players and actions, i.e. $O(m.n)$. It is therefore efficiently verifiable if an outcome is a PNE or not, as aspected from a NP-complete problem. We say that an outcome s is an approximation of a PNE according to some value $\varepsilon \geq 0$ (ε-PNE) [5–7] if and only if:

$$\forall i \in N, \forall s' \in S_i : u_i(s_i, s_{-i}) \geq u_i(s', s_{-i}) - \varepsilon \tag{14.2}$$

It is clear that if s is an ε-PNE then s is a μ-PNE as well, with $\mu \geq \varepsilon$. It follows from the definition that s is a PNE if and only if it is a 0-PNE. The approximation definition of PNE is appealing in practice when there is a cost associated every time a player alters his/her action. If ε is small enough, the approximation can prove to be useful, for example, when it would cost a lot of money and/or effort to change to a better action while it would only lead to a very small payoff increase.

14.3 Game Representations

In this section, we describe the game representations of the problem instances (i.e., games) used in our experiments. A general non-cooperative game can be represented in a *normal-form*, in which the payoff values of each player are given for every combination of the actions of all players. Our experiments are run on a number of games representated in this form with a limited numbers of players and actions, due to the fact that games with larger sizes would require extremely huge amount of memory to store. For games with larger sizes, we consider a subset of non-cooperative games called *succinct games*, in which a game can be described using a less costly representation. In particular, we choose a subclass of succinct games named *polymatrix games*.

14.3.1 Normal Form Games

A normal form description of a game describes the game by way of a multidimensional array. Each dimension of the array represents a player. The size of each dimension equals the number of actions available to the dimension's corresponding player. Each element of the multidimensional array is an outcome which has a payoff value for each player. The total number of payoff values in a normal form game is exponential in the number of players. More specifically, it is equal to $n \cdot \prod_{i \in N} m_i$ or $O(nm^n)$ in asymptotic notation.

As an example, Table 14.1 shows the normal form representation of a three-player game in which $N = \{1, 2, 3\}$; $\forall i \in N : S_i = \{s_i^1, s_i^2\}$ and $U = \{u_1(s), u_2(s), u_3(s)\}$. The values of $u_1(s)$, $u_2(s)$ and $u_3(s)$ are respectively represented by the first, second and third value in Table 14.1 in the relevant outcome. The outcome $s = \{s_1^1, s_2^2, s_3^1\}$ is a PNE since no player can unilaterally alters his action to gain a better payoff value for himself.

Table 14.1 A three player normal form game.

	s_3^1		s_3^2	
	s_2^1	s_2^2	s_2^1	s_2^2
s_1^1	$(-1, -1, -1)$	$(2, 1, 2)$	$(2, 2, 1)$	$(0, 1, 1)$
s_1^2	$(1, 2, 2)$	$(1, 1, 0)$	$(1, 0, 1)$	$(1, 1, 1)$

14.3.2   Polymatrix Games

By definition, all non-cooperative games can be described by the normal form description. Some games, however, can be described more efficiently using other representations. As an example, consider a game in which each player plays two-player games with a subset of the other players. Such a game is called a polymatrix game and can be represented by a graph. Each node of the graph represents a player. An edge is presented between two players if they play a two-player game with each other. The payoff for a player in a polymatrix game is the sum of payoffs from each two-player game the player is involved in. The number of two-player games is $O(n^2)$ and each game has $O(m^2)$ payoff values. Therefore, a polymatrix game described by the graph description has $O(n^2m^2)$ payoff values which is more efficient than the normal form description. Games, such as polymatrix games, which have polynomial size are called succinct games. They are of interest to us as we can not generate non-succinct games with a lot of players and/or actions. For example, consider a normal form game with 10 players and 10 actions. This game will have 10^{11} payoff values. If each payoff value is a 64 bit number, then the game would be 800 gigabytes big! A polymatrix game with the same amount of players and actions in which every player plays games with all the other players will be 304 kilobytes big using the graph description. If, in the polymatrix game, the number of players and actions is 100, we would still use only 792 megabytes.

14.4   Computing Approximate Pure Nash Equilibria Using Local Search-Based Metaheuristics

Using the definition of approximate PNEs described in Sect. 14.2, we convert the problem of finding a pure Nash equilibrium into an optimization problem by defining an objective function f that need to be minimized:

$$f : S \to \mathbb{R}$$
$$s \mapsto f(s) = \max_{i \in N, s' \in S_i} (u_i(s', s_{-i}) - u_i(s_i, s_{-i}))$$

The function represents the highest payoff increase that one of the players can gain by unilaterally switching his action, and can be computed in polynomial time $O(m.n)$. A solution $s \in S$ is a PNE if and only if $f(s) = 0$. s is an approximation for the PNE otherwise. We implement three local search-based metaheuristics for solving the problem:

- Random Restart Hill Climbing (RRHC): Hill Climbing is repeatedly restarted from a random solution. The First Improvement strategy, which accepts the first improved solution in the neighborhood, is used inside the Hill Climbing. Pseudo-code of the algorithm is presented in Algorithm 15.
- Simulated Annealing (SA) [11]: we implement a basic version of SA that allows reheating. The initial temperature is set as 2000. After t_{length} algorithm iterations, the temperature is decreased by a factor of α. When the temperature reaches a lower bound of ε, it is reset back to the initial temperature. We set $t_{length} = 100$, $\alpha = 0.9$ and $\varepsilon = 10^{-6}$. Pseudo-code of the algorithm is presented in Algorithm 16.
- Chained Local Optimization (CLO) [15]: this is a combination of the Simulated Annealing and the Hill Climbing algorithms. The algorithm resembles the Iterated Local Search framework [14] with the Hill Climbing algorithm as the local search component and the acceptance criteria taken from the Simulated Annealing. Algorithm 17 shows the pseudo-code of the algorithm implemented in this paper. Values of SA's parameters are the same as the ones for SA described above.

All metaheuristics use the same neighborhood structure. The neighborhood $\eta_k(s)$ of an outcome s is the set of all outcomes in which at most k players altered their action. The size of a neighborhood is then $O(m^k)$. The parameter k is set differently among the three metaheuristics: $k = 1$ for the Hill Climbing component inside RRHC and CLO, $k = 3$ for SA and the step of generating a random neighbor solution in the while loop of CLO. Random solutions are used for the initialization. All the three algorithms also share the same stopping criteria: either when a maximum running time limit is reached or when a PNE has been found.

Algorithm 15: Random Restart Hill Climbing

Require: Game $G = (N, S, U)$, running time limit T, objective function f, neighborhood structure $\eta_k(s)$
1: $s \leftarrow$ random solution $\in S$
2: $best \leftarrow s$
3: **while** time elapsed $\leq T$ **do**
4:     $s \leftarrow$ hillClimbing(s)
5:     **if** s is PNE **then return** s
6:     **if** $f(s) < f(best)$ **then**
7:         $best \leftarrow s$
8:     $s \leftarrow$ random solution $\in S$
    return $best$

Algorithm 16: Simulated Annealing

Require: Game $G = (N, S, U)$, running time limit T, objective function f, neighborhood structure
$\eta_k(s)$, initial temperature t_0, decreasing factor α, temperature stage length t_{length}
1: $s \leftarrow$ random solution $\in S$
2: $best \leftarrow s$
3: $t \leftarrow t_0$
4: **while** time elapsed $\leq T$ **do**
5:     **for** $i = 1..t_{length}$ **do**
6:         $s' \leftarrow$ random solution $\in \eta_k(s)$
7:         **if** s' is PNE **then return** s'
8:         $\Delta \leftarrow f(s') - f(s)$
9:         **if** $\Delta \leq 0$ **then**
10:            $s \leftarrow s'$
11:            **if** $f(s') < f(best)$ **then**
12:                $best \leftarrow s'$
13:         **else**
14:            $r \leftarrow$ random value $\in [0, 1]$
15:            **if** $r < e^{-\frac{\Delta}{t}}$ **then**
16:                $s \leftarrow s'$
17:     $t \leftarrow \alpha t$
18:     **if** $t < 10^{-6}$ **then**
19:         $t \leftarrow t_0$
    return best

Algorithm 17: Chained Local Optimization

Require: Game $G = (N, S, U)$, running time limit T, objective function f, neighborhood structure
$\eta_k(s)$, initial temperature t_0, decreasing factor α, temperature stage length t_{length}
1: $s \leftarrow$ random solution $\in S$
2: $s \leftarrow$ hillClimbing(s)
3: $best \leftarrow s$
4: $t \leftarrow t_0$
5: **while** time elapsed $\leq T$ **do**
6:     **for** $i = 1..t_{length}$ **do**
7:         $s' \leftarrow$ random solution $\in \eta_k(s)$
8:         $s' \leftarrow$ hillClimbing(s')
9:         **if** s' is PNE **then return** s'
10:         $\Delta \leftarrow f(s') - f(s)$
11:         **if** $\Delta \leq 0$ **then**
12:            $s \leftarrow s'$
13:            **if** $f(s') < f(best)$ **then**
14:                $best \leftarrow s'$
15:         **else**
16:            $r \leftarrow$ random value $\in [0, 1]$
17:            **if** $r < e^{-\frac{\Delta}{t}}$ **then**
18:                $s \leftarrow s'$
19:     $t \leftarrow \alpha t$
20:     **if** $t < 10^{-6}$ **then**
21:         $t \leftarrow t_0$
    return best

14.5 Results

In this section, the performance of the three metaheuristics on different problem instance sizes is presented. The size of a problem instance depends on two parameters: the number of players (n) and the number of possible actions for each players (m). For each pair of n and m, we generate ten random instances using GAMUT [18], an extensive library of different types of games with publicly available game generators. We first compare the performance of the metaheuristics with a MILP method on instances that are small enough so that the MILP approach can solve them within a reasonable time and memory limit. We then show how the metaheuristics perform on instances of succinct games that MILP can not solve due to time and/or memory limitations.

14.5.1 Comparison of the Metaheuristics with a Mixed Integer Linear Programming Approach

We first compare the metaheuristics with the exact method based on Mixed Integer Linear Programming (MILP) proposed in [20]. To the best of our knowledge, this MILP is the only method (except brute-force) that exists in the literature to solve the pure Nash equilibrium problem in normal form games. The comparison is done on a dataset of 28 problem instance sizes, leading to a total number of 280 instances. The payoff values are randomly selected from the range of [−100, 100]. Therefore, the value of the objective function f belongs to the range of [0, 200].

In this experiment, we set a maximum running time limit of 60 seconds for the metaheuristics. For the MILP, we use CPLEX 12.6 with the limits of 4GB memory and one hour of running time. Results are shown in Table 14.2. For each instance size, we report the number of instances in which a PNE has been found and the average running time (in milliseconds). For instances where a PNE is not found, we normalize the objective function values of the approximate solutions returned by the metaheuristics to the range of [0, 1] and report the mean over those instances. Additionally, the number of instances that the MILP cannot solve within the given memory and time limits are also listed.

We can see the clear outperformance of the metaheuristic approaches over the MILP:

- For problem sizes whose every instance is solvable by the MILP ($n \leq 5$ and $m \leq 6$), a PNE is always found by the metaheuristic approaches for every instance that has at least one PNE. Moreover, the time required by the metaheuristics to find a PNE is smaller with an order of magnitude when compared to the MILP. For instances where a PNE does not exist, the small objective function values given by the metaheuristics suggest good approximate PNEs. In fact, since the values returned by all the metaheuristics for each instance are exactly the same (except

Table 14.2 Random Normal Form Game results

n	m	Found				Time Found (ms)				Approximation			Unsolvable
		RRHC	SA	CLO	MILP	RRHC	SA	CLO	MILP	RRHC	SA	CLO	MILP
3	3	6	6	6	6	1.79	0.72	0.15	141.67	0.033	0.033	0.033	0
3	4	8	8	8	8	1.52	0.56	0.26	279.77	0.11	0.11	0.11	0
3	5	4	4	4	4	3.23	0.59	0.13	3696.24	0.057	0.057	0.057	0
3	6	6	6	6	6	2.60	0.389	0.44	2240.06	0.052	0.052	0.052	0
3	8	5	5	5	5	1.05	0.39	0.57	2654.57	0.026	0.026	0.026	0
3	9	5	5	5	5	0.58	1.013	0.68	8122.44	0.032	0.032	0.032	0
3	10	8	8	8	8	4.86	2.01	2.47	3731.40	0.011	0.011	0.011	0
3	15	6	6	6	6	16.34	5.82	12.05	1850.26	0.014	0.014	0.014	0
3	20	5	5	5	5	19.18	36.81	19.15	5929.46	0.011	0.011	0.011	0
3	25	6	6	6	6	67.70	29.78	133.63	42524.34	0.0061	0.0061	0.0061	0
3	30	3	3	3	3	108.31	55.99	68.93	49762.25	0.0065	0.0065	0.0065	0
3	35	5	5	5	5	432.00	92.37	285.10	215015.91	0.0066	0.0066	0.0066	0
3	40	6	6	6	6	670.20	173.30	165.02	324330.39	0.0066	0.0066	0.0066	0
3	45	7	7	7	7	409.89	247.53	293.42	469457.35	0.0060	0.0060	0.0060	0
3	50	8	8	8	8	921.45	297.51	788.01	818496.07	0.0012	0.0012	0.0012	0
5	3	8	8	8	8	1.92	0.82	0.51	10712.12	0.026	0.026	0.026	0
5	4	6	6	6	6	2.86	7.11	0.84	10423.76	0.021	0.021	0.021	0
5	5	7	7	7	7	3.16	2.27	5.64	87200.74	0.024	0.024	0.024	0
5	6	9	9	9	9	24.38	16.08	31.87	633587	0.012	0.012	0.012	0
5	7	5	5	5	4	55.80	129.75	50.72	1700474.00	0.23	0.23	0.23	6
5	8	6	6	6	2	78.73	34.41	110.84	2545067.13	0.013	0.013	0.013	8
5	9	4	4	4	0	348.69	82.07	352.03	/	0.009	0.009	0.009	10
5	10	6	6	6	0	592.63	362.79	1052.44	/	0.012	0.012	0.012	10
7	3	9	9	9	9	10.39	2.62	5.26	313190.35	0.028	0.028	0.028	0
7	4	9	9	9	1	50.40	37.74	298.16	3621348.80	0.022	0.022	0.022	9
7	5	4	4	4	0	452.33	395.58	467.50	/	0.018	0.018	0.018	10
9	3	5	5	5	0	57.58	42.95	45.19	/	0.015	0.015	0.015	10
9	5	7	8	7	0	9496.63	2863.37	10015.87	/	0.015	0.0082	0.020	10

the last problem size with $n = 9$ and $m = 5$), we suspect that these approximate solutions are actually the optimal ones, although we can not prove this.

- For problem sizes with $n \geq 5$ and $m \leq 7$ (except the case where $n = 7$ and $m = 3$), the MILP runs out of either memory or time in most instances. Those are the cases where the metaheuristics are dominant not only on the running time, but also on the ability of finding a PNE.

14.5.2 Results of the Metaheuristic Approaches on Large Games

We continue by running the metaheuristics on instances with more players and/or actions. As explained in Sect. 14.3.2, we need to use instances of succinct games. In particular, we chose polymatrix games in this experiment. We generate 10 instances of 25 different problem sizes, resulting in a total of 250 instances. In our n-player polymatrix game instances, each player plays $n - 1$ two-player games with the other players. The payoff values of the two-player games are randomly selected from the range of $[-100, 100]$. Consequently, since the payoff for a player is the sum of all payoff values he gets in the subgames, the payoff for a player lies in the range of $[100(1 - n), 100(n - 1)]$. Therefore, the value of the objective function f belongs to the range of $[0, 200(n - 1)]$.

We report results of the metaheuristics on those problem instances. The MILP approach is unable to solve those instances, due to the fact that the linear programs generated are exponential in problem sizes even for those succinct games. For the metaheuristics, we set a maximum running time limit of 10 minutes. For each instance size, we only report the approximations, since the metaheuristics do not find a PNE for almost all instances except a few ones with $n = 10$ and $m = 10$. Similarly to the small instances, we normalize the approximations to the range of $[0, 1]$. We observed that for the largest instances, the metaheuristics did not perform well. That is, within the time limit, not even one iteration of RRHC or CLO was finished. In other words, both algorithms only performed Hill Climbing until the time limit was reached. We reran some of those instances with an increased time limit of one hour but even then not one iteration was finished.

The reported approximations in Table 14.3 look similar to those for the small instances. While we can say that the approximations for the small instances are good, it would be short-sighted to say the same about the large instances. In a game with a lot of actions, the difference between the smallest payoff value to which a player can switch that is bigger than the current payoff and the current payoff is generally smaller than in a game with fewer actions. Therefore, the optimal approximation value is expected to be smaller.

Table 14.3 Random polymatrix game results

n	m	Approximation		
		RRHC	SA	CLO
10	10	0.0054	0.0040	0.0040
10	50	0.055	0.069	0.042
10	100	0.077	0.10	0.070
10	200	0.10	0.13	0.089
10	500	0.12	0.17	0.12
10	1000	0.15	0.20	0.15
10	2000	0.20	0.21	0.18
50	10	0.061	0.051	0.043
50	50	0.093	0.091	0.080
50	100	0.11	0.11	0.10
50	200	0.12	0.11	0.11
50	400	0.15	0.13	0.15
100	10	0.060	0.054	0.045
100	50	0.087	0.081	0.080
100	100	0.10	0.090	0.099
100	150	0.11	0.096	0.11
100	200	0.12	0.10	0.12
200	10	0.056	0.052	0.048
200	25	0.067	0.061	0.063
200	50	0.078	0.071	0.076
200	75	0.088	0.075	0.085
500	10	0.045	0.042	0.045
500	20	0.056	0.054	0.056
500	30	0.060	0.056	0.060
500	40	0.068	0.059	0.062

14.6   Conclusion and Future Work

The main aim of the paper is to show that existing metaheuristics are promising methods to solve the problem of finding a pure Nash equilibrium in game theory. From the experimental results on normal form games, it is clear that the considered metaheuristics, despite being implemented in a quite basic and general form, perform much better than the exact MILP method [20]. No other exact methods have been considered and there are none as far as we are aware other than enumerating all the possible outcomes (brute-force).

For larger instance sizes, the normal form description causes the memory usage of the game to be too big to be generated. Despite the fact that succinct games are polynomial in size in function of the number of players and actions, the MILP approach still reduces the problem to a linear program that is exponential in size. This means that not only the solving time of the MILP will take exponential time, but the reduction as well. We therefore deem the MILP approach as entirely unfit for these problem sizes. On the other hand, because our metaheuristic approaches do not require an exponential amount of memory, we can find ε-PNE in instances of succinct games that have large numbers of players and or actions.

All the problem instances and detailed results of the solving approaches used in this paper are publicly available at https://github.com/ElgersNiels/Metaheuristic-approach-to-PNE. We hope that it could serve as a first benchmark for applying metaheuristics on the problem of finding a pure Nash equilibrium in general non-cooperative games. For future work, we plan to improve the current metaheuristic approaches by exploiting knowledge from specific game types, such as the graph structure of polymatrix games. In that sense, one may design game specific heuristics that could make the metaheuristics more powerful. Moreover, in the current experiments, no real intensive work has been done to tune the parameters of the metaheuristics, so the application of an automated parameter tuning tool such as irace [13] will be interesting.

A second line of future work is to improve the current benchmark dataset. It should be noted that games generated by GAMUT are random and are not necessarily hard to find a PNE in. It might be the case that there exist specific generator settings that produce significantly harder problems than in the average case, which has been found in SAT community. In SAT, there exist specific settings so that the generated propositional formula has a 50% chance to be (un)satisfiable. These instances are shown to be the hardest to solve [16]. Perhaps such parameter settings exist for games as well so that finding a PNE, if it exists, is hard. It might prove to be fruitful to reduce hard SAT instances – or other hard instances of other problems – to games and evaluating their hardness in an attempt to create benchmarks.

Acknowledgements This work is funded by COMEX (Project P7/36), a BELSPO/IAP Programme. The authors would like to thank Prof. Danny Weyns and Dr. Rahul Savani for providing comments on how to improve the quality of this paper.

References

1. Bhalgat, A., Chakraborty, T., & Khanna, S. (2010). Approximating pure nash equilibrium in cut, party affiliation, and satisfiability games. In *Proceedings of the 11th ACM Conference on Electronic Commerce* (pp. 73–82). New York: ACM.
2. Caragiannis, I., Fanelli, A., Gravin, N., & Skopalik, A. (2011). Efficient computation of approximate pure nash equilibria in congestion games. In *2011 IEEE 52nd Annual Symposium on Foundations of Computer Science (FOCS)* (pp. 532–541). New Jersey: IEEE.
3. Ceppi, S., Gatti, N., Patrini, G., & Rocco, M. (2010). Local search methods for finding a nash equilibrium in two-player games. In *2010 IEEE/WIC/ACM International Conference on Web Intelligence and Intelligent Agent Technology (WI-IAT)* (Vol. 2, pp. 335–342).
4. Daskalakis, C., Goldberg, P. W., & Papadimitriou, C. H. (2009). The complexity of computing a nash equilibrium. *SIAM Journal on Computing, 39*(1), 195–259.

5. Daskalakis, C., Mehta, A., & Papadimitriou, C. (2007). Progress in approximate nash equilibria. In *Proceedings of the 8th ACM Conference on Electronic Commerce* (pp. 355–358). New York: ACM.
6. Even-Dar, E., Kesselman, A., & Mansour, Y. (2003). Convergence time to nash equilibria. In *International Colloquium on Automata, Languages, and Programming* (pp. 502–513). Berlin: Springer.
7. Fabrikant, A., Papadimitriou, C., & Talwar, K. (2004). The complexity of pure nash equilibria. In *Proceedings of the Thirty-Sixth Annual ACM Symposium on Theory of Computing* (pp. 604–612). New York: ACM.
8. Fearnley, J., Goldberg, P. W., Savani, R., & Sørensen, T. B. (2012). Approximate well-supported nash equilibria below two-thirds. In *Proceedings of the 5th International Symposium on Algorithmic Game Theory, SAGT 2012, Barcelona, Spain, October 22–23, 2012* (pp. 108–119). Berlin: Springer.
9. Gottlob, G., Greco, G., & Scarcello, F. (2005). Nash equilibria: hard and easy games. *Journal of Artificial Intelligence Research* (pp. 357–406).
10. Jiang, A. X. & Leyton-Brown, K. (2009). A tutorial on the proof of the existence of nash equilibria. *University of British Columbia Technical Report TR-2007-25. pdf.*
11. Kirkpatrick, S. & Vecchi, M. P. et al. (1983). Optimization by simulated annealing. *Science, 220*(4598), 671–680.
12. Lemke, C. E., & Howson, J. T, Jr. (1964). Equilibrium points of bimatrix games. *Journal of the Society for Industrial and Applied Mathematics, 12*(2), 413–423.
13. López-Ibáñez, M., Dubois-Lacoste, J., Cáceres, L. P., Birattari, M., & Stützle, T. (2016). The irace package: iterated racing for automatic algorithm configuration. *Operations Research Perspectives, 3*, 43–58.
14. Lourenço, H. R., Martin, O. C., & T. Stützle. (2003). *Iterated local search.* Berlin: Springer.
15. Martin, O. C., & Otto, S. W. (1996). Combining simulated annealing with local search heuristics. *Annals of Operations Research, 63*(1):57–75.
16. Mitchell, D., Selman, B., & Levesque, H. (1992). Hard and easy distributions of sat problems (pp. 459–465).
17. Nash, J. F. (1950). *A dissertation on non-cooperative game theory.* Princeton: Princeton university.
18. Nudelman, E., Wortman, J., Shoham, Y., & Leyton-Brown, K. (2004). Run the gamut: A comprehensive approach to evaluating game-theoretic algorithms. In *Proceedings of the Third International Joint Conference on Autonomous Agents and Multiagent Systems* (Vol. 2, pp. 880–887). Washington: IEEE Computer Society.
19. Tsaknakis, H., & Spirakis, P. G. (2007). An Optimization Approach for Approximate Nash Equilibria. In *Proceedings of the Third International Workshop on Internet and Network Economics, WINE 2007, San Diego, December 12–14, 2007* (pp. 42–56). Berlin: Springer.
20. Wu, Z., Dang, C., Karimi, H. R., Zhu, C., & Gao, Q. (2014). A mixed 0-1 linear programming approach to the computation of all pure-strategy nash equilibria of a finite n-person game in normal form. *Mathematical Problems in Engineering.*

Chapter 15
Hybrid Genetic Algorithms to Solve the Multidimensional Knapsack Problem

Abdellah Rezoug, Mohamed Bader-El-Den and Dalila Boughaci

Abstract This paper introduces solutions to deal with the Multidimensional Knapsack Problem (MKP), which is a NP-hard combinatorial optimisation problem. Two hybrid heuristics based on Genetic Algorithms (GA) are proposed: the Memetic Search Algorithm (MSA) and the Genetic Algorithm Guided by Pretreatment information (GAGP). MSA combines sequentially GA with the Stochastic Local Search-Simulated Annealing algorithm (SLSA). GAGP is composed of two steps, in the first, a ratio-based greedy algorithm extracts useful information and the core concept is utilised to decompose items according to their ratios; In the second, these information are integrated to the operators of a GA allowing to reach the best solutions faster. An operator is added to the GA to dynamically update the ratio values of the items. Two groups of data were used to examine the proposed approaches. A group of simple instances of MKP has been used to examine MSA and a group of complex MKP has been used to examine GAGP. The obtained results indicate that MSA and GAGP have the capability to give solutions of high quality.

Keywords Genetic algorithm · Multidimensional knapsack problem · Core concept · Guided genetic algorithm · Hybrid genetic algorithm · Memetic algorithm · Heuristics

A. Rezoug (✉)
Department of Computer Science, University of Boumerdes, Boumerdes, Algeria
e-mail: abdellah.rezoug@gmail.com

M. Bader-El-Den
School of Computing, University of Portsmouth, Portsmouth, UK
e-mail: mohamed.bader@port.ac.uk

D. Boughaci
Department of Computer Science, University of Science
and Technology Houari Boumediene, Algiers, Algeria
e-mail: dalila_info@yahoo.fr

© Springer International Publishing AG, part of Springer Nature 2019
E.-G. Talbi and A. Nakib (eds.), *Bioinspired Heuristics for Optimization*,
Studies in Computational Intelligence 774,
https://doi.org/10.1007/978-3-319-95104-1_15

15.1 Introduction

The MKP is composed of n items and a knapsack with m different capacities c_i ($i \in \{1, \ldots, m\}$). Each item j ($j \in \{1, \ldots, n\}$) has a weight w_{ij} on each capacity i of the knapsack and a value p_j. The goal is to pack the items in the knapsack so as to maximise the overall value without exceeding the capacities of the knapsack. The MKP model can be represented by the following integer program:

$$\text{Maximise}: \quad \sum_{j=1}^{n} p_j x_j \tag{15.1}$$

$$\text{Subject to}: \quad \sum_{j=1}^{n} w_{ij} x_j \leq c_i \quad i \in \{1 \ldots m\} \tag{15.2}$$

$$x_j \in \{0, 1\} \qquad j \in \{1 \ldots n\}$$

The MKP has received the attention of researchers for many years regarding to its different applications. Several methods have been proposed to deal with this problem that are divided into two categories: Exact and non-exact methods. Exact methods give the optimal solution however, they are slow when addressing data of large size. Methods such as Branch and Bound [1], Linear programming, Dynamic Programming [2] etc. are generally applied on small size MKP. Non-exact methods are used with large size data to obtain high-quality solutions within a reasonable time. Generally heuristics are hybridised in such a way to improve each other's performance and then guaranty outputs of good quality.

Hybridisation of different types have been suggested such as: combine two or more heuristics, analyse of data and annotation of heuristics with information of guidance, other hybridisation types exist such as heuristics and exact methods, relaxation and heuristics, hyper-heuristics etc. This study interests to the first two types of hybridisation especially the GA-based methods. Several approaches combine many heuristics have been proposed [3–6]. Many other heuristics integrate knowledge about the MKP as annotation in such a way to reduce the search time. This requires methods and mechanics to extract relevant information. To solve the Course Timetabling Problem, the approaches in [7, 8] use a memory denoted MEM to record useful information to guide the GA process and improve its performance. Other researchers used an external structure to guide GA such as [9, 10].

This study presents two non-exact approaches of different types to solve the MKP. Firstly, an improved Memetic Search Algorithm MSA is proposed. MSA is based on the hybridisation of a modified Genetic Algorithm (GA) and a Stochastic Local Search-Simulated Annealing (SLSA). SLSA is a local search algorithm that combines two local search heuristics: the Stochastic Local Search (SLS) [11] and the simulated annealing (SA) [12]. After offspring is created using the GA operators, it is improved by SLSA with local movements between the neighbour solutions. MSA is compared with GA, and with two hybrid heuristics: GA-SLS composed of GA and

SLS and GA-SA composed of GA and SA. The comparison is performed on small-size MKP instances MKP from the OR-Library.[1] Secondly, the second approach aims to reinforce the GA performance using the useful information about the items. To this purpose, the Genetic Algorithm Guided by Pretreatment information (GAGP) is proposed. First, GAGP applies the primal greedy with the core concept decomposition to extract a useful information about the subset of important items. Second, specific population initialisation, fitness function and update efficiency measurement operators augment a standard GA by exploiting the pretreatment information. In GAGP, an important rang of solutions are avoided and the process does not consider the non relevant solutions.

The paper is structured as follows: Sect. 15.2 explains the MSA components. Section 15.3 gives the experiments, the obtained results and comparisons. The proposed algorithm GAGP is introduced in Sect. 15.4. Section 15.5 presents the conducted experiments and the obtained results. The conclusions and final remarks are drawn in Sect. 15.6.

15.2   Memetic Search Algorithm (MSA)

MSA is a hybrid method composed of two algorithms: the Genetic Algorithm (GA) and the Stochastic Local Search-Simulated Annealing algorithm (SLSA). In MSA, the operators of the GA have been modified and adapted to the MKP and its efficiency has been improved by SLSA. In this section, we describe all the operators of GA and the SLSA algorithm.

15.2.1   Stochastic Local-Simulated Annealing Algorithm (SLSA)

The Stochastic Local-Simulated Annealing algorithm (SLSA) is a new approach based on two local search algorithms, the stochastic local search (SLS) [11] and the Simulated Annealing (SA) [12]. SLSA is applied on offspring X'_1, X'_2 a certain number of local steps that consists, first to create a neighbour solution by selecting an item to be added, second to remove all conflicting items.

With a probability $wp \in [0, 1]$, the item to be packed is selected according to one of the two following criteria:

- An item I is chosen randomly. If I increases the objective function $f(X)$ then it will be packed in the knapsack, otherwise it will be accepted if the following expression is true $r_1 < \exp^{(-\Delta f/T)}$ where r is a random value, $\Delta f = f(X') - f(X)$ and T is a temperature value initially equal to T_0 relatively high.
- Choose the best item to be accepted.

---

[1] http://people.brunel.ac.uk/~mastjjb/jeb/orlib/files/.

The first step may cause a conflict. To eliminate all conflicts, the worst item is removed from the knapsack. This process is repeated until no conflict remains. After that, the temperature value is updated. In our case, the decreasing rule is found empirically.

15.2.2 The Genetic Algorithm Step

The GA steps and some modifications concerning selection, crossover, mutation, replacement operations are presented in following.

- MSA begins by the creation of the initial population P of individuals of a population size (PS). The Random Key method (RK) [13] is used to create the individuals of P.
- The MSA selection operator selects items according to their fitness values. The individuals with higher values are most likely be selected to reproduce, whereas, those with low values will be discarded. In MSA, the two individuals X_1, X_2 with the first and the second best fitness values are designated as parents for the crossing step.
- X_1, X_2, are used to perform a uniform crossover. The resulting offspring X_1', X_2' may be unfeasible solutions, so a repair process removes the worst item until they become feasible. Offspring X_1' and X_2' represent a feasible solutions, these lead to calculate their objective functions $F(X_1')$ and $F(X_2')$. A List Q is used to store the individuals that have already participated in a crossover. The purpose of this list is to prevent parents to be selected more than once during a number of iterations. The size of Q is defined according to the population size PS so that it turns the major part of its individuals $WT = F * PS$, with $F < 0.5$ is factor coefficient to calculate the size of Q.
- A mutation number of items (MNI) from offspring X_1' and X_2' are replaced by other items selected randomly from the best individual found so far X_{best}. This principle is inspired by the adjustment procedure of the improvised harmony in the Self-adaptive Global best Harmony Search (SGHS) [14]. Substitute items must not be already included in the concerned offspring. The mutation operator can be summarised by the following pseudo code:

Algorithm 18: The Mutation.

1: **for** $Cpt = 1$ to MNI **do**
2:   $I_x = RandItem()$; $I_x \in X$
3:   $I_{(xb)} = RandItem()$; $I_{xb} \in X_{best}$, $I_{xb} \notin X$ and $I_x \neq I_{xb}$
4:   $X = X - \{I_x\}$
5:   $X = X \cup \{I_{xb}\}$
6: **end for**

Table 15.1 The values of MSA parameters

Parameter	Description	Value
NI	Number of iteration	50000
PS	Population size	200
WT	Wait time	70
N	Number of iteration of SLSA	100
wp	Probability of SLSA	0,93
T_0	Initial temperature	30
CT	Coefficient T update	0.0105
Nrun	Number RUN	29

- The offspring X'_1, X'_2, after been repaired, a local search improvement is performed using the SLSA (Algorithm of Sect. 15.2.1).
- The offspring will be compared to the worst individuals in the population in terms of fitness, they replace them in the population if they are better.

The optimisation process is repeated until the stopping criterion is checked. The criterion for stopping the optimisation process is a limited number of iteration NI determined empirically according to the size of the studied problem.

15.3   Experimental Results

MSA is coded in C++ and executed on a PC with 2 GHz Intel Core 2 Duo processor and 2 GB RAM. To evaluate the performance of our MSA algorithm, it was initially tested on 54 standard test problems (divided into six different sets) which are available at the OR-Library[2] maintained by Beasley. These problems are real-world problems widely used for the validation of the effectiveness of algorithms in the optimisation community. These problems consisting of $m = 2$–30 and $n = 6$–105. After several experiments, we set the parameters for the MSA as in Table 15.1.

15.3.1   Results on SAC-94 Data

Table 15.2 shows the results of MSA application on the 54 instances. Here, columns n and m represent the number of items, and the number of constraints (the number of dimensions) respectively, column $Optimum$ is the value of Z optimal, column $Time$ is the average time of 30 runs. Column AVI is the AVerage number of Iteration of all the 30 runs. Column DFO is the Distance From the Optimum and finally NSR is the Number Successful Runs i.e. number of times MSA reaches the optimal solution.

---

[2]http://people.brunel.ac.uk/~mastjjb/jeb/orlib/files/.

Table 15.2 Results of MSA applied on the 54 OR-Library instances

	n	m	Optimum	Time	AVI	DFO	NSR
HP	28	4	3418	5,020	24291	0	30/30
	35	4	3186	2,382	11237	0	30/30
PB	27	4	3090	1,133	5289	0	30/30
	34	4	3186	1,718	7717	0	30/30
	29	2	95168	0,335	1582	0	30/30
	20	10	2139	5,508	23963	0	30/30
	40	30	776	0,403	860	0	30/30
	37	30	1035	5,424	10746	0	30/30
PET	10	10	87061	0,013	1	0	30/30
	15	10	4015	0,010	2	0	30/30
	20	10	6120	0,013	13	0	30/30
	28	10	12400	0,905	3652	0	30/30
	39	5	10618	37,801	148	0,0922	9/30
	50	5	16537	13,059	47317	0,0026	29/30
SENTO	60	30	7772	0,773	1767	0	30/30
	60	30	8722	24,457	48509	0	30/30
WEING	28	2	141278	0,028	82	0	30/30
	28	2	130883	0,014	19	0	30/30
	28	2	95677	0,009	4	0	30/30
	28	2	119337	0,217	990	0	30/30
	28	2	98796	0,109	542	0	30/30
	28	2	130623	0,019	63	0	30/30
	105	2	1095445	58,112	121421	0	30/30
	105	2	624319	16,681	61761	0	30/30
WIESH	30	5	4554	0,010	10	0	30/30
	30	5	4536	0,057	216	0	30/30
	30	5	4115	0,063	248	0	30/30
	30	5	4561	0,013	28	0	30/30
	30	5	4514	0,013	35	0	30/30
	40	5	5557	0,647	2798	0	30/30
	40	5	5567	0,659	2863	0	30/30
	40	5	5605	0,893	3795	0	30/30
	40	5	5246	0,105	437	0	30/30
	50	5	6339	0,619	2491	0	30/30
	50	5	5643	0,774	3251	0	30/30
	50	5	6339	0,706	2810	0	30/30
	50	5	6159	0,782	3222	0	30/30
	60	5	6954	3,623	13939	0	30/30
	60	5	7486	0,613	2326	0	30/30
	60	5	7289	2,767	10593	0	30/30
	60	5	8633	9,375	31597	0	30/30

(continued)

Table 15.2 (continued)

n	m	Optimum	Time	AVI	DFO	NSR
70	5	9580	15,670	53919	0	30/30
70	5	7698	4,035	15589	0	30/30
70	5	9450	3,713	12805	0	30/30
70	5	9074	3,445	11963	0	30/30
80	5	8947	7,761	26499	0	30/30
80	5	8344	32,683	112813	0	30/30
80	5	10220	60,773	192505	0	30/30
80	5	9939	29,397	96748	0	30/30
90	5	9584	14,307	47672	0	30/30
90	5	9819	20,250	67586	0	30/30
90	5	9492	23,620	78641	0	30/30
90	5	9410	18,233	61025	0	30/30
90	5	11191	232,352	683614	0	30/30

From these results, we have identified various remarks. Firstly, MSA has succeeded to reach the optimal value once at least for all instances and 52 of the 54 in all 30 runs. The total average deviation of optimum is 0.0017%. Secondly, in 27 of 54 instances, the average execution time of MSA is less than one second, however, some instances required more time than the rest as WEISH 23–29, PET 5, WEING 7 and 8 and, in particular, the WEISH30, but in general, the time is relatively little with global average of 11,7 s. 7,4 s if the WEISH30 is ignored. Thirdly, the number of required average iteration varies from 1 to 600 thousand iterations (for PET1 and WEISH30 respectively) or an average of 35895 iterations and 23194 iterations without the WEISH30.

15.3.2 MSA Versus GA, GA-SA and GA-SLS

In this experiment the MSA approach was compared to three other approaches: the Genetic Algorithm (GA), its hybridisation with the Simulated Annealing (GA-SA) [5] and by the Stochastic Local Search (GA-SLS)[5]. In all tests, same parameters were used. We applied the GA, GA-SA, GA-SLS and MSA algorithms on the WEISH27 instance (optimum = 9819). The value of fitness was noted every second for 15 s. With the average of fitness obtained in 10 runs, the Fig. 15.1 was drawn.

The curves of the Fig. 15.1 represent the evolution of the four algorithms for 15 s. Algorithms start using population generated according to RK. The algorithms MSA and GA-SLS curves are too close together except at the end where only MSA reaches the optimum.

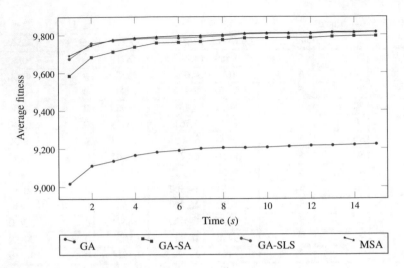

Fig. 15.1 Comparing MSA to GA, GA-SA, GA-SLS through the WEISH27

These two algorithms exceeded slightly that of the GA-SA algorithm and largely that of the GA algorithm. We can deduce that MSA is faster and more efficient than the other three algorithms.

15.3.3   MSA Versus CRGA and SRGA

Table 15.3 shows the comparison of MSA with CRGA and SRGA algorithms [15] in terms of effectiveness. Where CRGA and SRGA are two algorithms based on GA. The comparison was done according to the calculated means fitness. The values of results presented in Table 15.3 are the same published by the authors.

We can observe that MSA was able to reach the optimum in almost all instances, whereas CRGA and SRGA were not able to reach the optimum in any instance. MSA is more effective than CRGA and SRGA.

15.3.4   MSA Versus DPHEU

Table 15.4 shows the comparison of MSA with DPHEU algorithm [16] in terms of effectiveness. The comparison was done according to the calculated average percentage of deviation (A.P.O.D) and the number of problems for which the optimal solution was reached (N.O.P.T). The values of results presented in Table 15.4 are the same published by the authors.

Table 15.3   Mean fitness obtained by MSA compared to CRGA and SRGA

Instance	Optimum	CRGA	SRGA	MSA
		Mean	Mean	Mean
PET1	3800	3782	**3800**	**3800**
PET2	8706	8562	8662	**8706**
PET3	4015	3878	3941	**4015**
PET4	6120	5476	5630	**6120**
PET5	12400	11203	12240	**12400**
PET6	10618	10107	9953	**10608,2**
PET7	16537	15184	14915	**16536,5**
HP1	3418	3259	3214	**3418**
HP2	3186	2921	2864	**3186**
Weing1	141278	130885	131409	**141278**
Weing2	130883	113289	116883	**130883**
Weing4	119337	107535	106950	**119337**
Weing5	98796	79038	75109	**98796**
Weing6	130623	116773	115671	**130623**
Weing7	1095445	975269	783196	**1095445**
PB1	3090	2953	2936	**3090**
PB2	3186	2965	2907	**3186**
PB4	95168	83483	81412	**95168**
PB5	2139	1984	2016	**2139**
Weish1	4554	3774	3777	**4554**

Table 15.4   Average fitness and number of optimal solutions obtained by MSA compared to DPHEU

Data	Number of instances	DPHEU		MSA	
		A.P.O.D	N.O.P.T	A.P.O.D	N.O.P.T
HP	2	0.0	2	0.0	2
PB	6	0.04	5	**0.0**	**6**
PET	6	**0.0**	**6**	0,0158	4
SENTO	2	0.0	2	0.0	2
WEING	8	0.0	8	0.0	8
WEISH	30	0.03	28	**0.0**	**30**

Both MSA and DPHEU obtained similar results in three of the six sets (HP, SENTO and WEING). MSA is more effective than DPHEU in two sets PB and WEISH. DPHEU is more effective than MSA in the PET set. We can say that MSA and DPHEU are quite similar in terms of effectiveness.

15.4 Genetic Algorithm Guided by Pretreatment Information (GAGP)

The GAGP algorithm is motivated by the observation that in may optimisation real-world problem, we may have some prior information about the components/patterns that are likely to appear in the good solutions. For example, in MKP, it is possible using linear relaxation or the "optimal fractional solution" [17, 18] to predict some of the items that are likely or unlikely to appear in the good solutions. This study proposes a method for using such prior information as an additional guide for the GA evolutionary process for the MKP problem. By guide, we mean any structure external to GA, which maintains its original composition and is used to drive its search process. This can be through a subset of operators, in order to accelerate the search process and improve the speed of convergence. This section aims to describe the GAGP components.

15.4.1 Pretreatment

The guiding information is based on the work by [19]. The items are sorted in decreasing order according to a statistical efficiency e_j based on the profit and the cost. In simple words, the items are sorted based on how likely each item to appear in high performing individuals, the item at the top of this list are the items that are likely to be selected while the items at the bottom of the list are the items that are unlikely appear in good solutions. However, it is important to note here that this list is just an estimate and not a predefined part of the solution. It should be noted also that the Greedy heuristic as by [20] only based on the efficiency sorting is not an effective solution for the strongly correlated problem instances of the MKP [21].

$$e_j^{st} = \frac{p_j}{\sum_{j=1}^{m} w_{ij}\left(\sum_{l=1}^{n} w_{il} - c_i\right)} \tag{15.3}$$

The sorting operation allows favouring items that have a good compromise (i.e. efficiency) between the average profit and overall capacity. The efficiency of an item is high if its profit is high while its required global capacity is low. The sorted items are split into three sets where the value of each variable is assigned as follows:

- $X_1 : x_j = 1$ The variables have the best efficiency e_j. These variables are most likely to build the best solutions even the optimal solution.
- $Core : x_j =?$ The variables have the values of the efficiency e_j very close. In this group, it is difficult to determine the best.
- $X_0 : x_j = 0$ The variables have a very low efficiency e_j, in other words, the profit is low or the capacity is large or both.

The guide is represented by the items of $X_1 \cup Core \cup X_0$. The sizes of X_1, $Core$ and X_0 are determined as follows: Construct a feasible solution by adding the items in the order. The item that makes the solution unfeasible represents the centre of $Core$. The size of each part of the guide depends on the size of $Core$. Set the size of $Core$ defines the size of the other parts.

15.4.2   Guided Genetic Algorithm Optimisation

The GAGP chromosome consists of the set of the items to be added to the knapsack. GAGP uses the integer representation, where each gene presents an item ID. The items are coded as integer numbers. A chromosome is formed only by the number of items that it contains. This representation allows reducing the size of the processed data.

1. *Initial population.* GAGP algorithm uses a special initialisation process which allows the GA to make use of the prior information available about the items, and in the same time generates a diverse initial population to ensure exploration of the search space. A chromosome is generated from the items of X_1 completed by items generated randomly. In each chromosome, X_1 is integrated with a probability α. If α is set to zero this means that all the items in each individual are selected randomly, while $\alpha = 1$ means that each individual in the initial population contains all the items in X_1. This method allows having an initial population of good quality by integrating X_1 and ensures the diversification by adding the rest randomly.

2. *Fitness evaluation.* Besides the population initialisation, the guidance by the pre-treatment information is integrated in the GA by this operator. The fitness function $f(j)$ is evaluated according to Eq. 15.1. The efficiency e_j is introduced in its evaluation according to Eq. 15.4. Each generation, the fitness value of each chromosome is calculated. The fitness formula allows giving more chance to the chromosome that has a high efficiency to be selected more than the others.

$$f(j) = \sum_{j=0}^{n} e_j p_j x_j \qquad (15.4)$$

3. *Genetic operators.* GAGP uses standard genetic crossover and mutation operators. A tournament selection of size 5 is used as the selection method, and the random single point method is applied with a probability p_c as a crossover method. For the mutation operator, the random multiple point bit flip with the probability p_m is adopted. And finally, a reproduction operator copies a subset of individuals with the probability p_r such as $p_c + p_m + p_r = 1$.

4. *Update efficiency.* The Sorting efficiency is not always efficient especially for the problems with strong correlation. A step of efficiency update is proposed that aims to make a perturbation in the items ratios. Two items j, j', $j \in X_1$ and

are selected and their efficiency is permuted. Rather than maintaining the same guidance, the search process diversify the guide with the items in X_1 and $Core$. This modification has an impact on the fitness evaluation and so on the whole process of GA.

5. *Stopping condition.* The process of optimisation is repeated until a specific number of iteration is reached.

15.5   Experimental Results

The experiments aims to compare the proposed GAGP with the state-of-art results reported in the literature (Sect. 15.5.2). For an experimental purpose, and because the chosen sorting method concerns MKP, it is natural to use data from this problem. The test platform is a Toshiba laptop with 4GB RAM capacity and an Intel Core (TM) i5-4200 M 2.5 Ghz CPU. The Java language is used to implement the approach. As for the test data, two well known benchmarks from the OR-Library[3] are used.

15.5.1   Analytical Study of the Guidance

This analytical study compares firstly, between the items of the optimal solution and the two main parts of the guide (i.e. X_1 and $Core$) and secondly, between the optimal solution and the solution obtained by GAGP. The aim is on the one hand, for understanding how significant is the sort and measure its effective impact on the GA; On the other hand, to find if the GAGP does effectively follow the guide and what happens to drop the GA in the wrong solution whether because of the guide or because of the optimisation process itself.

The composition of the optimal solution S^*, calculated using the deterministic CPLEX optimiser 12.5, is compared to a feasible solution S obtained by GAGP. Also, the items of the X_1 and $Core$ and the placement of the items of the S^* in the three parts of the guide are given (where $+$, $*$ and $-$ corresponds to item in X_1, in $Core$ and in X_0 respectively). The first four instances (OR5x100.0.25 1-4) are used to conduct this analysis. Finally, the Distance From the Optimum D.F.O of the solution calculated by GAGP is given. The obtained results of the comparison are reported in Table 15.5.

A percentage of 75−90% of the items in S^* are included in X_1 or $Core$. Similarly, S contained 75−90% of the items of S^*. Almost the same items initially contained in X_1 and $Core$ are maintained in S. Some (3–7) items of the excluded part X_0 appear in S^* at the same time some were introduced in S by the mutation operator. In the first three instances, at most one item form X_1 has not been contained in S^*. GAGP could be more effective by introducing a better mutation operator. The efficiency

---

[3]http://people.brunel.ac.uk/~mastjjb/jeb/orlib/files/.

Table 15.5 Analytical comparison of GAGP feasible solution composition to the composition of the optimal solution obtained by CPLEX and the composition of the parts of the guidance information using the OR5x100.0.25 1–4 instances. S: Items of the solution obtained by GAGP. S*: Items of the optimal solution obtained by CPLEX. G: Group of S* item in the guide (item is: $+ \in X_1$, $- \in X_0$ or $* \in Core$). A.D.F.O: Average Distance Form the Optimum in % of the GAGP solution

S	S*	G	X_1	S	S*	G	X_1	S	S*	G	X_1	S	S*	G	X_1
1	1	+	1	1	3	+	3	7	4	+	4	0	0	*	3
3	3	*	4	3	10	+	10	10	11	+	11	1	1	*	5
6	6	+	6	10	18	*	27	11	13	+	13	3	3	+	11
8	8	+	8	20	20	*	28	13	18	+	18	5	5	+	12
10	10	-	23	27	27	+	34	18	19	-	21	6	8	-	22
12	18	*	26	34	28	+	45	21	21	+	28	8	11	+	27
17	23	+	31	36	34	+	49	28	26	-	37	11	13	-	28
23	25	-	43	39	36	-	56	32	28	+	44	12	22	+	30
26	26	+	49	41	41	*	57	34	32	*	48	22	24	-	34
28	28	*	56	42	42	*	61	36	34	-	55	24	27	+	35
29	29	*	62	45	45	+	62	37	37	+	69	26	30	+	42
31	31	+	65	48	48	*	73	42	42	*	72	27	34	+	53
41	43	+	68	49	49	+	90	44	44	+	74	30	35	+	63
43	49	+	76	53	53	-	91	48	48	+	84	34	42	+	69
49	56	+	78	56	56	+	93	51	51	-	87	35	49	*	70
56	61	-	85	57	57	+	95	53	55	+	92	42	53	+	78
62	62	+	92	58	58	-	99	55	59	*		49	54	-	86
65	65	+		61	61	+		59	60	-		53	55	-	94
68	68	+	Core	62	62	+	Core	60	64	*	Core	54	56	*	
76	70	-	3	70	64	*	18	64	72	+	10	55	58	*	Core
78	73	*	15	73	73	+	20	72	74	+	32	56	61	*	0
84	76	+	18	74	74	*	41	74	78	*	36	61	63	+	1
85	78	+	28	81	81	*	42	78	79	-	42	63	74	-	15
91	84	-	29	88	88	*	48	79	84	+	59	68	78	+	49
92	85	+	34	90	90	+	64	84	87	+	64	70	79	-	56
94	91	-	66	91	91	+	74	87	92	+	78	76	86	+	58
95	92	+	73	92	93	+	81	92	93	*	93	79	94	+	61
98	95	-	81	93	95	+	88	93	96	*	96	86	95	+	65
	98	+	98	95	99	+	92	96	99	*	99	95			68
			99	99											95
A.D.F.O = 0.82				A.D.F.O = 0.51				A.D.F.O = 0.23				A.D.F.O = 0.50			

measurement function would be more effective if *Core* contained a slightly more items of X_0. Most items of *Core* were components of S*, that supports the efficiency update operator proposed in GAGP.

15.5.2 Comparison with the Literature

As with most optimisation problems, MKP heuristics could be classified in two groups: the first is*constructive* heuristics, that aim to construct a solution. The second is *improvement* heuristics which aim to improve a given initial solution normally generated first by a *constructive* heuristic. The proposed method is considered as a *constructive* heuristic. However, in order to demonstrate the performance of the proposed method, the performance of the GAGP is compared with both *constrictive* and *improvement* approaches. The following is short description of the methods (*constructive* and *improvement*) used in the comparison presented in this section. GAGP is compared with the standard GA algorithm and other state-of-the-art optimisation methods reported in the literature. GAGP is compared to the following construc-

Table 15.6 Comparison of results obtained by GAGP with GA, constructive and improvement heuristics

n	m	α	Constructive							Improvement		
			GAGP	GA	PECH	MAG	VZ	PIR	SCE	CB	NR(P)	MCF
5	100	0.25	**0.35***	2.17	7.3	13.6	10.3	1.6	3.5	0.99	0.94	1.09
		0.50	**0.48**	0.86	3.4	6.7	6.9	0.77	2.6	0.45	**0.44***	0.57
		0.75	**0.21***	0.42	2.02	5.1	5.6	0.48	1.1	0.32	**0.22**	0.38
	250	0.25	0.58	4.03	7.1	6.6	5.8	**0.53**	4.3	**0.23***	0.46	0.41
		0.50	0.36	1.15	3.2	5.2	4.4	**0.24**	3.3	**0.12***	0.17	0.22
		0.75	0.23	0.58	1.8	3.5	3.5	**0.16**	1.5	**0.08***	0.1	0.14
	500	0.25	0.51	4.27	6.4	4.9	4.1	**0.22**	4.6	1.56	**0.15***	0.21
		0.50	0.36	1.45	3.4	2.9	2.5	**0.08**	3.6	0.79	**0.06***	0.1
		0.75	0.22	0.65	1.7	2.3	2.41	**0.06**	1.8	0.48	**0.03***	0.06
10	100	0.25	**1.0**	2.40	8.2	15.8	15.5	3.4	6.8	**0.09***	2.05	1.87
		0.50	**0.53**	1.53	3.7	10.4	10.7	1.8	5.1	**0.04***	0.81	0.95
		0.75	**0.27**	0.53	1.8	6.1	5.67	1.1	2.4	**0.03***	0.44	0.53
	250	0.25	**0.75**	3.56	5.8	11.7	10.5	1.1	6.9	**0.51***	0.88	0.79
		0.50	0.48	1.35	2.5	6.8	5.9	**0.57**	5.4	**0.25***	0.39	0.41
		0.75	**0.27**	0.66	1.5	4.4	3.7	0.33	2.8	**0.15***	0.19	0.24
	500	0.25	0.71	3.61	5.1	8.8	7.9	**0.52**	6.8	**0.24***	0.34	0.44
		0.50	0.4	1.44	2.4	5.7	4.1	**0.22**	5.8	**0.11***	0.14	0.2
		0.75	0.29	0.71	1.2	3.6	2.9	**0.14**	3.4	**0.07***	0.1	0.13
30	100	0.25	**1.56***	2.27	6.8	17.3	17.2	9.1	8.6	2.91	2.24	3.61
		0.50	**1.07***	1.72	3.2	11.8	10.1	3.51	6.6	1.34	1.32	1.6
		0.75	**0.36***	0.78	1.9	6.58	5.9	2.03	3.6	0.83	0.8	0.97
	250	0.25	**1.66**	3.20	4.8	13.5	12.4	3.7	8.3	**1.19***	1.27	1.75
		0.50	**1.0**	1.46	2.1	8.6	7.1	1.5	6.9	**0.53***	0.75	0.79
		0.75	**0.5**	0.73	1.2	4.4	3.9	0.84	3.8	**0.31***	0.38	0.43
	500	0.25	4.07	3.50	3.7	9.8	9.6	**1.89**	8.6	**0.61***	0.89	1.05
		0.50	2.14	1.45	1.7	7.1	5.7	**0.73**	7.4	**0.26***	0.36	0.44
		0.75	0.51	0.69	0.9	3.7	3.5	**0.48**	4	**0.17***	0.23	0.27

tive approaches: PECH (Primal Effective Capacity Heuristic) [22]; MAG [23]; VZ [2]; PIR (Pirkul 1987) and SCE (Shuffled Complex Evolution) [24]. GAGP is also compared to the following improvement approaches: CB [25]; NR (P) (New Reduction (Pirkul)) [26] and MCF (Modified Choice Function - Late Acceptance Strategy) [27]. The comparison is shown in Table 15.6. As shown in Table 15.6, GAGP is competitive with both construction and improvement methods and has managed to outperform both group of methods on a few instances.

15.6 Conclusions

In this paper the multidimensional knapsack problem (MKP) is addressed. Two GA-Based approaches were introduced: the first is denoted Memetic Search Algorithm (MSA) and the second Genetic Algorithm Guided by Pretreatment information (GAGP). MSA is a sequential hybridisation of GA and the Stochastic Local Search-Simulated Annealing algorithm (SLSA). GAGP is a two steps algorithm that uses a greedy method the extract information about the items and then integrates these knowledge in the operators of a GA. The conducted experiments used two groups of MKP instances of small and large size. The obtained results from applying MSA on small-size data and GAGP on large-size data indicated that both algorithms are able to obtain encouraging results. The use of SLSA in MSA allows to improve each offspring. Similarly, the use of knowledge, the guidance by fitness function and the dynamic strategy to update the efficiency of items allow to reduce significantly the processing time in GAGP.

References

1. Fukunaga, A. S. (2011). A branch-and-bound algorithm for hard multiple knapsack problems. *Annals of Operations Research, 184*(1), 97–119.
2. Volgenant, A., & Zoon, J. (1990). An improved heuristic for multidimensional 0–1 knapsack problems. *Journal of the Operational Research Society, 41*(10), 963–970.
3. Deane, J., & Agarwal, A. (2013). Neural, genetic, and neurogenetic approaches for solving the 0–1 multidimensional knapsack problem. *International Journal of Management & Information Systems (Online), 17*(1), 43.
4. Yoon, Y., & Kim, Y. H. (2013). A memetic lagrangian heuristic for the 0–1 multidimensional knapsack problem. Discrete Dynamics in Nature and Society, *2013*.
5. Rezoug, A., Boughaci, D., & Badr-El-Den, M. (2015). Memetic algorithm for solving the 0–1 multidimensional knapsack problem. In *Portuguese Conference on Artificial Intelligence* (pp. 298–304). Berlin: Springer.
6. Valls, V., Ballestin, F., & Quintanilla, S. (2008). A hybrid genetic algorithm for the resource-constrained project scheduling problem. *European Journal of Operational Research, 185*(2), 495–508.
7. Jat, S. N., & Yang, S. (2009). A guided search genetic algorithm for the university course timetabling problem.

8. Yang, S., & Jat, S. N. (2011). Genetic algorithms with guided and local search strategies for university course timetabling. *IEEE Transactions on Systems, Man, and Cybernetics, Part C (Applications and Reviews), 41*(1), 93–106.
9. Acan, A., & Tekol, Y. (2003). Chromosome reuse in genetic algorithms. In *Genetic and Evolutionary Computation Conference* (pp. 695–705). Berlin: Springer.
10. Louis, S., & Li, G. (1997). Augmenting genetic algorithms with memory to solve traveling salesman problems. In *Proceedings of the Joint Conference on Information Sciences* (pp. 108–111).
11. Boughaci, D., Benhamou, B., & Drias, H. (2010). Local search methods for the optimal winner determination problem in combinatorial auctions. *Journal of Mathematical Modelling and Algorithms, 9*(2), 165–180.
12. Kirkpatrick, S., Gelatt, C. D., Vecchi, M. P., et al. (1983). Optimization by simulated annealing. *science, 220*(4598), 671–680.
13. Bean, J. C. (1994). Genetic algorithms and random keys for sequencing and optimization. *ORSA Journal on Computing, 6*(2), 154–160.
14. Pan, Q. K., Suganthan, P. N., Tasgetiren, M. F., & Liang, J. J. (2010). A self-adaptive global best harmony search algorithm for continuous optimization problems. *Applied Mathematics and Computation, 216*(3), 830–848.
15. Yang, H., Wang, M., & Yang, C. (2013). A hybrid of rough sets and genetic algorithms for solving the 0–1 multidimensional knapsack problem. *Int. J. Innovative Comput Inf. Control, 9*(9), 3537–3548.
16. Veni, K. K., & Balachandar, S. R. (2010). A new heuristic approach for large size zero-one multi knapsack problem using intercept matrix. *International Journal of Computing Science and Mathematics, 4*(5), 259–263.
17. Dantzig, G. B. (1957). Discrete-variable extremum problems. *Operations Research, 5*(2), 266–288.
18. Shih, W. (1979). A branch and bound method for the multiconstraint zero-one knapsack problem. *Journal of the Operational Research Society, 30*(4), 369–378.
19. Puchinger, J., Raidl, G. R., & Pferschy, U. (2006). The core concept for the multidimensional knapsack problem. In *European Conference on Evolutionary Computation in Combinatorial Optimization* (pp. 195–208). Berlin: Springer.
20. Senju, S., & Toyoda, Y. (1968). An approach to linear programming with 0–1 variables. *Management Science, 15*, B196–B207.
21. Huston, S., Puchinger, J., & Stuckey, P. (2008). The core concept for 0/1 integer programming. In *Proceedings of the Fourteenth Symposium on Computing: The Australasian Theory-Volume 77* (pp. 39–47). South Melbourne: Australian Computer Society, Inc.
22. Akçay, Y., Li, H., & Xu, S. H. (2007). Greedy algorithm for the general multidimensional knapsack problem. *Annals of Operations Research, 150*(1), 17–29.
23. Magazine, M., & Oguz, O. (1984). A heuristic algorithm for the multidimensional zero-one knapsack problem. *European Journal of Operational Research, 16*(3), 319–326.
24. Baroni, M. D. V., & Varejão, F. M. (2015). A shuffled complex evolution algorithm for the multidimensional knapsack problem. In *Iberoamerican Congress on Pattern Recognition* (pp. 768–775). Berlin: Springer.
25. Chu, P. C., & Beasley, J. E. (1998). A genetic algorithm for the multidimensional knapsack problem. *Journal of Heuristics, 4*(1), 63–86.
26. Hill, R. R., Cho, Y. K., & Moore, J. T. (2012). Problem reduction heuristic for the 0–1 multidimensional knapsack problem. *Computers & Operations Research, 39*(1), 19–26.
27. Drake, J. H., Özcan, E., & Burke, E. K. (2015). Modified choice function heuristic selection for the multidimensional knapsack problem. In *Genetic and Evolutionary Computing* (pp. 225–234). Berlin: Springer.

Chapter 16
Solving the Home Health Care Problem with Temporal Precedence and Synchronization

Sophie Lasfargeas, Caroline Gagné and Aymen Sioud

Abstract Home health care (HHC) services aim to improve patients' living condition by providing different services at patients home. This chapter presents a medium-term home health care problem (HHCP) model. This model considers qualification requirements, possible patients/nurses exclusions, temporal precedences, synchronized services, routing with time windows, and working time constraints. As building a proper HHC planning is a challenging task, we proposed a two-stages approach including a new construction heuristic and an enhance basic VNS. Various alternatives in the construction heuristic and tested neighborhoods are exposed and discussed. The obtained results of the solving approach are compared to the literature, on challenging publicly available short-term datasets with temporal precedences and synchronized services. Amongst the 40 instances used, we obtained better results than the literature for 37 of them.

16.1 Introduction

In most of the developed countries, health systems tend to struggle to solve economic, organizational, and sanitary problems. The aging population, growing number of new chronic pathologies, overcrowding in hospitals, and risks of nosocomial infections are the main cause behind them [1]. Recently, Home health care (HHC) services have emerged as a viable solution to meet the challenges of these issues. They seek to improve patients living conditions by providing medical, paramedical, and social services to their homes, thereby avoiding or reducing hospitalizations.

S. Lasfargeas · C. Gagné (✉) · A. Sioud
Université du Québec à Chicoutimi, Chicoutimi, QC, Canada
e-mail: Caroline.Gagne@uqac.ca

S. Lasfargeas
e-mail: Sophie.Lasfargeas1@uqac.ca

A. Sioud
e-mail: Aymen_Sioud@uqac.ca

© Springer International Publishing AG, part of Springer Nature 2019
E.-G. Talbi and A. Nakib (eds.), *Bioinspired Heuristics for Optimization*,
Studies in Computational Intelligence 774,
https://doi.org/10.1007/978-3-319-95104-1_16

However, building a proper HHC planning is a complex task. Indeed, a wide range of caregivers is required to perform the different services such as nurses, physicians, social workers, etc. Each of them travels using a particular transportation mode from one patient's home to the next one in their tours, to deliver needed care within the desired time windows. Furthermore, multiple constraints must be considered like workload constraints, qualification, routing, interdependencies between services, etc. Interdependent services can be defined as having temporal dependencies between their starting times or as synchronized services (also called shared visits or simultaneous services), see [2]. It is thus possible for a single patient to need pairs of services that must be either separated from a given time interval or take place simultaneously. Interdependent services appear to be rarely considered despite being able to improvement significantly the planning quality or utility [3].

The home health care problems (HHCP) firstly distinguish themselves regarding the short, medium or long-term planning horizon, ranging from one day to several months. Short-term (or daily) problems have received the most attention. Less addressed, medium-term problems consider a planning horizon from a week to up to a month. The constraints above-mentioned, presumably essential to reflect the scope and complexity of real world problems, are rarely met. Furthermore, they change depending on the planning horizon. Indeed, according to the review of Fikar and Hirsch [4], short-term problems usually consider time windows, qualification requirements and working time regulations but interdependent services or breaks are barely borne in mind. In the case of medium-term problems, however, workload balance maybe considers but temporal precedence constraints are not. Additionally, synchronized services are weak in the HHC literature. Finally, it appears that many aspects should be taken into account together.

This chapter proposes a modeling of the medium-term HHCP problem with temporal precedence-constraints, synchronized services, multi-modality and working time requirements. A two-step solving approach is presented. It includes a new construction heuristic specifically studied for the model, and a metaheuristic, the variable neighborhood search (VNS) [5], designed to solve the HHCP efficiently. Emphasis has been placed on the study and comparison of various possible neighborhoods to highlight appropriate ones. While no method comparison, to the best of our knowledge, has yet been made in the domain, neither on medium-term nor short-term, our approach effectiveness is exemplified by the results obtained on short-term HHC existing datasets.

The organization of this chapter is as follows. Section 16.2 reviews the related literature. Section 16.3 presents the medium-term HHCP model by defining the key aspects. The construction heuristic and VNS are detailed, focusing on alternatives and neighborhoods in Sect. 16.4. Section 16.5 shows and compares their performances on HHC datasets. Section 16.6 makes some concluding comments and gives an outlook on future work.

16.2   Literature Review

This section presents a short overview of the HHC literature, and papers that deal with HHC planning focusing on their models, constraints, and resolution methods, on medium-term horizon first, then short-term. The extended definition of HHC defined by Fikar and Hirsch [4] is used. Therefore, the terms nurse and patient may set forth any staff member performing a service or customer receiving one.

In [6], various challenges dealt in the HHC literature has been reported such as partitioning the territory, dimensioning human resources, allocating resources to districts, assigning nurses to patients (or visits), scheduling and routing optimization. Basically, the objective of the assignment problem is to find an ideal pairing of nurses and patients, whereas the routing problem aims to design nurses tours. Analogously, the scheduling problem generates timetables to visit patients. These issues, covered by the HHCP, have received the most attention [7]. The HHCP is an NP-Hard problem [8]. Since proposed models, knowing that different objectives and set of constraints are considered, are evaluated on homemade datasets or derived ones from protected data [4], make a comparison is difficult. Indeed, there is no method comparison in the HHC domain and no commonly accepted benchmark. A systematical evaluation using publicly available benchmarks should be determined.

Nickel et al. [9] proposed a two-stage approach for solving the medium-term HHCP (i.e., with a weekly planning horizon). Initially, a feasible solution is generated using a constraint programming (CP) heuristic. Afterward, it is enhanced with an adaptive large neighborhood search (ALNS). The ALNS significantly alters solutions with task removal operations to overcome local optima. The resulting infeasible solutions are then repaired. This approach penalizes but allows not assigned visits. Additionally, interdependent services, transport mode diverseness and working time requirements are not considered. Duque et al. [10] focused on a flexible decision support system for the planning of a social profit organization providing HHC services in Belgium. The resolution, which aims to optimize the service level and minimize the total travel time, is performed using a two-phase algorithm. Precisely, a CP is chained with a randomized local search algorithm. The employed model is more restrictive to the extent that all visits of a patient must be the same duration and needed the same qualification from the nurses. While interdependent services are not taken into account, working time requirements are restricted to a maximum weekly working hours for each nurse. Trautsamwieser and Hirsch [11] specifically focused on this aspect of working time requirements and proposed a mathematical model formulation for the medium-term HHC planning problem without interdependent services. The solution approach used to solve their generated test instances to optimality is composed by a VNS providing upper bounds to a Branch-Price-and-Cut algorithm. The VNS used two neighborhoods that either move a sequence of patients from a nurse schedule to another or swap two sequences of patients between two nurses schedules.

Hiermann et al. [12] presented a framework to solve the short-term HHC scheduling problem for a major Austrian HHC provider. Its formal definition includes patients and nurses preferences, a maximal and minimal amount of working time for nurse, pre-assign job, and multi-modality. Infeasible solutions and tardiness are allowed in the resulting planning. The resolution approach described is two-phased. First, a random construction heuristic or a CP based approach generates the initial solution. Then, one of four metaheuristics improves it: a general VNS algorithm, simulated annealing, memetic, and a scatter search. The VNS neighborhoods are a swap of two nurses, a repositioning of a job in the best feasible position in the current tour and a shift job from one position to another in a different tour. Despite this short-term model completeness, it does not consider interdependent services or detailed working time requirements. On the contrary, Bredström and Rönnqvist [3] emphasized the importance of the temporal synchronization and precedence. The proposed mathematical programming model is a generalization of the vehicle routing and scheduling problem with time-windows. A local branching heuristic is used to solve generated instances of a daily planning problem for HHC staff members. Services are organized in three sets: simple services, services with temporal precedence and synchronized services. However, to ensure feasible solutions without hiring a new nurse, time windows are flexible for synchronized services. Finally, Mankowska et al. [2] proposed a refine model for the short-term routing and scheduling problem (e.g., without nurses working time windows) but considering interdependent services. Several datasets, publicly released, are presented and used to test the solving approach design for their model. An adaptive VNS is presented along with its neighborhoods. Highly dependent on the daily solution representation chosen, those neighborhoods are unsuitable for the medium-term HHCP.

16.3 Description of the Medium-Term HHCP Problem

The HHC planning on a weekly horizon plan aims to assign and schedule all jobs (or services) required by patients to the proper nurses. In the following description, the HHCP has no initial assignment.

Over a given planning horizon, i.e., a set of days $\mathcal{T} = \{1, \ldots, T\}$, the HHCP has essential features to consider such as the jobs, interdependent jobs, nurses, legal working requirements, multiple transportation modes, constraints, and its objectives.

16.3.1 Jobs

Within this modeling, a patient is represented by its needed job or jobs. The jobs $\mathcal{J} = \{1, \ldots, J\}$ are cares that must be performed at patients home, and have different nature: toileting, providing medication, dressing, making an injection, etc. A single patient may relate to one or more jobs. However, a job $j \in \mathcal{J}$ belongs to a unique patient.

To perform a job j, a nurse must possess the required qualification $r_j \in \mathscr{Q}$. Let $\mathscr{Q} = \{1, \ldots, Q\}$ be the set of qualifications required by jobs. In the event of a job need not only one but several qualifications then a new one is introduce in the set as a combination of others.

A job j has to start in a time window $[b_j, e_j] \subseteq [0, M]$ with M corresponding to the maximal amount of minutes in a day. If the starting time of a job j exceeds e_j a day $t \in \mathscr{T}$, a tardiness value z_{jt} penalized the objective function. Then, a job j has a processing time p_j (independent from whom will perform or receive it) and a frequency f_{jt} specifying if it must take place a day t (1 if it does, 0 otherwise). Therefore, a job j must be schedule $\sum_{t \in \mathscr{T}} f_{jt}$ many times during the planning horizon. Each occurrence is called a task.

Finally, some patients might want to avoid particular nurses for various reasons such as gender, that can be inconvenient for some patients. Thereby, for each job j, rejected nurses by the patient are stored in R_j.

16.3.2 Interdependent Jobs

A job j can have a prerequisite job stated in pre_j. This feature contains the unique identifier of the prerequisite job (0 if j does not have one). Additionally, a minimal time distance δ_j^{min} and maximal time distance δ_j^{max} that must be observed between their starting times, are defined. These three elements help characterized synchronized jobs and temporal precedence. Indeed, if $\delta_j^{min} = \delta_j^{max} = 0$ then j and its prerequisite job must be synchronized (i.e., start at the same exact time) which means that two nurses are needed simultaneously at the patients home. Synchronized jobs are often meet in case of overweight patients or palliative cares. Temporal precedence becomes handy with some jobs (e.g., take a blood sample or administrate drugs), that must be done a certain amount of time before another (e.g., providing a meal).

This notation, inspired by the double services in [2], differs by considering interdependent jobs as two distinct jobs in their own rights, and assignable on a medium-term planning. Moreover, interdependent jobs are never assigned both to the same nurse because synchronized jobs need in essence, two distinct nurses.

16.3.3 Nurses

Let $\mathscr{N} = \{1, \ldots, N\}$ be the set of nurses. Then, a nurse $n \in \mathscr{N}$ is a qualified staff member able to perform jobs at patients home. Each nurse can accomplish a set of jobs. Her qualifications $q_n \subseteq \mathscr{Q}$ must fit the jobs requirements. However, she can deliver a job only within her time window. Thus, $[b_{nt}, e_{nt}]$ represents the lower bound and upper bound of availability for the nurse n during the day t.

Nurses may refuse some patients (e.g., if they have dogs) and hence refuse to perform any of these patients jobs. In such case, the list of excluded jobs for a nurse

n is stored in E_n. The binary matrix Q_{jn} combines these hard constraints of job/nurse compatibilities by taking into account the adequacy between the skill r_j required by the job j, the qualifications q_n of the nurse n and the mutual exclusions R_j and E_n. Therefore, Q_{jn} equals 1 if the job j can be offered by the nurse n, 0 otherwise. An example is provided in Sect. 16.4.

All nurses tours must start and finish at the depot (e.g., the HHC office). A nurse n might have a different departure and arrival such as her home but this specificity, for simplicity, is considered only when the traveled distances are calculated.

16.3.4   Legal Working Requirements

Beyond the daily planning, legal working requirements are important to consider. Here, these are taken into account similarly to [11]. To do so, the nurses have a contractual number of hours h_n. A break P of a duration p must be planned for a nurse n a day t if her working time is longer than B minutes. The nurses are not authorized to take more than one break per tour. This break can, however, be taken at any time during the day as the successive working time does not exceed B. It is planned before or after a job. Therefore, it does not alter the already planned tours.

The number of daily hours that a nurse is required to work must not exceed a threshold L (in hours). If $|\mathscr{T}| > 1$, in other words the planning is done on multiple days, then the specified total weekly working time, noted W, is considered. Additionally, compliant resting times must be planned. Indeed, a nurse needs a period Rd of rest for recovery between two working days. In order to ensure proper rest periods per week, a minimal of $Rw + M$ is required.

16.3.5   Multi-modality

Since the model comprises a unique depot, which constitutes the first job, denoted by 0, let the set of jobs including the depot be $\mathscr{J}_0 = \mathscr{J} \cup \{0\}$. In the same way as Hiermann et al. [12] defined the concept of multi-modality, a pairwise distance matrix D^{m_n} of travel time is available. The travel time $d_{ij}^{m_n}$, with i and $j \in \mathscr{J}_0 \cup \mathscr{N}$, expressed in minutes, therefore represents the average time taken by the nurse n to travel from the location of i to the location of j using a certain transportation mode m_n defined as her preferred. A location can be the depot, a patient or nurse home.

16.3.6   Constraints

Constraints include satisfaction, precedence, synchronization between services, routing with time windows, and working time constraints.

Above all, the demand must be satisfied. It is unacceptable that services required by patients are not supported. Each task of each job must be planned, and only one unique nurse is assigned by task. The assignment is only made when the nurse is available in the required time window, she has the expertise to carry it, and there is no mutual exclusion ($Q_{jn} = 1$). Exactly two nurses must be assigned by interdependent jobs, and begin their respective task at the required time. Each tour starts and ends at the depot. Assigning a nurse to a given task is applicable only if the daily, or weekly ($\mathcal{T} > 1$), workload allows it.

The tours of the nurses are modeled by the decision variable x_{ijnt}, S_{jnt}, and z_{jt}. The binary variable x_{ijnt} is 1 if the nurse n performs the job j, or take a break P, after the job i on the day t, 0 otherwise. The real variable S_{jnt} indicates the beginning of a job j, or a break P, performed by the nurse n, on the day t. In this way, S_{0nt} and $S_{(J+1)nt}$ respectively represent the start and end of the tour of a nurse n, on the day t. The last real variable, z_{jt}, contains the potential delay of S_{jnt} on e_j if the job j was performed by the nurse n on the day t.

All above-mentioned constraints can be formalized in the following manner.

- A job j assigned to the nurse n at the day t ($x_{ijnt} = 1$) has to start in its time window $[b_j, e_j]$. In a case of delay, a penalty z_{jt} is calculated:

$$b_j \leq S_{jnt} \leq e_j + z_{jt} \quad \forall j \in \mathcal{J} : x_{ijnt} = 1, n \in \mathcal{N}, t \in \mathcal{T} \qquad (16.1)$$

- A job j assigned to the nurse n at the day t ($x_{ijnt} = 1$) must begin in $[b_{nt}, e_{nt}]$

$$b_{nt} \leq S_{jnt} \leq e_{nt} \quad \forall j \in \mathcal{J}_0 : x_{ijnt} = 1, n \in \mathcal{N}, t \in \mathcal{T} \qquad (16.2)$$

- Each task of a job j must be assigned to an unique nurse n at the desired day t

$$\sum_{i \in \mathcal{J}_0} \sum_{n \in \mathcal{N}} x_{ijnt} = f_{jt} \quad \forall j \in \mathcal{J}, t \in \mathcal{T} \qquad (16.3)$$

- When the jobs i and j are successive on the nurse n planning a day t, ($x_{ijnt} = 1$), j must begin at least $p_i + d_{ij}$ after the beginning of i:

$$S_{int} + p_i + d_{ij}^{mn} \leq S_{jnt} \quad \forall i, j \in \mathcal{J}_0 : x_{ijnt} = 1, n \in \mathcal{N}, t \in \mathcal{T} \qquad (16.4)$$

- The nurse n must have the expertise to carry a job j, and the mutual exclusion Q_{jn} conditions must be fulfilled:

$$x_{ijnt} \leq Q_{jn} \quad \forall i \in \mathcal{J}_0, j \in \mathcal{J}, n \in \mathcal{N}, t \in \mathcal{T} \qquad (16.5)$$

- A tour must start at the depot when the nurse is available ($e_{nt} - b_{nt} > 0$):

$$\sum_{j \in \mathcal{J}} x_{0jnt} = 1 \quad t \in \mathcal{T}, n \in \mathcal{N} : e_{nt} - b_{nt} > 0 \qquad (16.6)$$

- A nurse n who arrives at the patient's residence to perform the job j must leave it at the end of the task. Similarly, the nurse leaves the depot at the beginning of a tour and arrives at it at the end. This constraint ensures the routing coherence.

$$\sum_{i \in \mathscr{J}_0} x_{ijnt} = \sum_{k \in \mathscr{J}_0} x_{jknt} \quad \forall j \in \mathscr{J}_0, n \in \mathscr{N}, t \in \mathscr{T} \qquad (16.7)$$

- The time interval between the beginning of two interdependent jobs k and j, where k is the prerequisite of j $(k = pre_j)$, must be in the time window $[\delta_j^{min}, \delta_j^{max}]$:

$$\delta_j^{min} \leq S_{jnt} - S_{knt} \leq \delta_j^{max} \quad \forall k, j \in \mathscr{J} : k = pre_j, n \in \mathscr{N}, t \in \mathscr{T} \quad (16.8)$$

- Two distinct nurses must be assigned to interdependent jobs. Considering the respect of Eq. (16.3):

$$\sum_{i \in \mathscr{J}} x_{ijnt} = \sum_{m \in \mathscr{N} \setminus \{n\}} \sum_{i \in \mathscr{J}_0} x_{ikmt} = 0 \quad \forall k, j \in \mathscr{J} : k = pre_j, n \in \mathscr{N}, t \in \mathscr{T}$$

$$(16.9)$$

- Constraints on decision variables:

$$S_{jnt} \in \mathbb{R}^+ \quad \forall j \in \mathscr{J}_0 \cup P, n \in \mathscr{N}, t \in \mathscr{T} \qquad (16.10)$$

$$z_{jt} \in \mathbb{R}^+ \quad \forall j \in \mathscr{J}_0, n \in \mathscr{N}, t \in \mathscr{T} \qquad (16.11)$$

$$x_{ijnt} \in \{0, 1\} \quad \forall i, j \in \mathscr{J}_0 \cup P, n \in \mathscr{N}, t \in \mathscr{T} \qquad (16.12)$$

The definition of the working time requirements constraints for this problem is identical to the ones formulated by Trautsamwieser and Hirsch [11].

16.3.7 Objective Function

The model objective function incorporates the minimization of the working time, of delays and of the maximum delay. The duration of jobs being constant regardless the nurse assigned, the minimization of the total working time refers to the minimization of travel time and waiting time. The minimization of delays reflects the need for the provider to offer a good quality service. The minimization of the maximum delay avoids the over-penalization of a patient. As a result, the retained objectives aim to enhance the satisfaction whether of patients, nurses or service providers and can be expressed as follows:

$$\min\left(\alpha_1 * \sum_{n\in\mathcal{N}}\sum_{t\in\mathcal{T}}\left(S_{(J+1)nt} - S_{0nt} - p * \sum_{i\in\mathcal{J}} x_{iPnt}\right) + \right.$$
$$\left. \alpha_2 * \sum_{j\in\mathcal{J}}\sum_{t\in\mathcal{T}} z_{jt} + \alpha_3 * \max z_{jt}\right)$$

$$(16.13)$$

where each $\alpha_k = 1/K$ and K is the number of objectives. The flexibility of Eq. (16.13) easily allows the nullification of a given sub-objective or the adding of constraints. In this way, the presented model can be adapted and tested on datasets like the one provided by Mankowska et al. [2].

16.4 Solving Approach

The proposed method for resolving the HHCP is two stages approach. A feasible solution is firstly generated using a novel heuristic. It is afterward transmitted to a VNS derived from the basic VNS, presumably firstly proposed by Hansen and Mladenović [5], to enhance it. Fleszar and Hindi [13] have shown that the VNS produces better and faster results using an improved initial solution, careful consideration will thus be given to the construction heuristic in the following.

In this section, we firstly present the solution representation. We then describe the construction heuristic. Finally, we explain the VNS main elements such as its neighborhoods used during the shaking phase, its neighborhood order and its local search.

16.4.1 Solution Representation

A solution is a potential global planning. It is represented using a matrix, where each entry contains the assigned job list for a specific nurse on a particular day among the planning horizon. The starting and ending time of jobs are also encoded, it is called a slot. Figure 16.1 gives a solution example for a specific problem with $N = 3$ nurses, $T = 5$ days and $J = 10$ jobs. It illustrates that jobs can include multiple tasks on the planning horizon. Indeed, the 5th job has a task each day, and the 6–10th jobs respectively have two tasks. The 7th and 8th jobs are synchronized. Jobs 9 and 10 share a temporal precedence constraint. As a reminder, the distance between their starting time is always between δ_j^{min} and δ_j^{max} ($j = 10$). It may be observed that the nurse 1 has no assignment on the 5th day, which means that she does not work that precise day.

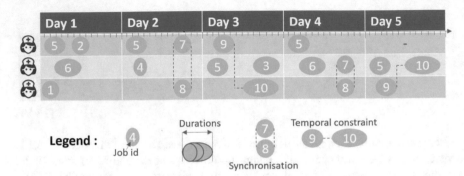

Fig. 16.1 Representation of a solution for a problem with $N = 3$ nurses, $T = 5$ days and $J = 10$ jobs

16.4.2 Construction Heuristic

To generate a feasible solution with the proposed heuristic, it is required to explain three rules of list ordering beforehand.

The first rule is applied to the job list. Many alternatives are available such as a purely random order (R) to improve the solution diversity, an increasing order of time windows (TW) given priority to the urgent jobs, and an increasing order of the number of possible job/nurse associations (PN). The last one aims to prioritize the jobs that can be assigned to a limited amount of nurses. It has the benefit of considering the rarity of the required qualifications and the excluded job/nurse association (previously defined in the model as Q_{jn}), as illustrated in Fig. 16.2. It exemplifies a compatibility matrix, Q_{jn}, summarizing that the nurse 1 excludes the job 3, and she has the required skill. During the jobs sequencing, if a job is tied with another, a random order will be applied to them. Regardless of the chosen order, if there are any interdependent jobs, they must be adjacent to assign in the first place the prerequisite job right before the job that dependent on him. It simplifies the processing and maximizes the feasible solutions.

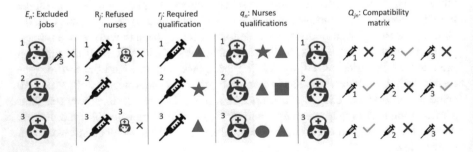

Fig. 16.2 An example of a compatibility matrix summarizing requirements and exclusions

The second order is applied to the nurse list. Nurses can be ordered randomly (R) for diversity purpose, by increasing number of qualification (Q), or by increasing nurse workload (WL) to balance workload between nurses. The choice of a workload ordering implies to update the nurse list at each iteration of the heuristic.

The last order is applied to the slot list that is generated before a job assignment. Slots can be ordered randomly (R), by earliest possible starting times (E), by increasing induced traveling distances (D) or waiting times (W) at patient home.

Now that orders have been determined, the heuristic can run properly. Basically, for each task of each job from the ordered job list, a qualified nurse is searched in the ordered nurse list. Then, all possible slots (fulfilling hard constraints) in the selected nurse planning are gathered for the given task. The previously chosen order is used to sort them. Then, the resulting first slot is used to assign the current task to the current nurse. All the tasks of the current job are dealt in the same way, one after the other, then the next job and its tasks, and so on. If no possible slots have been found earlier in the current qualified nurse' planning, then the heuristic would have searched for another qualified nurse in the nurse list. If none is available, an exception is thrown. This construction mode improves the generation of good feasible solutions, see Algorithm 1. All the jobs and their tasks must be assigned to a nurse with respect to the problem hard constraints.

Algorithm 19: Construction Heuristic

$jobList \leftarrow$ Sort J;
$nurseList \leftarrow []$;
for job in $jobList$ **do**
    for t in $frequency(job)$ **do**
        Update($nurseList$);
        $assigned \leftarrow false$;
        $currentNurse \leftarrow empty$;
        while $assigned! = true$ or $currentNurse$ is not empty **do**
            $currentNurse \leftarrow$ next qualified nurse in $nurseList$;
            $slotList \leftarrow$ find compatible slots in $s[currentNurse][t]$;
            if $Size(slotList) > 0$ **then**
                Sort $slotList$;
                $s[currentNurse][t][slotList[1]] \leftarrow job$;
                $assigned \leftarrow true$;
            end
        end
    end
end

The inherent possibilities make the heuristic more robust depending on the datasets nature. The experimentation results are shown in Sect. 16.5.

16.4.3 Neighborhoods

Various neighborhoods have been investigated for the HHCP problem. These are moves on the current solution, capable of improving it. In the case where an unfeasible solution is made, it is discarded. Here, a limited number of attempts $max_{iteration}$ to generate a feasible neighbor has been established.

Only two types of moves are used, a move (M) and a swap (S). Since adding or suppressing operations tend to generate unfeasible solutions all too often, they have not been considered. The moves concern either slots (SL) or shifts (i.e., a nurse daily planning, SH), one nurse (1N) or more (2N), and one day (1D) or two (2D). Therefore, the proposed neighborhoods on short-term are:

- M/SL/1N/1D moves a random slot to a random position on the same shift.
- M/SL/2N/1D moves a random slot to a random position in another nurse shift on the same day.
- S/SL/1N/1D exchanges two random slots from the same nurse shift.
- S/SL/2N/1D exchanges two random slots from random nurses on the same day.
- S/SH/2N/1D exchanges two random shifts from random nurses on the same day.

These neighborhoods can be transposed to a medium-term planning, where 1D would be replaced by 2D. Depending on the dataset, swapping shifts concerning two different days could make hard to respect the qualification constraint. For instance, jobs of the first picked shift that must have a task a day t will not necessarily need to have a task the day $t + 1$, the day of the second picked shift. All proposed neighborhoods also work with interdependent jobs. Indeed, just after its creation, the neighbor is updated along with the starting and ending times of each slot impacted by the move performed. The slots of interdependent jobs that have been moved are mutually updated. Thereby, the fulfillment of hard constraints is maximized. The neighborhoods' effectiveness is discussed in Sect. 16.5.

16.4.4 Neighborhood Order and Local Search

Neighborhoods are used in a determined sequence. This sequence is known to bring solution diversity avoid local optima. Here, the following order is used during the first VNS iteration: M/SL/2N/1D, S/SL/2N/1D, S/SL/1N/1D, M/SL/1N/1D. The first three neighborhoods are applied during the shaking phase, while the last one is only used in the local search. After each VNS iteration, this neighborhood sequence is shuffle, thus possibly introducing new improvements. The last neighborhood in the current neighborhood sequence is always used in the local search, known as an intensification phase.

Two types of local search have been considered, the first improvement and the best improvement algorithms. They both stop when they have reached a local optimum, but regarding the number of hard constraints and the complexity of the HHCP problem, only the first improvement method have been found suitable.

16.5   Results and Discussion

In the following, the datasets ([2]) employed for testing the proposed methods are described, along with the experimental setup. Subsequently, the obtained results are compared and discussed. Particular attention will be given to the different neighborhood and configurations of the construction heuristic.

As no medium-term dataset with interdependent services was publicly available by the time we searched, we used the datasets proposed by Mankowaska et al. [2]. They represent short-term (i.e., daily) HHC planning problems challenging on interdependent services. These are seven datasets, where each one comprises ten instances. Datasets complexity ranges from 10 to 300 patients and from 3 to 40 nurses. There are as many jobs as patients; 15% of the jobs are synchronized jobs, 15% are jobs with temporal precedence. These datasets intentionally overestimate the frequency of interdependent jobs that might occur in a real world case. To compare our model and methods, and to avoid constraint relaxation, the first four datasets (namely A, B, C, D) has only been considered.

All components of the proposed model and methods have been implemented in C++. The experimentation has been carried out on a computer equipped with an Intel Core i7 processor (4.00 GHz) and 32 GB of RAM. The construction heuristic has been run 10,000 times on each datasets instances with every possible configuration (3 job orders*3 nurse orders*4 slot orders, i.e., 36 possible configurations). Each configuration on any instance takes less 2.0 s to run 10,000 times. For the sake of consistency, the VNS has been run 40 times on each datasets instances. The VNS runtime was limited to 5 minutes.

Table 16.1 reports the results presented in [2] for comparison purpose, then our VNS results and our construction heuristic results. The 2th to the 4th column contains the best found solution by their construction heuristic, their Cplex and their best heuristics. Then, the 5th to the 12th column indicate our VNS results, respectively our best solutions found, the gap with the ones found in [2], our average solution (with the standard deviation), the maximal and minimal runtimes in seconds to obtain our best solutions, and the average initial solutions used (with the standard deviation). The construction heuristic results can be found in the last two columns. The last one shows the best heuristic configuration for the given instance, as a reminder, the orders chosen for the job list (TW, PN, R), nurse list (Q, WL, R) and slot list (D, W, E, R). If more than one configuration has been able to generate the best found solution, only the best one has been retained in Table 16.1, judging by average solutions and the number of time the best solution has been found.

As it can be noticed in Table 16.1, while in the dataset A, multiple optima have been reached by the construction heuristic, better solutions, in the great majority of cases, have been generated on the other datasets than the ones found by the construction heuristic in [2]. Then, amongst the 40 instances used, our VNS obtained better results than those in [2] for 37 of them.

Our study of the possible combination orders, for the construction heuristic, highlighted three interesting behaviors. Some combinations (e.g., TW/Q or TW/WL

Table 16.1 Experimental results on datasets A–D

Instance	Mankowska et al. [2]		AVNS/TL	VNS results for 40 runs								CH results for 10,000 runs	
	Init	Cplex		Best found	GAP (%)	AVG	STD	Max CPU	Min CPU	AVG Init	STD	Best found	Best orders
A1	464.4	218.2	218.2	218.2	0.0	224.3	14.6	0.2	0.0	279.4	31.4	218.2	TW/R/D
A2	1034	246.6	248.1	246.6	0.0	258.1	31.6	0.2	0.0	352.1	49.0	247.6	R/WL/W
A3	569.1	305.9	305.9	305.9	0.0	358.8	38.5	0.9	0.0	569.8	129.9	305.9	R/WL/E
A4	676.3	186.9	186.9	186.9	0.0	196.4	11.1	0.1	0.0	286.1	66.5	189.6	TW/R/W
A5	421.2	189.5	192	189.5	0.0	216.5	36.6	0.3	0.0	266.2	31.8	189.5	R/R/R
A6	1282	200.1	200.1	200.1	0.0	200.1	0.0	0.0	0.0	217.3	11.1	200.1	R/WL/D
A7	933.6	225.4	225.4	225.4	0.0	232.4	24.2	0.3	0.0	305.5	92.1	225.4	PN/R/E
A8	296	232	232	232.0	0.0	281.4	47.6	0.0	0.0	319.6	47.5	232.0	R/WL/R
A9	781.2	222.3	222.3	222.3	0.0	225.4	4.0	0.1	0.0	358.0	79.9	222.3	R/WL/E
A10	1540	225	225	225.0	0.0	225.0	0.0	0.0	0.0	295.0	41.8	230.8	PN/Q/D
B1	2215	1135	458.9	434.1	-5.6	552.8	93.4	53.1	0.0	940.6	136.3	475.5	TW/R/E
B2	1229	476.2	580.9	476.0	0.0	561.3	61.4	27.7	0.0	812.8	126.5	509.9	R/Q/E
B3	3664	399.2	431.4	399.1	0.0	527.6	72.5	63.5	0.0	967.2	127.7	543.1	R/R/E
B4	1402	576	587.3	414.0	-32.7	509.7	74.5	66.8	0.0	852.8	156.2	482.5	TW/R/D
B5	2469	599.4	391.1	385.6	-1.4	496.9	98.1	13.7	0.0	809.1	105.5	469.6	TW/R/E
B6	1059	1358	545.9	447.8	-17.7	611.8	129.9	43.7	0.0	1064.5	229.3	521.8	R/WL/E
B7	1184	432.3	356.6	328.7	-7.8	398.8	64.8	61.5	0.1	820.5	139.8	400.5	TW/R/D
B8	504.2	357.8	410.9	359.7	0.5	488.7	116.2	79.3	0.0	945.4	179.8	447.9	R/R/E
B9	782.7	403.8	487.9	404.1	0.1	483.4	60.3	62.1	0.0	1152.6	183.8	474.6	TW/WL/R

(continued)

Table 16.1 (continued)

Instance	Mankowska et al. [2]			VNS results for 40 runs								CH results for 10,000 runs	
	Init	Cplex	AVNS/TL	Best found	GAP (%)	AVG	STD	Max CPU	Min CPU	AVG Init	STD	Best found	Best orders
B10	3217	629.9	500.4	**462.7**	−7.8	616.8	147.7	8.7	0.0	961.8	138.6	533.3	PN/WL/E
C1	2264	–	1123.6	**974.2**	−14.2	1350.4	365.3	96.2	1.9	2792.6	414.6	1582.6	R/WL/E
C2	1063	–	677	**605.1**	−10.7	685.5	55.6	106.4	2.1	1724.5	247.0	959.4	TW/WL/D
C3	1360	–	642.4	**562.9**	−13.2	698.2	82.7	109.8	3.7	2043.7	254.6	953.3	TW/WL/R
C4	1460	–	580.4	**521.9**	−10.6	630.4	101.8	112.4	1.9	1726.4	249.1	814.0	TW/WL/D
C5	1412	–	754.6	**683.1**	−10.0	822.6	119.3	114.9	0.2	1606.2	268.7	948.7	PN/WL/E
C6	2313	–	951.6	**854.6**	−10.7	1010.6	146.4	115.9	4.0	2504.6	372.9	1321.1	R/WL/E
C7	772	4754	577.4	**529.2**	−8.7	572.5	29.7	109.4	0.9	1321.1	142.4	876.9	TW/WL/R
C8	949.2	–	540.6	**471.0**	−13.8	522.8	29.8	110.8	0.4	1168.8	181.7	765.4	TW/Q/E
C9	1073	–	608.7	**551.1**	−9.9	642.7	77.6	115.4	3.3	1903.2	240.7	945.2	TW/WL/D
C10	1368	–	679.3	**608.9**	−10.9	653.0	35.6	99.0	0.6	1647.2	268.2	929.5	TW/R/D
D1	4073	–	1321.8	**1278.2**	−3.4	1498.8	199.0	143.0	11.9	2898.4	279.0	1982.4	R/WL/E
D2	1362	–	892.7	**746.9**	−17.8	914.3	97.9	168.7	40.5	2525.4	318.0	1482.2	TW/WL/D
D3	1482	–	819.4	**678.6**	−18.8	817.8	8C.9	155.4	3.1	2177.5	213.9	1421.5	TW/WL/D
D4	2781	–	877.4	**809.7**	−8.0	1073.1	197.5	148.5	29.0	2319.4	286.3	1509.8	R/WL/E
D5	2039	–	872.1	**777.0**	−11.5	924.9	120.8	150.3	39.1	2220.7	292.1	1284.8	TW/WL/D
D6	1985	–	835.2	**768.6**	−8.3	886.6	97.4	154.6	18.3	2600.2	288.6	1361.1	TW/WL/R
D7	1026	–	706.3	**600.1**	−16.3	680.4	31.6	168.1	18.0	1796.3	179.3	1241.0	TW/WL/E
D8	1178	–	811.4	**715.5**	−12.6	775.8	31.1	149.8	8.3	1865.7	191.5	1262.3	R/WL/E
D9	1325	–	860.3	**741.0**	−12.8	818.2	46.5	156.0	15.8	2324.4	236.5	1314.6	TW/WL/D
D10	6592	–	**1306.6**	1424.6	8.6	1867.7	258.6	173.1	30.4	3805.4	356.6	2483.7	R/WL/E

regardless of the slot order) bring no solution diversity on all instances of the dataset A (except for 9), and on B4. Indeed, on small instances, deterministic orders tend to always generate the same solution. On the other instances, cases of a tie are met during the ordering, which allows applying a random order creating diversity. Similarly, the same combinations TW/Q and TW/WL generate only unfeasible solutions on C1 and for TW/Q also on D9. The last thing to notice is that amongst all the combinations that obtain the best found solutions, the average solutions is always better if slots are ordered by E.

The VNS uses initial solutions generated with an R/WL/R setup because it offers a compromise which not necessarily aims to found the best solution but rather a good feasible solution in average, regardless the considered dataset instance. Starting from these initial solutions, our VNS found better solutions than those in [2] and also obtains optima on the dataset A. However, on instances B8 and B9, the Cplex [2] finds slightly better solutions than ours, and on instance D10, their heuristics beats ours. It also seems useful to precise that their Cplex requires no less than 10 hours and their heuristics at most 8 s.

Our neighborhoods tests on the datasets revealed that the M/SL/2N/1D is the one generating the greatest number of solution improvements. By contrast, the M/SL/1N/1D is the one with the best ratio of feasible neighbors generated upon unfeasible neighbor, fostering the exploration. The other neighborhoods can also bring a solution diversification and find couple solution improvements. It might be possible for some neighborhoods to be ineffective therefore we kept the four instead of eventually less. For instance, S/SL/1N/1D generates almost no feasible solutions on the instance A4. We considered additional neighborhoods, not retained, capable of making the same moves as ours but not at random, in the best positions. They had two major drawbacks. They increased the runtime significantly and covered less the solution space. After our tests, we chose the sequence M/SL/2N/1D, S/SL/2N/1D, S/SL/1N/1D, and M/SL/1N/1D, causing in average the largest number of solution improvements. However, it is changed at random each iteration of the VNS to find new improvements if it seems stuck in a local optimum. Amongst the five bests neighborhood sequences, we observe that S/SL/1N/1D is in the last position, used in the local search. The neighborhoods M/SL/2N/1D and S/SL/2N/1D are predominantly adjacent and in this order.

16.6 Conclusion

This chapter proposes a medium-term HHCP model and a two-stages approach to solve an HHCP problem, including a new construction heuristic and an enhance basic VNS. Experimentations were carried out on challenging short-term HHC datasets, where interdependent jobs have been considered. A better solution quality has been obtained by our VNS than in the existing literature on these datasets. One of the construction heuristic strengths, other than its runtime, is that it could be tailored to each dataset, thanks to all its possible configurations. Indeed, before choosing our setup and running the heuristic, we could consider a pre-analyze instance phase. Since our model and solving approach have been tested on short-term, and can effectively deal with interdependent constraints, a systematical evaluation for medium-term should be established. In this sense, it will bring a unified approach to solve HHCP problem.

References

1. Bashir, B., Chabrol, M., & Caux, C. (2012). Literature review in home care. In *9th International Conference of Modelling and Simulation (MOSIM)*. Citeseer.
2. Mankowska, D. S., Meisel, F., & Bierwirth, C. (2014). The home health care routing and scheduling problem with interdependent services. *Health Care Management Science, 17*(1), 15–30.
3. Bredström, D., & Rönnqvist, M. (2008). Combined vehicle routing and scheduling with temporal precedence and synchronization constraints. *European Journal of Operational Research, 191*(1), 19–31.
4. Fikar, C., & Hirsch, P. (2017). Home health care routing and scheduling: A review. *Computers & Operations Research, 77*, 86–95.
5. Hansen, P., & Mladenović, N. (2001). Variable neighborhood search: Principles and applications. *European Journal of Operational Research, 130*(3), 449–467.
6. Lanzarone, E., Matta, A., & Sahin, E. (2012). Operations management applied to home care services: the problem of assigning human resources to patients. *IEEE Transactions on Systems, Man, and Cybernetics-Part A: Systems and Humans, 42*(6), 1346–1363.
7. Benzarti, E., Sahin, E., & Dallery, Y. (2013). Operations management applied to home care services: Analysis of the districting problem. *Decision Support Systems, 55*(2), 587–598.
8. Bertels, S., & Fahle, T. (2006). A hybrid setup for a hybrid scenario: combining heuristics for the home health care problem. *Computers & Operations Research, 33*(10), 2866–2890.
9. Nickel, S., Schröder, M., & Steeg, J. (2012). Mid-term and short-term planning support for home health care services. *European Journal of Operational Research, 219*(3), 574–587.
10. Duque, P. M., Castro, M., Sörensen, K., & Goos, P. (2015). Home care service planning. the case of landelijke thuiszorg. *European Journal of Operational Research, 243*(1), 292–301.
11. Trautsamwieser, A., & Hirsch, P. (2014). A branch-price-and-cut approach for solving the medium-term home health care planning problem. *Networks, 64*(3), 143–159.
12. Hiermann, G., Prandtstetter, M., Rendl, A., Puchinger, J., & Raidl, G. R. (2015). Metaheuristics for solving a multimodal home-healthcare scheduling problem. *Central European Journal of Operations Research, 23*(1), 89–113.
13. Fleszar, K., & Hindi, K. S. (2004). Solving the resource-constrained project scheduling problem by a variable neighbourhood search. *European Journal of Operational Research, 155*(2), 402–413.

Chapter 17
Automatically Generating Assessment Tests Within Higher Education Context Thanks to Genetic Approach

R. Ciguené, C. Joiron and G. Dequen

Abstract In educational context (online or face-to-face), with increasing cohort size and the need for individualization, the question of partial or full automation of assessments is growing. This paper deals with preliminary works that tackle the question of automatic generation of assessment *Tests* that could guarantee fairness and reasonable difference, by the content and the structure. A structural metric characterizing the distance between two given *Tests* is presented. This metric provides a dedicated fitness function that leads to define a Genetic Algorithm (GA) technique. The original use of GA allows optimizing this structural differentiation and thus guarantees the generation of collections of *Tests* with the largest distance possible while involving the smallest items source database. Preliminary experiments and results on the basis of multiple choice questions items are discussed.

Keywords Genetic algorithm · Individual assessment · Automatic generation
Structural distance

R. Ciguené
Laboratoire SITERE, Ecole Supérieure d'Infotronique d'Haiti, Port-au-Prince, Haiti
e-mail: richardson.ciguene@esih.edu

C. Joiron · G. Dequen (✉)
Laboratoire MIS, Université de Picardie Jules Verne, 33 Rue Saint-Leu,
80039 Amiens Cedex 1, France
e-mail: gilles.dequen@u-picardie.fr

C. Joiron
e-mail: celine.joiron@u-picardie.fr

© Springer International Publishing AG, part of Springer Nature 2019   269
E.-G. Talbi and A. Nakib (eds.), *Bioinspired Heuristics for Optimization*,
Studies in Computational Intelligence 774,
https://doi.org/10.1007/978-3-319-95104-1_17

17.1 Introduction

The work described in this paper is at the crossroads of two main research fields: the higher educational learning assessments and the Constraint Satisfaction Problems (CSP) (consisting in finding a solution sometimes optimal to a problem with constraints [1, 2]). Summative assessments [3–5] are done at the end of a curricula in order to evaluate learning outcomes of learners [6]. When this kind of assessment aims at delivering a certificate or a diploma, it is called *certificative*. For a certificative assessment, *Test* collections must guarantee fairness between all learners. However, teachers might have to create differentiated assessment *Tests* within various contexts: for distinct cohorts of learners, related to the same topics, and in various university sections or from a session to another; for the same cohort with multisession assessments; for a single assessment session with the aim to limit fraud (e.g. in a large amphitheater).

As a consequence, the question is then: *How to guarantee fairness of a given assessment if we suppose it consists of distinct Tests ?.*[1] Thus, we tackle in this paper the problem of assisting, thanks to an automatic process, the teacher in designing and generating assessments where each *Test* is structurally different from another. This work proposes a metric with the aim of characterizing the distance between two given *Tests*. This metric provides a dedicated fitness function that leads to define a Genetic Algorithm (GA) technique. At first, we especially focus automatic generation of Multiple Choice Questionnaires.

As a requirement, we suppose that the teacher has at his disposal a set of questions where each of them is coupled to a set of possible choices of answer (named *source database* in the following). The original use of GA allows to optimize the structural differentiation and thus guarantees the generation of collections of *Tests* with the largest distance possible while involving the smallest *source database*.

The next section presents the problem definition, how are structured entities involved. Section 17.3 presents our progression toward genetic approach, especially trough a metric that measures the distance between two *Tests*, and the random approach performances. Section 17.4 describes our genetic approach and its results regarding to the random one. Finally, Sect. 17.5 gives some preliminary conclusions about this work and presents prospects.

17.2 Problem Description

The goal of the teacher is to construct collection of differentiated *Tests*, starting from a source database. Then, these *Tests* are structurally different regarding their contents and/or the places of this contents. This research aims at proposing generating algorithms that maximize this differentiation even if the source database size is short.

---

[1]The notion of distinction between two *Tests* is then considered from a structural angle and as a first step, fairness of the assessments is out of the scope of this paper.

To achieve this, it is essential to ensure that each couple of *Tests* of the collection is the most distant as possible.

Therefore, in order to precise the complexity of this problem, we describe the structure of a *Test* and the process of its construction in Sect. 17.2.1. The Sect. 17.2.2 describes how a *Test* collection is built.

17.2.1 Test Structure

The automatic generation process aims to build a collection C of m assessment *Tests*. $Test_i$ consists of a set of p elements called *ItemTests*. An *ItemTest* is constituted of a statement and a list of choices that the student has to select when answering. Through the generation process, each *ItemTest* is constructed starting from an *ItemTestPattern*, and available *ItemTestPatterns* are stored in a *source database* containing n elements.

An *ItemTestPattern* contains a statement (called *StatementItem*) and two sets of answers, the first one contains $nbITP_1$ right answers (ra_i) for $(nbITP_1 \geq 1)$, and the second one contains $nbITP_2$ wrong answers (wa_i) for $(nbITP_2 \geq 1)$. It can be defined as the following:

- *StatementItem*
- $ra_i, i = 1, \ldots, nbITP_1$
- $wa_i, i = 1, \ldots, nbITP_2$

Moreover, in an *ItemTest* we ca find a statement (called *StatementItem*) and a set of answers (ans_i) consisting of a subset of $nbIT_1$ right answers (ra_i) and $nbITP_2$ wrong answers (wa_i). It is defined as the following:

- *StatementItem*
- $ans_i=\{ra_i, i = 1, \ldots, nbIT1(nbIT1 \leq nbITP1)\} \cup \{wa_i, i = 1, \ldots, nbIT2$
  $(nbIT2 \leq nbITP2)\}$

Then, each p *ItemTest_x* is constructed starting form an *ItemTestPattern_y* as the following:

- *ItemTest_x*.*StatementItem* = *ItemTestPattern_y*.*StatementItem*
- *ItemTest_x*.*ans*   = $\{a \in$ *ItemTestPattern_y*.*ra*$\} \cup \{rep_i \in$ *ItemTestPattern_y*. *ra*\$\{a\} \cup$ *ItemTestPattern_y*.*wa*$\}$

To constitute *ItemTest_x*.*ans*, one right answer is pulled at first and the others answers are randomly pulled from the sets of right and wrong answers. This makes it possible to have at least one right answer among the pulled choices.

The construction of each *Test* collection is done by following three conditions presented bellow:

- $nbIT_1 + nbIT_2 \leq nbITP_1 + nbITP_2$ (less or equal answers in the *ItemTest* than in the source *ItemTestPattern*)
- $nbIT_1 + nbIT_2 = x$, x is fixed by the teacher when deciding the number of choices per question required in the *Test* collection.

- $p \leq m$ (less or equal elements in each *Test* than in the source database).

Thus, each *ItemTest* contains a set of x answers, extracting from the associated source *ItemTestPattern* answers, including at least a positive one. The following paragraph presents more precisely the collection construction process based on random approach.

17.2.2 Random Construction of a Test Collection

To generate each *Test* collection the teacher ensures the configuration of the generation by fixing the *Tests* collection size, the number of questions per test and the number of choices per question. With a source database of n *ItemTestPatterns*, the generator can automatically generate a collection of m differentiated *Tests*. It proceeds to a random sampling of p *ItemTestPatterns* without replacement. The initial drawing order of constituents is the order of *ItemTests* sequence appearing on the *Tests*.

For each *ItemTestPattern$_i$*, the generator selects a set of answers following a similar process. First, an element *ans* is randomly selected without replacement from ra_{ITP_i}. This first random sampling guarantees that at least one right answer is associated to the *ItemTest$_i$*. At the end of this first step, a second random sampling without replacement of $x - 1$ elements is computed in the set $\{wa_{ITP_i} \cup ra_{ITP_i} \setminus ans\}$. This corresponds to the completion of the set of multiple choices linked to *ItemTestPattern$_i$*. Finally, the generator shuffles the set of multiple choices thanks to the Fisher–Yates algorithm popularized by Knuth [7].

Its from this method of generation, the issue: *[How can you guarantee that each exam script is unique?]* is raised. This is conditional upon the minimum requirements mentioned above (i.e. $p \leq n$). Thus, from successive random samples of p *ItemTests* among the n available *ItemTestPatterns*, the size of the ordered set of *ItemTests* that you will be able to generate corresponds to a *p-permutation* of n (denoted $_nP_p$) and equal to $\frac{n!}{(n-p)!}$.

For each of these *ItemTests$_i$*, it exists $|ra_{IT_i}|$ possibilities of drawing *ans*. Thus, the number of ordered sets of multiple choices, excluding *ans*, is then equal to a permutation of $x - 1$ elements among $|ra_{IT_i} \cup wa_{IT_i}| - 1$ and so $_{|ra_{IT_i} \cup wa_{IT_i}|-1}P_{x-1}$. Finally, following the sequence described previously, this means our generator can generate a number of distinct *Test* equals to

$$_nP_p \cdot |ra_{IT_i}| \cdot {}_{|ra_{IT_i} \cup wa_{IT_i}|-1}P_{x-1} \quad (1 \leq i \leq p) \tag{17.1}$$

It is interesting to notice that for a special case, which consists in exactly p *ItemTests* where exactly x possible choices are associated each and among which exactly one is right (which is the minimum requirements), the number of distinct exam scripts represented in the above equation is then simplified in

$$_pP_p \cdot 1 \cdot {}_{x-1}P_{x-1} \equiv p! \cdot (x-1)! \tag{17.2}$$

The next section presents how this process is experimented and its differentiation performance is measured.

17.3 Toward Genetic Generation

Starting from the generation process presented in previous section, we aim at measuring the differentiation in *Test* collections. We defined a metric that calculates the distance between two *Tests*. Then, Sect. 17.3.1 presents this metric, called *DTest()*, and Sect. 17.3.2 presents results about random generation differentiation performances.

17.3.1 The DTest Metric

A *Test* can be considered as a string if we associate a unique identifier to each *ItemTest* which constitutes it. Since the goal is to be able to estimate the distance between two *Tests*, related works on structural distances between strings have been studied. Hamming [8, 9] measures the distance between two strings that supposed to have the same length. It considers the number of positions where the strings are different. By example, considering *String1* = "*aklm*", *String2* = "*amkl*" and *String3* = "*abcd*", the distance $d(String1, String2) = d(String1, String3) = 3$. In a pedagogical context (if the strings is considered as *Tests* assessment and each character is an *ItemTest*), $d(String1, String2)$ should be less than $d(String1, String3)$ since in the first case *StatementItems* are the same but in the second case *StatementItem* are strongly disparate. On another hand, Levenshtein estimates the distance between two strings by calculating the number of insertions, deletions and substitutions that should be achieved in one of them to make it identical to the other [10, 11]. So, Levenshtein allows the measure of disparity and substitution between them. In our case, Levenshtein is more suitable than Hamming to measure the differentiation between two *Tests* because it offers the possibility to weithing separately each operation to make a string becomes identical to the other one.

Thus, DTEST is a normalized metric (scored within the range [0, 1]) that estimates the structural distance between a couple of *Tests* (t_i and t_j). From this range, 0 corresponds to $t_i \equiv t_j$, meaning each *StatmentItem* from t_i belongs to t_j at the same rank and is associated to the same set of answers with the same order. On the other hand, 1 is the maximal distance and corresponds to $t_i \neq t_j$ meaning each *StatmentItem* from t_i does not belong to t_j. Else, others values represent a measure of 2 amounts applied to the sets of *StatmentItems* and answers respectively *Disparity* and *Permutation*. More precisely, *Disparity* (17.3) estimates the amount of distinct elements from two sets, by analogy with the Levenshtein distance. *Disparity* measures the amount of insertions and deletions from the two sets, whereas *Permutation* (17.4) expresses, within the framework of ordered sets, the mean rank difference between identical elements that represent the substitution for the Levenshtein distance. a and b are ordered sets and $\mathscr{I}_i^j$ denotes the rank of the element i in the set j. Then, #a and #b represent respectively the cardinal of a and b.

$$Disparity(a, b) = \frac{max(\#a, \#b) - \#(a \cap b)}{max(\#a, \#b)} \qquad (17.3)$$

$$Permutation(a, b) = \frac{\sum_{x \in (a \cap b)} | \mathscr{I}_x^a - \mathscr{I}_x^b |}{2 \left\lfloor \frac{max(\#a, \#b)}{2} \right\rfloor \left\lceil \frac{max(\#a, \#b)}{2} \right\rceil} \qquad (17.4)$$

Figure 17.1 presents the scale of our final metric which consists in combining the 4 following scores:

(i) $qd(t_1, t_2) = Disparity(EnonceItem_{t_1}, EnonceItem_{t_2})$: $t1$ and $t2$ have almost disparate *ItemTests*. *DTest(t1, t2)* is the range of high disparity in the *ItemTests* and greater than 0.75

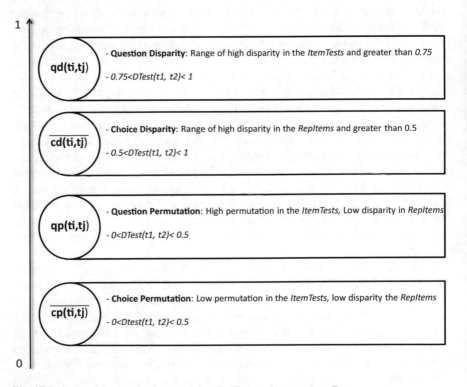

Fig. 17.1 Scale of the metric characterizing the distance between two *Tests*

(ii) $\overline{cd(t_1, t_2)} = \forall_{r\in(EnonceItem_{t_1} \cap EnonceItem_{t_2})} Disparity(RepItem_{t_1}^r, RepItem_{t_1}^r)$:
t1 and *t2* have almost the same *ItemTests* where almost all *RepItems* are disparate. *DTest(t1, t2)* is the range of high disparity in the *RepItems* and greater than 0.5.

(iii) $qp(t_1, t_2) = Permutation(EnonceItem_{t_1}, EnonceItem_{t_2})$: Then, *t1* and *t2* have almost identical *ItemTests*, but in a very different order with slightly different *RepItems*. *DTest(t1, t2)* is the range of high permutation in the *ItemTests* and Low disparity in *RepItems* and $0 < DTest(t1, t2) < 0.5$

(iv) $cp(t_1, t_2 = \forall_{r\in(EnonceItem_{t_1} \cap EnonceItem_{t_2})} Permutation(RepItem_{t_1}^r, RepItem_{t_1}^r))$: Then, *t1* and *t2* have almost the same *ItemTests* in a slightly different order with slightly different *RepItems*. So *DTest (t1, t2)* is in a range of low permutation in the *ItemTests* and low disparity the *RepItems* and $0 < DTest(t1, t2) < 0.5$

Scores *(cp, qp, cd and qd)* are weighted respectively from the lowest *(cp)* to the strongest *(qd)* and then combined. The result is normalized to be fit between 0 and 1. Thanks to DTEST, the differentiation performances through a random generation process have been measured. Results are presented in next sub-section.

17.3.2 Differentiation in Random Generation

To generate randomly a *Test* collection, the process presented in Sect. 17.2.2 is followed. More specifically, below is presented the related algorithm 20.

Algorithm 20: randomGenerateTest()

Require:
  db : {source database}
  p : {number of ItemTests per Test}
  x : {number of answers per ItemTest}
  m : {number of Tests}
  c : {a collection}
Ensure:
  while $\#c \le m$ **do**
    repeat
      {——— Pulling of statement item from the source database ———}
      ItemTest.StatementItem ← *randomPullStatementItem(db)*
      {——— pulling of set of right and wong answers ———}
      ItemTest.ans ← *randomPullAnswer(db, StatementItem, x)*
      test ← *test* ∪ {*StatementItem + Answers*}
    until $\#test = p$
    $c \leftarrow c \cup \{test\}$
  end while
  return *c*

The performances of the random generation regarding to the differentiation constraint is measured by generating various collection of *Tests*. For each couple of *Tests* of a collection, the average distance, the minimal distance, the maximal distance, and the standard deviation are calculated. We considered following parameters:

- m: size of *Test* collections, ($m \in \{2, 5, 10, 20, 30, 40, 50, 60, 70, 80, 90, 100, 150, 200, 250, 300\}$);
- p: quantity of *ItemTests* per *Tests*, ($p \in \{3, 5, 10, 20, 30, 40, 50, 60, 70, 80, 90, 100, 150, 200, 250\}$);
- n: size of source databases of *ItemTestPatterns*, ($n \in \{10, 20, 30, 40, 50, 60, 70, 80, 90, 100, 150, 200, 250\}$);
- x: quantity of *Answers* per *ItemTest* fixed at 4 for the experiment.

The results demonstrate that the average distance between *Tests* in a same collection strictly depends on the ratio of *ItemTestPatterns* in source database by *ItemTests* in *Tests*.

$$Ratio = \frac{\#ItemTestPattern \in Source\ Database}{\#ItemTest \in Test_i}. \tag{17.5}$$

Even if the databases of *ItemTestPatterns* has different sizes, the average distance stays the same for a same ratio. For example, generating *Tests* with 10 *ItemTests* chosen among a set of 20 *ItemTestPatterns* provides the same average distance than generating *Tests* with 30 *ItemTests* chosen among a set of 60 *ItemTestPatterns*. When the ratio is greater than 3.00, the random generation provides an average distance greater than 0.8 between couples of generated *Tests*, which means there is a high rate of disparity in the *ItemTests* of the collection. But, when the ratio is less or equal to 2 it provides an average distance less than 0.7, which means almost the same *ItemTests* appears in the *Tests* but in different places (high rate of permutation and poor rate of disparity in the *ItemTests*).

The genetic algorithm makes it possible to obtain a close solution of a problem when one is not able to find the exact solution. This approach can be relevant to the problem of generation of collections of differentiated *Tests* because there is no way to generate a collection of *Tests* with a total differentiation between all couple of *Tests*. In order to maximize the differentiation, we decided to explore the genetic pathway which will enable us to obtain collections of *Tests* increasingly differentiated from population to population. More specifically, we tackle the possibility to generate *Test* collections with the greatest differentiation possible in the ratios less than 3. This approach is developed in the next section.

17.4   Genetic Algorithm

Genetic Algorithms are type of algorithm used to obtain approximate solutions when the exact method to bring a solution in a reasonable time is unknown [12, 13]. We are not the first to use GA to resolve problems relative to higher education context. For example, [14] explains the application of data mining in higher education through algorithms as genetic algorithm, to predict the paths of leaners; [15] implements an hybrid algorithm constituted by a genetic algorithm and some local methods to resolve the problem of exam scheduling in universities. However, we don't find any trace of use of genetic algorithms in the automatic generation of *Tests*.

17.4.1   The Algorithm Features

The process starts from a first population which it ameliorates after several iterations by considering a fitness function and by applying operations such as selection by crossovers and mutations [16], until obtaining a population close to that expected. As the random generation, the teacher sets the generation by specifying all necessary parameters as a number of iterations, a timeout, a size of the population (where individuals are *Test* collections), and the percentages of deletion and mutation. Then, a first population of *n* *Test* collections is randomly generated and the average DTEST of each collection is calculated. So, a collection is an individual of the population. The population is ordered from the weakest collection to the strongest collection related to their average DTESTs.

The deletion of a percentage of weakly individuals is done. As Fig. 17.2 illustrates it, we consider a population containing *n* individuals and the deletion percentage is empirically fixed at 8%. Then, sample crossovers are done between couple of individuals by exchanging each other half of their constituents (Fig. 17.3) to create a new individual. Crossovers are done until the population becomes complete after deletion.

At the end of the process, a mutation is done one time on three. To do so, we delete randomly the predefined percentage of mutation from the population and we generate randomly this same percentage to complete the population. The whole process is repeated until the given time out or the given quantity of iterations is reached. The algorithm 21 illustrates it and shows that the solution expected in our

Percentage of weakest Tests to delete

C1	C1	...	$C_{(8\% \, of \, n)}$	C81	C82	...	C(n-2)	C(n-1)	Cn

Classified by their average DTests →

Fig. 17.2 Example of deletion operator (8%)

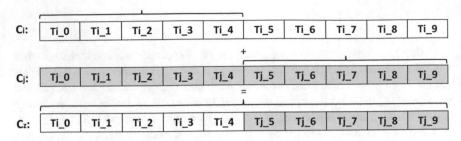

Fig. 17.3 Example of crossover operator

case is a collection of *Tests* with great differentiation (better than random) between each pair of *Tests*. *Test* collections represent individuals of our population and the metric *DTest* is used as the fitness function and applied on each collection in order to evaluate its performance in term of differentiation. However, the execution time can go up to 2 hours more than the random algorithm when the ratio is less than 3 and the expected differentiation threshold is quite high. These results have been presented in META-2016 [17].

17.4.2   Performance Tests

The GA has been experimented with a population of 1000 individuals where each of them is initially a collection of *Tests* randomly sampled among those belonging to a database of *ItemTestPatterns*. The percentage of deletion and mutation was empirically set to 8 and 3% respectively. We consider 1000 generations or a timeout at 3600 s. For parameters m, n, p and x, they are fixed in order to take into account only small ratios (≤ 3) because random generation has already demonstrated a sufficient differentiation performance in higher ratios. So, parameters m, n, p and x are fixed as follow:

- m *Test* collections, ($m \in \{2, 3, 5, 7, 10, 15, 30\}$);
- p quantity of *ItemTests* per *Tests*, ($p \in \{10, 20, 30, 40\}$);
- n size of source databases of *ItemTestPatterns*, ($n = \{40\}$);
- x quantity of *Answers* per *ItemTest* fixed at 4.

Some details on the experimental contribution of genetic generation compared to the random generation are given in Fig. 17.4. It shows comparative average distances between them according to DTEST. For instance this guarantees to the teacher that if he has at disposal n *ItemTestPatterns* in the database to build *Tests* of p *ItemTests* (i.e. ratio $= 1.00$), DTEST enhances the classical random shuffle generation, according to the global metric DTEST, by 10%.

Algorithm 21: Genetic generation

Require:
  nbG : {number of iteration}
  szP : {population size}
  percentOfDeletion : {% of deletion}
  percentOfMutation : {% of mutation}
  mutationVar{} : {probability to mutate}
  timeOut : {maximal process duration}
  db : {source database}
  p : {number of ItemTests}
  x : {number of Answers}
  m : {number of Tests}
  P{} : {population of collection}
  c{} : {a collection}
  test : {a test}
  i : {counter}
Ensure:
  {————— Generation of the first population of Test collections —————}
  repeat
    $c \leftarrow randomGenerateTest(db, m, p, x)$
    $P \leftarrow P \cup \{c\}$
  until $\#p = szP$
  $i \leftarrow 0$
  while $i < nbG$ **or** $!timeOut$ **do**
    for all $c \in P$ **do**
      averageDtest(c) {The FITNESS function calculates the average distance between each
      couple of Tests}
    end for
    $P.sort()$
    {—— Deletion of a predefined % and crossover to recomplete the population ——}
    deletion$(P, percentOfDeletion)$
    repeat
      $P \leftarrow P \cup \{crossover(c1, c2)\}$
      $P.shuffle()$
    until $\#p = szP$
    {————— Mutation of a predefined % one time on three —————}
    mutationVar $\leftarrow range(0, 1000)$
    mutationVar.shuffle()
    if *mutationVar*$[0] > 300$ **then**
      deletion$(P, percentOfMutation)$
      repeat
        $c \leftarrow randomGenerateTest(db, m, p, x)$
        $P \leftarrow P \cup \{c\}$
      until $\#P = szP$
    end if
    $i \leftarrow i + 1$
  end while
  $P.sort()$
  $c \leftarrow max(P)$ {the must differentiated collection}
  return c

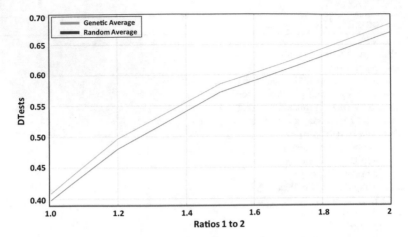

Fig. 17.4 Average distance on the random and the genetic generation according to ratios

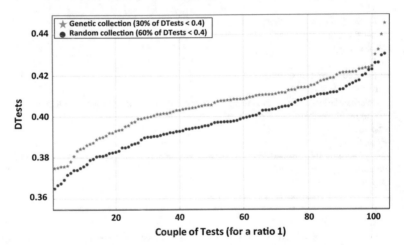

Fig. 17.5 Comparative distance for couple of Tests between the random and the genetic algorithm

Moreover, Fig. 17.5 presents the distribution of DTEST metric for each couple of *Tests* belonging to a given assessment at a ratio 1.00. According to our scale (see Fig. 17.1), and within a practical angle this means that DTEST can simulate a high disparity in *EnonceItems* while the ratio structurally cannot allow it.

The following section concludes this paper and presents prospects of these research works.

17.5 Conclusion

This paper has presented an approach that can assist the teacher in automatically generating collections of *Tests* through MCQ. As a contribution, we propose a metric partially inspired from Levenshtein distance [10, 11] that characterizes the difference between two *Tests* of a given collection. Since this metric can be used as a fitness function, we adapted a genetic approach to automatically generate collections that maximizes the distance between each couple of *Tests* of a given collection and that even if the teacher has few patterns in its source database. The approach of Genetic Algorithm with DTEST, can be compared to a random selection and allows some substantial improvements on a difficult problem. Its improvements are more significant to generate small collections. This is why, we are currently experimenting other approaches with the aim to optimize distribution of differentiated *Tests* in classrooms by using techniques of graph coloring.

Others aspects of this work consist in taking into account the difficulty level of a *Test* as a new constraint for our generation algorithms. The final goal is to be able to maximize the differentiation by keeping the equity between all *Tests* in a collection. To perform it, a method in order to calculate the level of difficulty of a *Test* has to be developed, starting from the difficulty levels of *ItemTests*, themselves starting from statistic and semantic information available in the database of the corrected *Tests* and in the source database of *ItemTestPatterns*. At more long term, others types of *ItemTestPatterns* than MCQ will be integrated and combined in our algorithms.

References

1. Kumar, V. (1992). Algorithms for constraint-satisfaction problems: A survey. *AI Magazine*, *13*(1), 32.
2. Mackworth, A. K. (1992). Constraint satisfaction problems. *Encyclopedia of AI*, 285–293.
3. Garrison, C., & Ehringhaus, M. (2007). Formative and summative assessments in the classroom.
4. Harlen, W. (2006). On the relationship between assessment for formative and summative purposes. *Assessment and Learning*, 103–118.
5. Taras, M. (2005). Assessment-summative and formative-some theoretical reflections. *British Journal of Educational Studies*, *53*(4), 466–478.
6. Harlen, W., & James, M. (1997). Assessment and learning: Differences and relationships between formative and summative assessment. *Assessment in Education*, *4*(3), 365–379.
7. Knuth, D. (1998). *The art of computer programming, volume 2: Seminumerical algorithms* (3rd ed.). Boston: Addison-Wesley Publishing Company.
8. Hamming, R. W. (1950). Error detecting and error correcting codes. *Bell Labs Technical Journal*, *29*(2), 147–160.
9. Norouzi, M., Fleet, D. J., & Salakhutdinov. R. R. (2012). Hamming distance metric learning. In *Advances in Neural Information Processing Systems* (pp. 1061–1069).
10. Levenshtein, V. I. (1966, February). *Binary codes capable of correcting deletions, insertions, and reversals, 10*.
11. Yujian, L., & Bo, L. (2007). A normalized levenshtein distance metric. *IEEE Transactions on Pattern Analysis and Machine Intelligence*, *29*(6), 1091–1095.
12. Melanie, M. (1988). *An introduction to genetic algorithms*. London: MIT press.

13. Mitsuo, G., & Runwei, C. (2000). *Genetic algorithms and engineering optimization*. New York: Wiley.
14. Luan, J. (2002). Data mining and its applications in higher education. *New Directions for Institutional Research, 2002*(113), 17–36.
15. Burke, E., Newall, J., & Weare, R. (1996). A memetic algorithm for university exam timetabling. *Practice and Theory of Automated Timetabling*, 241–250.
16. Srinivas, M., & Patnaik, L. M. (1994). Adaptive probabilities of crossover and mutation in genetic algorithms. *IEEE Transactions on Systems, Man, and Cybernetics, 24*(4), 656–667.
17. Ciguene, R., Joiron, C., & Dequen, G. (2016). Automatically generating assessment tests within higher education context thanks to genetic approach. In *International Conference on Metaheuristics and Nature Inspired Computing, Morocco* (Vol. 3, pp. 168–170). Marocco: Marrakech.

Chapter 18
Ant Colony Optimization for Optimal Low-Pass Filter Sizing

Loubna Kritele, Bachir Benhala and Izeddine Zorkani

Abstract In analog filter design, discrete components values such as resistors (R) and capacitors (C) are selected from the series following constant values chosen. Exhaustive search on all possible combinations for an optimized design is not feasible. In this chapter, we present an application of the Ant Colony Optimization (ACO) technique for optimal filter design considering different manufacturing series for both the resistors and capacitors. Three variants of the Ant Colony Optimization are applied, namely, the AS (Ant System), the MMAS (Min-Max AS) and the ACS (Ant Colony System), for the optimal sizing of the Low-Pass Butterworth filter. Different optimal designs of the filter are provided depending on the preference between two conflicting factors, namely the cutoff frequency and selectivity factor. SPICE simulations are used to validate the obtained results/performances. A comparison with published works is also highlighted.

Keywords Optimization · Metaheuristic · Ant Colony Optimization · Low-Pass Butterworth Filter

18.1 Introduction

The optimal sizing of analog circuits is very challengin task, due to the number of variables involved, to the number of required objectives to be optimized and to the

L. Kritele (✉) · B. Benhala
University of Sidi Mohamed Ben Abdellah, Faculty of Sciences Dhar el Mahraz,
Fez, Morocco
e-mail: loubnakritele@gmail.com

B. Benhala
University of Moulay Ismail, Faculty of Sciences, Meknes, Morocco
e-mail: b.benhala@fs-umi.ac.ma

© Springer International Publishing AG, part of Springer Nature 2019
E.-G. Talbi and A. Nakib (eds.), *Bioinspired Heuristics for Optimization*,
Studies in Computational Intelligence 774,
https://doi.org/10.1007/978-3-319-95104-1_18

constraint functions restrictions. The aim is to automate this task in order to make the design faster and more accurate. Recently, the used of the metaheuristics have proved a capacity to treat these problem efficiently, such as Tabu Search (TS) [13], Genetic Algorithms (GA) [1], Local search (LS) [2], Simulated Annealing (SA) [11], Ant Colony Optimization (ACO) [4–6] and Particle Swarm Optimization (PSO) [12].

Active analog filters are constituted of amplifying elements, resistors and capacitors; therefore, the filter design depends strongly on passive component values. However, the manufacturing constraints makes difficult an optimal selection of passive compenent values.

Indeed, the search on all possible combinations in preferred values for capacitors and resistors is an exhaustive process, because discrete components are produced according to a series of values constants such as the series: E12, E24, E48, E96 or E192.

Consequently, an intelligent search method requires short computation time with high accuracy, must be used. Ant Colony Optimization is one such technique which is increasingly used in the domains of optimization and has been applied successfully in the field of analog circuits [4, 5, 16, 19].

In this work, we propose to apply the AS, the MMAS and the ACS which are three variants of the ACO technique, for the optimal sizing of the Low-Pass Butterworth Filter.

This chapter consists of six parts. The second part presents an overview of the ant colony algorithm. The third part illustrates the application example, the fourth part deals with the simulations and comparison results, and the fifth part is devoted to the presentation of design varieties. Conclusions are made in the sixth part.

18.2 Ant Colony Optimization: ACO Technique: An over View

ACO has been inspired by the foraging behavior of real ant colonies [9]. Figure 18.1 shows an illustration of the ability of ants to find the shortest path between food and their nest [7, 8]. It is illustrated through the example of the appearance of an obstacle on their path. Every ant initially chooses path randomly to move and leaves a chemical substance, called pheromone in the path. The quantity of pheromone deposited will guide other ants to the food source. The indirect communication between the ants via the pheromone trail allows them to find shortest paths from their nest to the food source.

18.2.1 Ant System

The first variant of the ACO is "Ant System" (AS) which is used to solve combinatorial optimization problems such as the traveling salesman problem (TSP) [15],

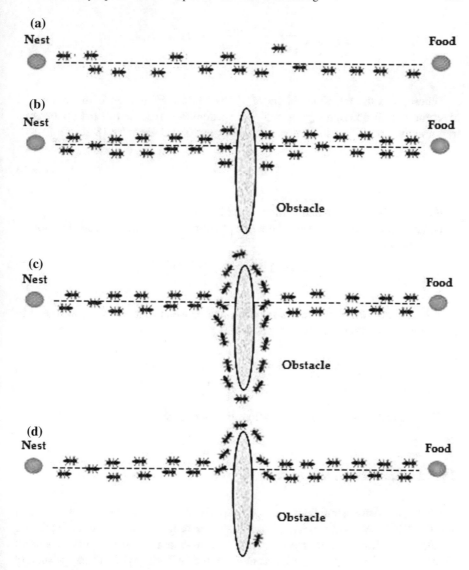

Fig. 18.1 Self-adaptive behavior of a real ant colony

vehicle routing problem [24]... For solving such problems, ants randomly select the vertex to be visited. When ant k is in vertex i, the probability of going to vertex j is given by expression 18.1:

$$P_{ij}^k = \begin{cases} \dfrac{(\tau_{ij})^\alpha * (\eta_{ij})^\beta}{\sum_{l \in J_i^k}(\tau_{ij})^\alpha * (\eta_{ij})^\beta} & if \ j \ \in \ J_i^k \\ \\ 0 & if \ j \ \notin \ J_i^k \end{cases} \qquad (18.1)$$

Where J_k^i is the set of neighbors of vertex i of the kth ant, τ_{ij} is the amount of pheromone trail on edge (i,j), α and β are weightings that control the pheromone trail and the visibility value, i.e. η_{ij}, which expression is given by 18.2:

$$\eta_{ij} = \frac{1}{d_{ij}} \qquad (18.2)$$

d_{ij} is the distance between vertices i and j.

Once all ants have completed a tour, the pheromone trails are updated. The update follows this rule 18.3:

$$\tau_{ij} = (1 - \rho)\tau_{ij} + \sum_{k=1}^{m} \Delta\tau_{ij}^k \qquad (18.3)$$

Where ρ is the evaporation rate, m is the number of ants, and $\Delta\tau_{ij}^k(t)$ is the quantity of pheromone laid on edge (i,j) by ant k 18.4 :

$$\Delta_{ij}^k = \begin{cases} Q/L_k & if \ ant \ k \ used \ edge \ (i, j) \ in \ its \ tour \\ 0 & otherwise \end{cases} \qquad (18.4)$$

Q is a constant and L_k is the length of the tour constructed by ant k.

18.2.2 Max-Min Ant System

The Max-Min Ant System is another variant of ACO, which was developed by Stutzle and Hoos [21]. Max-Min ant system has always been to achieve the optimal path searching by allowing only the best solution to increase the information and use a simple mechanism to limit the pheromone, which effectively avoid the premature stagnation [10, 17].

MMAS which based on the ant system does the following areas of improvement:

1. During the operation of the algorithm, only a single ant was allowed to increase the pheromone. The ant may be the one which found the best solution in the current iteration or the one which found the best solution from the beginning of the trial. Consequently the modified pheromone trail update rule is given by 18.5:

$$\tau_{ij} = (1 - \rho)\tau_{ij} + \Delta\tau_{ij}^{best} \qquad (18.5)$$

2. In order to avoid stagnation of the search, the range of the pheromone trail is limited to an interval $[\tau_{min}, \tau_{max}]$.
3. The pheromone is initialized to τ_{max} in each edge.

18.2.3   Ant Colony System

The ACS algorithm represents an improvement with respect to the AS [7, 14].

The ACS incorporates three main differences with respect to the AS algorithm:

1. ACS introduced a transition rule depending on a parameter q_0, which provides a direct way to balance between diversification and intensification. In the ACS algorithm, an ant positioned on node i chooses the city j to move to by applying the rule given by 18.6:

$$J = \begin{cases} argmax_{u \in J_i^k} = [(\tau_{ui}(t)).(\mu_{ij})^\beta] & if\ q \leq q_0 \\ (1) & if\ q > q_0 \end{cases} \tag{18.6}$$

Where q is a random number uniformly distributed in $[0, 1]$, q_0 is a parameter $0 \leq q_0 \leq 1$.
2. The global updating rule is applied only to edges which belong to the best ant tour. The pheromone level is updated as follows 18.7:

$$\tau_{ij} = (1 - \rho)\tau_{ij} + \rho\Delta\tau_{ij} \tag{18.7}$$

$$\begin{cases} \dfrac{1}{L} & if\ (i, j) \in best - global - tour \\ 0 & otherwise \end{cases} \tag{18.8}$$

3. While ants construct a solution a local pheromone updating rule is applied :

$$\tau_{ij} = (1 - \rho)\tau_{ij} + \rho\tau_{int} \tag{18.9}$$

18.3   Application to the Optimal Design of Low Pass Butterworth Filter

The three proposed variants of ACO algorithm were used to optimize the analog circuit, namely low-pass Butterworth filter.

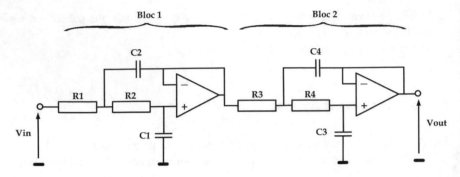

Fig. 18.2 Butterworth fourth order low-pass filter

Analog active Filters are important building blocks in signal processing circuits. They are widely used in the separation and demodulation of signals, frequency selection decoding, and estimation of a signal from noise [18].

Analog active filters are characterized by four basic properties: the filter type (low-pass, high-pass, bandpass, and others), the passband gain (generally all the filters have unity gain in the passband), the cutoff frequency (the point where the output level has fallen by 3 dB from the maximum level within the passband), and the quality factor Q (determines the sharpness of the amplitude response curve) [22].

Butterworth filters are termed maximally flat-espace filters, optimized for gain flatness in the passband. The transient response of a Butterworth filter to a pulse input shows moderate overshoot and ringing [20].

The fourth order low pass Butterworth filter formed by two operational amplifiers, four resistors and four capacitors. The schematic of this filter is given in Fig. 18.2.

The transfer function of this filter can be obtained as follows:

$$H(s) = \frac{\omega_{c1}^2}{s^2 + \frac{\omega_{c1}}{Q_1}s + \omega_{c1}^2} \times \frac{\omega_{c2}^2}{s^2 + \frac{\omega_{c2}}{Q_2}s + \omega_{c2}^2} \qquad (18.10)$$

The cutoff frequency (ω_{c1}, ω_{c2}) and the selectivity factor (Q_1, Q_2) of filter, which depend only on the values of the passives components, are given as follows [3]:

$$\omega_{c1} = \frac{1}{\sqrt{R_1 R_2 C_1 C_2}} \qquad \omega_{c2} = \frac{1}{\sqrt{R_3 R_4 C_3 C_4}} \qquad (18.11)$$

$$Q_1 = \frac{\sqrt{R_1 R_2 C_1 C_2}}{R_1 C_1 + R_2 C_2} \qquad Q_2 = \frac{\sqrt{R_3 R_4 C_3 C_4}}{R_3 C_3 + R_4 C_4} \qquad (18.12)$$

For comparison reasons, the specification chosen here is [22]:

- $\omega_{c1} = \omega_{c2} = 10000 rad/s (1591.55 Hz)$
- $Q_1 = 1/0.7654$
- $Q_2 = 1/1.8478$

In order to generate ω_{c1}, ω_{c2}, Q_1 and Q_2 approaching the specified values; the values of the resistors and capacitors to choose should be able to satisfy desired constraints. For this, we define the Total Error (TE) [22] which expresses the offset values, of the cut-off frequency and the selectivity factor, compared to the desired values, by:

$$Total - error = \delta_1 \Delta\omega + \delta_2 \Delta Q \qquad (18.13)$$

Where:

δ_1 and δ_2 are the weights for each objective, $(\delta_1 + \delta_2 = 1)$

$$\Delta\omega = \frac{|\omega_{c1} - \omega_c| + |\omega_{c2} - \omega_c|}{\omega_c} \quad and \quad \Delta Q = \left| Q_1 - \frac{1}{0.7654} \right| + \left| Q_2 - \frac{1}{1.8478} \right|$$
$$(18.14)$$

The objective function considered is the Total Error (15) which is calculated for equal weights ($\delta_1 = \delta_2 = 0.5$). The decision variables are the resistors and capacitors forming the circuit. Each component must have a value of the standard series (E12, E24, E48, E96, E192). The resistors have values in the range of 10^3 to $10^6 \Omega$. Similarly, each capacitor must have a value in the range of 10^{-9} to $10^{-6} F$.

The aim is to obtain the exact values of design parameters ($R_{1...4}$, $C_{1...4}$) which equate the Total-error to a very close value to 0.

$$Total - error = \delta_1 \cdot \frac{\frac{1}{\sqrt{R_1 R_2 C_1 C_2}} - \omega_c + \frac{1}{\sqrt{R_3 R_4 C_3 C_4}} - \omega_c}{\omega_c} +$$
$$\delta_2 \cdot \left(\left| \frac{\sqrt{R_1 R_2 C_1 C_2}}{R_1 C_1 + R_2 C_2} - \frac{1}{0.7654} \right| + \left| \frac{\sqrt{R_3 R_4 C_3 C_4}}{R_3 C_3 + R_4 C_4} - \frac{1}{1.8478} \right| \right)$$
$$(18.15)$$

18.4 Comparison Results

In this section we applied ACO algorithms to perform optimization of a low-pass Butterworth fourth order filter.

The studied algorithms parameters are given in Table 18.1 with a generation algorithm of 1000. The optimization techniques work on C codes and are able to link SPICE to measure performances.

The optimal values of resistors and capacitors forming the low-pass Butterworth filter and the performance associated with these values for the different series using the three variants of the ACO technique: The MMAS, the AS and the ACS are shown in Tables 18.2, 18.3 and 18.4 respectively.

From the results, we notice that the MMAS achieved a smaller design error.

Table 18.5 presents the ACO results for E192 series, compared to those of other already published metaheurisrtics (GA [23], ABC [23], PSO [22]).

Table 18.1 The ACO algorithms parameters

Evaporation rate (ρ)	0.1
Quantity of deposit pheromone (Q)	0.4
Pheromone factor (α)	1
Heuristics factor (β)	1
τ_{min}	0.5
τ_{max}	1.5
q_0	0.2

Table 18.2 Values of components and related butterworth filter performances using the "MMAS"

	R_1 $(K\Omega)$	R_2 $(K\Omega)$	C_1 (nF)	C_2 (nF)	R_3 $(K\Omega)$	R_4 $(K\Omega)$	C_3 (nF)	C_4 (nF)	$\Delta\omega$	ΔQ	Total error
Linear values	4.900	2.750	10.0	74.2	3.750	1.150	37.7	61.6	0.0008	0.0011	**0.0010**
E12	4.700	2.700	10.0	68.0	3.900	1.200	39.0	56.0	0.0874	0.0841	**0.0857**
E24	5.100	2.700	10.0	75.0	3.900	1.200	39.0	62.0	0.0759	0.0010	**0.0430**
E48	4.870	2.740	10.0	75.0	3.830	1.150	38.3	61.9	0.0218	0.0135	**0.0176**
E96	4.870	2.740	10.0	75.0	3.740	1.150	37.4	61.9	0.0025	0.0125	**0.0075**
E192	4.930	2.740	10.0	74.1	3.740	1.150	37.9	61.9	0.0049	0.0029	**0.0039**

Table 18.3 Values of components and related butterworth filter performances using the "AS"

	R_1 $(K\Omega)$	R_2 $(K\Omega)$	C_1 (nF)	C_2 (nF)	R_3 $(K\Omega)$	R_4 $(K\Omega)$	C_3 (nF)	C_4 (nF)	$\Delta\omega$	ΔQ	Total error
Linear values	4.100	2.750	11.2	79.3	1.050	2.100	58.6	77.5	0.0014	0.0031	**0.0022**
E12	3.900	2.700	12.0	82.0	1.000	2.200	56.0	82.0	0.0227	0.0410	**0.0318**
E24	4.300	2.700	11.0	82.0	1.100	2.200	56.0	75.0	0.0309	0.0269	**0.0289**
E48	4.020	2.740	11.0	78.7	1.050	2.150	59.0	78.7	0.0473	0.0078	**0.0276**
E96	4.120	2.740	11.3	78.7	1.050	2.100	59.0	76.8	0.0024	0.0173	**0.0098**
E192	4.120	2.740	11.1	79.6	1.050	2.100	58.3	77.7	0.0019	0.0081	**0.0050**

A comparison between GA, PSO, ABC-based design and the ACO techniques shows that the ACO algorithms achieved a smaller design error.

One can also notice that the MMAS can find the smaller Total Error value.

Table 18.6 presents the computation time of the GA, ABC, PSO and the three variants of the ant colony optimization for the optimal design of Butterworth Filter.

Table 18.4 Values of components and related butterworth filter performances using the"ACS"

	R_1 $(K\Omega)$	R_2 $(K\Omega)$	C_1 (nF)	C_2 (nF)	R_3 $(K\Omega)$	R_4 $(K\Omega)$	C_3 (nF)	C_4 (nF)	$\Delta\omega$	ΔQ	Total error
Linear values	4.450	4.250	8.80	60.1	2.250	1.300	52.1	65.7	0.0007	0.0004	**0.0005**
E12	4.700	3.900	8.20	56.0	2.200	1.200	56.0	68.0	0.0926	0.0201	**0.0564**
E24	4.300	4.300	9.10	62.0	2.200	1.300	51.0	68.0	0.0250	0.0181	**0.0215**
E48	4.420	4.220	8.66	59.0	2.260	1.270	51.1	64.9	0.0493	0.0021	**0.0257**
E96	4.420	4.220	8.87	60.4	2.260	1.300	52.3	64.9	0.0017	0.0069	**0.0043**
E192	4.420	4.270	8.76	60.4	2.260	1.300	52.3	65.7	0.0054	0.0077	**0.0066**

Table 18.5 Component values and performance of GA, ABC, PSO and ACO techniques for Butterworth filter design

	GA	ABC	PSO	ACO MMAS	ACO AS	ACO ACS
$R_1(K\Omega)$	6.80	4.70	4.58	4.93	4.12	4.42
$R_2(K\Omega)$	6.80	4.70	4.70	2.74	2.74	4.22
$C_1(nF)$	5.60	8.20	8.20	10.0	11.1	8.87
$C_2(nF)$	39.0	56.0	56.0	74.1	79.6	60.4
$R_3(K\Omega)$	39.0	1.00	1.10	3.74	1.05	2.26
$R_4(K\Omega)$	1.00	39.0	1.00	1.15	2.10	1.30
$C_3(nF)$	4.70	4.70	87.6	37.9	58.3	52.3
$C_4(nF)$	56.0	56.0	102.2	61.9	77.7	64.9
$\Delta\omega$	0.0179	0.0201	0.0135	0.0049	0.0019	0.0017
ΔQ	0.0153	0.0024	0.0018	0.0029	0.0081	0.0069
TE	0.0166	0.0113	0.0076	0.0039	0.0050	0.0043

Table 18.6 The execution time of GA, ABC, PSO and ACO techniques for Butterworth filter design

	GA	ABC	PSO	ACO MMAS	ACO AS	ACO ACS
Execution time (s)	246.0	0.7	222.0	98.61	84.0	127.1

In terms of execution time, ABC algorithm presents the shorter computation time, followed by the AS algorithm.

In order to check the validity of the results, the following figures show the PSPICE simulation in the filter gain for the optimal values of the E192 series. The practical cuts off frequency using the three variants of the ACO technique: The MMAS, the AS and the ACS are equal to 1.586, 1.596 and 1.589 KHz respectively (Figs. 18.3, 18.4 and 18.5).

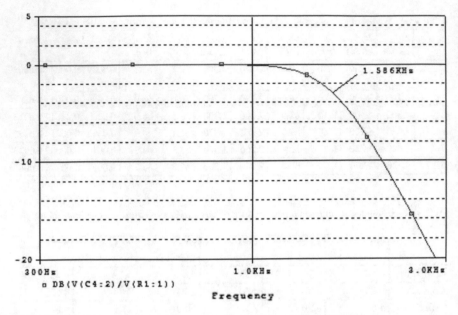

Fig. 18.3 Frequency responses of low-pass butterwoth filter using the MMAS

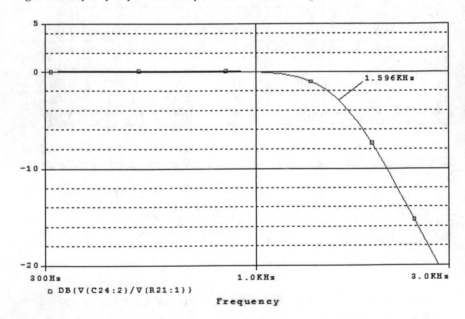

Fig. 18.4 Frequency responses of low-pass butterwoth filter using the AS

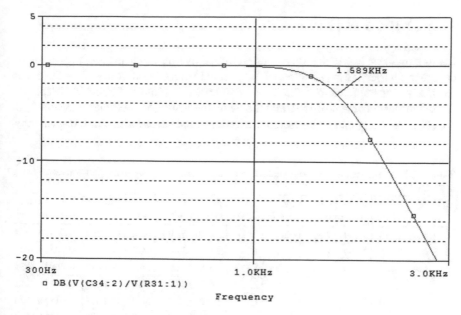

Fig. 18.5 Frequency responses of low-pass butterwoth filter using the ACS

Table 18.7 shows the comparison between the theoretical values and those practices for the error on the cut-off frequency for different series.

From the results presented in Table 18.7, we notice that simulation results are in
good agreement with those obtained using the ACO algorithm. The slight difference
between the two values is mainly due to imperfections of the op-amp which are
considered perfect in the theoretical calculations.

Table 18.7 Comparisons between the theoretical and practices for the error on the cut-off frequency

	MMAS		AS		ACS	
	$\Delta\omega$ theoretical	$\Delta\omega$ practical	$\Delta\omega$ theoretical	$\Delta\omega$ practical	$\Delta\omega$ theoretical	$\Delta\omega$ practical
E12	0.0874	0.1100	0.0227	0.0126	0.0926	0.0214
E24	0.0759	0.1307	0.0309	0.0126	0.0250	0.0100
E48	0.0218	0.0340	0.0473	0.0013	0.0493	0.0327
E96	0.0025	0.0100	0.0024	0.014	0.0017	0.0126
E192	0.0049	0.0013	0.0019	0.0100	0.0054	0.0013

18.5 Variety of Design

In this part, the values of the weights of each objective are varied in order to give for the designer a variety of possibilities of the optimal values in order to satisfy the needs for the various applications that favor one criterion (cut-off frequency/selectivity factor) to the detriment of the other. For this, the values of coefficients δ_1 and δ_2 in the objective function will be varied in ascending and descending order, respectively.

Table 18.8 Linear values of components and related Butterworth filter performances using the "MMAS"

δ_1	δ_2	R_1 $(K\Omega)$	R_2 $(K\Omega)$	C_1 (nF)	C_2 (nF)	R_3 $(K\Omega)$	R_4 $(K\Omega)$	C_3 (nF)	C_4 (nF)	$\Delta\omega$	ΔQ	Total error
0	1	4.570	3.800	9.00	62.0	2.090	2.210	52.0	61.0	0.1897	0.0004	**0.0004**
0.1	0.9	3.380	2.460	13.0	91.0	1.720	3.270	37.0	48.0	0.0088	0.0003	**0.0011**
0.2	0.8	3.100	3.850	11.0	76.0	1.120	3.070	44.0	66.0	0.0019	0.0009	**0.0011**
0.3	0.7	3.240	4.440	10.0	70.0	2.800	1.400	44.0	58.0	0.0037	0.0002	**0.0012**
0.4	0.6	3.650	2.240	13.0	94.0	1.330	3.510	38.0	56.0	0.0037	0.0018	**0.0026**
0.5	0.5	4.900	2.750	10.0	74.2	3.750	1.150	37.7	61.6	0.0008	0.0011	**0.0010**
0.6	0.4	3.570	3.400	11.0	75.0	3.980	3.740	24.0	28.0	0.0008	0.0027	**0.0016**
0.7	0.3	3.470	4.170	10.0	69.0	1.160	2.210	55.0	71.0	0.0013	0.0027	**0.0017**
0.8	0.2	3.880	4.620	9.00	62.0	1.350	2.080	54.0	66.0	0.0005	0.0019	**0.0008**
0.9	0.1	3.770	3.170	11.0	76.0	3.840	1.100	37.0	64.0	0.0006	0.0088	**0.0014**
1	0	3.430	1.780	26.0	63.0	4.310	2.800	36.0	23	0.0004	0.7189	**0.0004**

Table 18.9 values of components selected from E192 and related Butterworth filter performances using the "MMAS"

δ_1	δ_2	R_1 $(K\Omega)$	R_2 $(K\Omega)$	C_1 (nF)	C_2 (nF)	R_3 $(K\Omega)$	R_4 $(K\Omega)$	C_3 (nF)	C_4 (nF)	$\Delta\omega$	ΔQ	Total error
0	1	4.590	3.790	8.98	61.9	2.100	2.210	52.3	61.2	0.1964	0.0007	**0.0007**
0.1	0.9	3.360	2.460	13.0	90.9	1.720	3.280	37.0	48.1	0.0138	0.0007	**0.0020**
0.2	0.8	3.090	3.830	11.0	75.9	1.130	3.050	44.2	65.7	0.0064	0.0009	**0.0020**
0.3	0.7	3.240	4.420	10.0	69.8	2.800	1.400	44.2	58.3	0.0052	0.0015	**0.0026**
0.4	0.6	3.650	2.230	13.0	94.2	1.330	3.520	37.9	56.2	0.0030	0.0025	**0.0027**
0.5	0.5	4.930	2.740	10.0	74.1	3.740	1.150	37.9	61.9	0.0049	0.0029	**0.0039**
0.6	0.4	3.570	3.400	11.0	75.0	3.970	3.740	24.0	28.0	0.0018	0.0027	**0.0021**
0.7	0.3	3.480	4.170	10.0	69.0	1.170	2.210	54.9	70.6	0.0017	0.0032	**0.0022**
0.8	0.2	3.880	4.640	8.98	61.9	1.350	2.080	54.2	65.7	0.0004	0.0043	**0.0012**
0.9	0.1	3.790	3.160	11.0	75.9	3.830	1.100	37.0	64.2	0.0004	0.0087	**0.0013**
1	0	3.440	1.780	26.1	62.6	4.320	2.800	36.1	22.9	0.0002	0.7245	**0.0002**

Table 18.10 Linear values of components and related Butterworth filter performances using the"AS"

δ_1	δ_2	R_1 ($K\Omega$)	R_2 ($K\Omega$)	C_1 (nF)	C_2 (nF)	R_3 ($K\Omega$)	R_4 ($K\Omega$)	C_3 (nF)	C_4 (nF)	$\Delta\omega$	ΔQ	Total error
0	1	3.510	4.310	10.0	69.0	4.550	4.550	35.0	41.0	0.4410	0.00003	**0.00003**
0.1	0.9	3.430	4.200	10.0	69.0	1.690	1.390	60.0	71.0	0.0034	0.0003	**0.0006**
0.2	0.8	3.110	3.850	11.0	76.0	2.560	3.220	32.0	38.0	0.0017	0.0004	**0.0007**
0.3	0.7	4.230	3.430	10	69	1.670	3.460	36	48	0.0013	0.0004	**0.0007**
0.4	0.6	3.580	2.310	13.0	93.0	2.290	3.300	33.0	40.0	0.0013	0.0009	**0.0010**
0.5	0.5	4.100	2.750	11.2	79.3	1.050	2.100	58.6	77.5	0.0014	0.0031	**0.0022**
0.6	0.4	4.570	3.080	10.0	71.0	2.370	4.470	27.0	35.0	0.0009	0.0009	**0.0009**
0.7	0.3	3.840	4.670	9.00	62.0	2.240	4.160	29.0	37.0	0.0004	0.0028	**0.0011**
0.8	0.2	3.090	2.360	14.0	98.0	3.520	2.880	29.0	34.0	0.0005	0.0069	**0.0018**
0.9	0.1	5.000	3.530	9.00	63.0	4.250	4.480	21.0	25.0	0.0006	0.0076	**0.0013**
1	0	3.860	2.290	87.0	13.0	1.200	4.480	20.0	93.0	0.0002	1.4587	**0.0002**

Table 18.11 Values of components selected from E192 and related Butterworth filter performances using the"AS"

δ_1	δ_2	R_1 ($K\Omega$)	R_2 ($K\Omega$)	C_1 (nF)	C_2 (nF)	R_3 ($K\Omega$)	R_4 ($K\Omega$)	C_3 (nF)	C_4 (nF)	$\Delta\omega$	ΔQ	Total error
0	1	3.520	4.320	10.0	69.0	4.530	4.530	35.2	41.2	0.4441	0.0003	**0.0003**
0.1	0.9	3.440	4.220	10.0	69.0	1.690	1.380	59.7	70.6	0.0094	0.0003	**0.0012**
0.2	0.8	3.120	3.830	11.0	75.9	2.550	3.200	32.0	37.9	0.0064	0.0005	**0.0017**
0.3	0.7	4.220	3.440	10.0	69	1.670	3.480	36.1	48.1	0.0054	0.0009	**0.0022**
0.4	0.6	3.570	2.320	13.0	93.1	2.290	3.280	33.2	40.2	0.0024	0.0013	**0.0017**
0.5	0.5	4.120	2.740	11.1	79.6	1.050	2.100	58.3	77.7	0.0019	0.0081	**0.0050**
0.6	0.4	4.590	3.090	10.0	70.6	2.370	4.480	27.1	34.8	0.0013	0.0057	**0.0031**
0.7	0.3	3.830	4.700	8.98	61.9	2.230	4.170	29.1	37.0	0.0009	0.0045	**0.0020**
0.8	0.2	3.090	2.370	14.0	97.6	3.520	2.870	29.1	34.0	0.0006	0.0057	**0.0016**
0.9	0.1	4.990	3.520	8.98	63.4	4.270	4.480	21.0	24.9	0.0001	0.0052	**0.0006**
1	0	3.880	2.290	86.6	13.0	1.200	4.480	20.0	93.1	0.0006	1.4588	**0.0006**

The following tables show the variation of the error on the cut-off frequency and the error on the selectivity factor according to the δ_1 and δ_2 values for the linear values and E192 series (Tables 18.8, 18.9, 18.10, 18.11, 18.12 and 18.13).

The δ_1 and δ_2 variations offer a range of optimal solutions. The choice between solutions depends on the designer.

$\Delta\omega$ and ΔQ values versus δ_1 and δ_2 values are plotted in Figs. 18.6, 18.7 and 18.8 for the MMAS, the AS, and the ACS algorithms, respectively.

Table 18.12 Linear values of components and related Butterworth filter performances using the"ACS"

δ_1	δ_2	R_1 ($K\Omega$)	R_2 ($K\Omega$)	C_1 (nF)	C_2 (nF)	R_3 ($K\Omega$)	R_4 ($K\Omega$)	C_3 (nF)	C_4 (nF)	$\Delta\omega$	ΔQ	Total error
0	1	4.190	3.460	9.00	62.0	4.960	4.650	52.0	61.0	0.7421	0.0002	**0.0002**
0.1	0.9	3.450	3.480	11.0	75.0	1.270	1.540	66.0	78.0	0.0082	0.0011	**0.0018**
0.2	0.8	3.260	3.120	12.0	82.0	1.150	2.870	46.0	66.0	0.0014	0.0003	**0.0006**
0.3	0.7	3.950	2.420	12.0	87.0	1.140	2.630	49.0	68.0	0.0015	0.0005	**0.0008**
0.4	0.6	4.090	2.290	12.0	89.0	2.310	1.890	44.0	52.0	0.0007	0.0005	**0.0006**
0.5	0.5	4.450	4.250	8.80	60.1	2.250	1.300	52.1	65.7	0.0007	0.0004	**0.0005**
0.6	0.4	2.550	2.920	14.0	96.0	3.360	3.700	26.0	31.0	0.0014	0.0043	**0.0026**
0.7	0.3	2.240	4.730	11.0	86.0	1.100	2.030	59.0	76.0	0.0018	0.0014	**0.0017**
0.8	0.2	4.860	2.100	11.0	89.0	4.430	3.240	24.0	29.0	0.0009	0.0027	**0.0013**
0.9	0.1	3.840	3.830	10.0	68.0	3.310	4.680	23.0	28.0	0.0012	0.0050	**0.0016**
1	0	2.440	2.600	83.0	19.0	4.760	1.220	82.0	21.0	0.0002	1.4046	**0.0002**

Table 18.13 values of components selected from E192 and related Butterworth filter performances using the"ACS"

δ_1	δ_2	R_1 ($K\Omega$)	R_2 ($K\Omega$)	C_1 (nF)	C_2 (nF)	R_3 ($K\Omega$)	R_4 ($K\Omega$)	C_3 (nF)	C_4 (nF)	$\Delta\omega$	ΔQ	Total error
0	1	4.170	3.440	8.98	61.9	4.930	4.640	52.3	61.2	0.7503	0.0007	**0.0007**
0.1	0.9	3.440	3.480	11.0	75.0	1.270	1.540	65.7	77.7	0.0070	0.0010	**0.0016**
0.2	0.8	3.240	3.120	12.0	81.6	1.150	2.870	45.9	65.7	0.0075	0.0034	**0.0042**
0.3	0.7	3.970	2.430	12.0	86.6	1.140	2.640	49.3	68.1	0.0064	0.0046	**0.0051**
0.4	0.6	4.070	2.290	12.0	88.7	2.320	1.890	44.2	51.7	0.0050	0.0047	**0.0048**
0.5	0.5	4.420	4.270	8.76	60.4	2.260	1.300	52.3	65.7	0.0054	0.0077	**0.0066**
0.6	0.4	2.550	2.910	14.0	96.5	3.360	3.700	26.1	30.9	0.0026	0.0056	**0.0038**
0.7	0.3	2.230	4.750	11.0	85.6	1.100	2.030	59.0	75.9	0.0013	0.0061	**0.0027**
0.8	0.2	4.870	2.100	11.0	88.7	4.420	3.240	24.0	29.1	0.0011	0.0064	**0.0022**
0.9	0.1	3.830	3.830	10.0	68.1	3.320	4.700	22.9	28.0	0.0008	0.0051	**0.0012**
1	0	2.430	2.610	82.5	19.1	4.750	1.230	81.6	21.0	0.0009	1.4022	**0.0009**

From the Figs. 18.6, 18.7 and 18.8, we notice that is very difficult to make a compromise choice, except for the cases of [δ_1, δ_2] values for [0,1] and [0,1]. A judicious choice would be possible by treating the problem in multi-objective to have an efficient range of compromise values between the two conflicting criteria.

Fig. 18.6 $\Delta\omega$ and ΔQ values versus δ_1 and δ_2 values using the MMAS

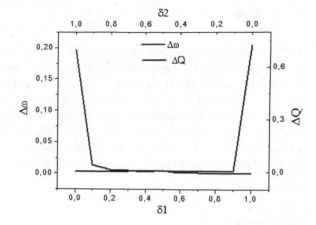

Fig. 18.7 $\Delta\omega$ and ΔQ values versus δ_1 and δ_2 values using the AS

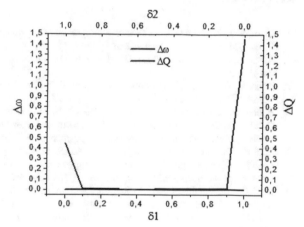

Fig. 18.8 $\Delta\omega$ and ΔQ values versus δ_1 and δ_2 values using the ACS

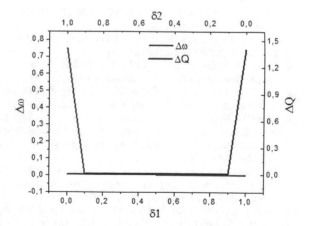

18.6 Conclusion

We presented in this chapter an application of the Ant Colony Optimization technique for optimal sizing of a fourth order Butterworth low pass analog filter. The design of the analog filter with high accuracy and short execution time is successfully realized using the ACO technique. SPICE simulation confirms the validity of the proposed approaches whose performances are also compared to already published evolutionary methods. We can argue that the ACO technique is a priori an adequate technique to use in the field of analog filter sizing. Now, we are focusing on transforming the proposed ACO mono-objective algorithms into multi-objective ones.

References

1. Grimbleby, J. B. (2000). Automatic analogue circuit synthesis using genetic algorithms. *IEE Proceedings - Circuits, Devices and Systems, 147* (6).
2. Aarts, E., & Lenstra, K. (2003). *Local search in combinatorial optimization.* Princeton: Princeton University Press.
3. Benhala, B. (2014). Ant colony optimization for optimal low-pass butterworth filter design. *WSEAS Transactions on Circuits and Systems, 13,* 313–318.
4. Benhala, B., Ahaitouf, A., Kotti, M., Fakhfakh, M., Benlahbib, B., Mecheqrane, A., et al. (2011). *Application of the ACO technique to the optimization of analog circuit performances. Analog circuits: Applications, design and performance.* NOVA Science Publishers.
5. Benhala, B., Ahaitouf, A., Mechaqrane, A., & Benlahbib, B. (2012). Multiobjective optimization of second generation current conveyors by the ACO technique. In *The International Conference on Multimedia Computing and Systems (ICMCS'12)* (pp. 1147–1151).
6. Benhala, B., Ahaitouf, A., Mechaqrane, A., Benlahbib, B., Abarkan, F. A. E., & Fakhfakh, M. (2011). Sizing of current conveyors by means of an ant colony optimization technique. In *The IEEE International Conference on Multimedia Computing and Systems (ICMCS'11)* (pp. 899–904).
7. Dorigo, M., DiCaro, G., & Gambardella, L. M. (1999). Ant algorithms for discrete optimization. *Artificial Life Journal, 5,* 137–172.
8. Dorigo, M., & Krzysztof, S. (2006). An introduction to ant colony optimization. In *Approximation algorithms and metaheuristics.*
9. Dorigo, M., Maniezzo, V., & Colorni, A. (1996). The ant system: Optimization by a colony of cooperating agents. *IEEE Transactions on Systems, Man and Cybernetics, 26,* 29–42.
10. Elzeki, O., Reshad, M., & Elsoud, M. (2012). Improved max-min algorithm in cloud computing. *International Journal of Computer Applications, 5,* 22–27.
11. Fakhfakh, M., Boughariou, M., Sallem, A., & Loulou, M. (2012). Design of low noise amplifiers through flow-graphs and their optimization by the simulated annealing technique. In *Advances in monolithic microwave integrated circuits for wireless systems: Modeling and design technologies.* IGI Global (pp. 69–88).
12. Fakhfakh, M., Cooren, Y., Sallem, A., Loulou, M., & Siarry, P. (2010). Analog circuit design optimization through the particle swarm optimization technique. *Journal of Analog Integrated Circuits & Signal Processing Springer, 63 N 1,* 71–82.
13. Glover, F. (1990). Tabu search-part ii. *ORSA Journal on Computing, 2,* 4–32.
14. Gmez, J. F., Khodr, H. M., Oliveira, P. M. D., Ocque, L., Yusta, J. M., Villasana, R., et al. Ant colony system algorithm for the planning of primary distribution circuits.
15. Jinhui, Y., Xiaohu, S., Maurizio, M., & Yanchun, L. (2009). An ant colony optimization method for generalized tsp problem. *Progress in Natural Science, e18,* 1417–1422.

16. Kotti, M., Benhala, B., Fakhfakh, M., Ahaitouf, A., Benlahbib, B., Loulou, M., et al. (2011). Comparison between pso and aco techniques for analog circuit performance optimization. In E.E.A.o.O.T. IEEE TN CEDAs (Ed.), *The International Conference on Microelectronics*.
17. Neumann, F., Sudholt, D., Witt., C. (2009). Analysis of different mmas aco algorithms on unimodal functions and plateaus. *Swarm Intelligence, 3*, 35–68.
18. Paarman, L. D. (2007). *Design and analysis of analog filters*. Norwell, MA: Kluwer.
19. Sallem, A., Benhala, B., Kotti, M., Fakhfakh, M., Ahaitouf, A., & Loulou, M. (2013). Application of swarm intelligence techniques to the design of analog circuits: Evaluation and comparison. In Springer (Ed.), *Analog integrated circuits and signal processing* (Vol. 75, pp. 499–516).
20. Schaumann, R., & Valkenburg, M. V. (2001). *Design of analog filters*. New York: Oxford University Press.
21. Sttzle, T., & Hoos, H. (2000). Max-min ant system. *Future Generation Computer System, 16*, 889–914.
22. Vural, R., Yildirim, T., Kadioglu, T., & Basargan, A. (2012). Performance evaluation of evolutionary algorithms for optimal filter design. *IEEE Transactions on Evolutionary Computation, 16*, 135–147.
23. Vural, R. A., Yildirim, T. (2010). Component value selection for analog active filter using particle swarm optimization. In *2nd ICCAE* (Vol. 1, pp. 25–28).
24. Yu, B., Yang, Z., & Yao, B. (2009). An improved ant colony optimization for vehicle routing problem. *European Journal of Operational Research, 196*, 171–176.

Chapter 19
Maximum a Posteriori Based Evolutionary Algorithm

Asmaa Ghoumari, Amir Nakib and Patrick Siarry

Abstract This work is dedicated to the presentation and the analysis of the performance of Maximum a posteriori based Evolutionary Algorithm (MEA). MEA allows a hybridization of set of operators to preserve the diversity during the search. This approach is based on a set of search strategies which are composed of one crossover with one mutation method, respectively. The algorithm uses the Maximum a Posteriori Principle (MAP) to select the most probable strategy from those available in the search set. Experiments were performed on well-known continuous optimization problems to observe the impact of population size and operators rates on MEA's behaviour, robustness and performance.

19.1 Introduction

Optimization algorithms struggle to perform in solving large scale problems. Although, the number of fitness evaluations increases with the dimension, the stopping criterion challenges limits of algorithms' effectiveness. Moreover, EAs lose diversity during the evolution, which makes exploration less efficient. In particular, Evolutionary Algorithms (EAs) have to deal with intensification and exploration, and their performance increases with their ability to respect the balance between

A. Ghoumari · A. Nakib (✉) · P. Siarry
Université Paris Est Laboratoire Images, Signaux et Systèmes
Intelligents (LISSI) - EA 3956, 122 rue Paul Armangot,
94400 Vitry-sur-Seine, France
e-mail: nakib@u-pec.fr

A. Ghoumari
e-mail: asmaa.ghoumari@u-pec.fr

P. Siarry
e-mail: siarry@u-pec.fr

© Springer International Publishing AG, part of Springer Nature 2019   301
E.-G. Talbi and A. Nakib (eds.), *Bioinspired Heuristics for Optimization*,
Studies in Computational Intelligence 774,
https://doi.org/10.1007/978-3-319-95104-1_19

these two phases. Thus, we propose an EA whose principle is to deal with a set of N search strategies (couples of mutation and crossover operators). The choice of the applied strategy is based on Bayesian theory, that is why this EA is called Maximum a posteriori based Evolutionary Algorithm (MEA).

This approach is a predictive since its uses past diversities to predict the best strategy the future to use. Diversity is measured thanks to the Euclidean distance between individuals of the population. Moreover, the choice of applied operators is questioned at regular intervals of Δ generations, called *decision moments'*. The purpose is to find the compromise between increasing the speed of convergence to reach the best solution, while maximizing the diversity to avoid local optima.

MEA is based on diversity's evolution, and it is a mono-population based, unlike others EAs based on diversity variations: the Diversity-Control-Oriented Genetic Algorithm [2], the Shifting-Balance Genetic Algorithm (SBGA) [1], the Forking Genetic Algorithm (fGA) [3], the Diversity-Guided Evolutionary Algorithm (DGEA) [4].

The flowchart includes two important steps: the first which consists in selecting the first strategy to apply, and the second an evolution loop which regularly change the selected strategy.

This paper presents the algorithm, but also experiments on sensitivities regarding the size of the population, the crossover and the mutation rates, and the study of establishing a dynamic Δ: value for each strategy.

MEA's process is described in the first section, then results are detailed and discussed in the next section. To finish this paper, conclusion and future work are presented.

19.2 Proposed Approach

The purpose is to solve two issues: the absence of rule for the choice of operators, and the loss of diversity during generations. This paper proposes the MEA as a solution and its description is in the following.

First, a working assumption is: a strategy is a couple of one crossover with one mutation. A set of N search strategies is available to the algorithm. After a random initialization, there are, on one hand, the population, and on the other hand, a set of search strategies. Thus, the first step consists in evaluating strategies' diversification potential. To do so, we use a diversity metric: the Euclidean distance, and the MAP principle.

During Δ generations, each of N strategies is applied on the initial population, diversities are measured, then probabilities deduced, and the MAP is used to predict the best strategy for the next Δ generations, this mechanism is illustrated in Fig. 19.1. This first step allows to evaluate fitnesses.

Then, during the evolving phase, which stops when the stopping criterion is reached, strategies are evaluated at regular intervals that is to say every Δ generations, those are called decision moments, illustration is in Fig. 19.2. The Δ genera-

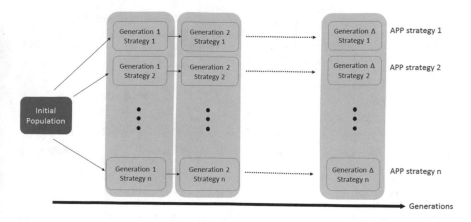

APP: A posteriori probability

Fig. 19.1 Illustration of the first step. Each of N strategies is applied on the initial population, in parallel, during Δ generations. Then a posteriori probabilities are found, and the MAP used

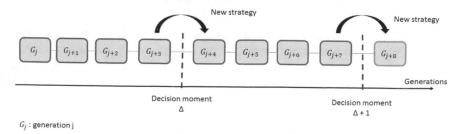

G_j : generation j

Fig. 19.2 Evolution of the population during generations. Only one population is considered, and the strategy applied is changed at decision moments

tions produced by the selected strategy are kept, and $N - 1$ strategies are respectively applied on them. The selection is based on the same process used before: diversities are measured and the MAP is used on probabilities deduced, see Fig. 19.3. Unlike the previous step, fitnesses are not evaluated, only diversities are measured in order to save fitness evaluations, a common stopping criterion.

The key point is the algorithm deals with only one population, the change is on the strategy applied on this population evolving over time, as shown in Fig. 19.2. Considering a use case of a set of $N = 20$ strategies, we have four crossovers and five mutations, so fifteen strategies possible. We use the BLX-α, the discrete, the one-point and the linear crossovers. Moreover we have the Levy, Gaussian, Scramble and the DE/RAND/1/BIN mutations.

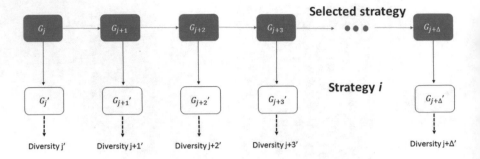

G_j : generation j

Fig. 19.3 Illustration of the method used to find the next applied strategy. In gray, the population produced by the selected strategy. Each $N - 1$ strategy left is applied on each generation of the population from the selected strategy, and diversities are measured

19.3 Performance Evaluation and Analysis

The objective of experiments performed in this section is to study the MEA's behavior by sensitivity of two parameters: the size of the population and operators' rates.

All simulations are obtained using an Intel (R) Core (TM) i7-3770 with 3.40 GHz speed and 8-GB RAM on Windows 7.

19.3.1 Sensitivity Against the Population Size

The purpose is to study the sensitivity of MEA regarding the size of the population. To do so, the experiment is carried out for values of the population size equal to 50, 100, 300, 500 and 800 individuals.

The experiment is performed on Rastrigin and Griewank problems and results are obtained for $\Delta = 10, 20, 50, 100, 200, 500$ and dimension $D = 20, 50, 100, 200$. Other parameter settings are the crossover rate $P_c = 0.3$, the mutation rate $P_m = 0.7$ and the size of the search set $N = 20$. The stopping criterion is set to 5,000 generations, if the final solution is inferior to $10E-8$ then it is considered equal to 0. Tables 19.1, 19.2, 19.3, 19.4, 19.5, 19.6, 19.7, 19.8, 19.9 and 19.10 detail final average results and Figs. 19.4, 19.5, 19.6, 19.7, 19.8 and 19.9 are illustrated them.

Concerning Rastrigin problem, Tables 19.1, 19.2, 19.3, 19.4 and 19.5 show that $\Delta = 500$ never reaches the optimum whatever the dimension, the Δ value or the population size. Then, from 300 individuals MEA reaches optimum for 200 dimension which chose that it is more efficient in large scale with a important population, the illustration to Fig. 19.4. Another point might be the number of Δ values getting the best solution from a population size to another: indeed, more N increases and more Δ can be increased. The phenomenon is presented in Fig. 19.5 more the population

Table 19.1 Final results (average final best fitness and standard deviation on 100 executions in brackets) depending on Δ and dimension values for Rastrigin function with a population of 50 individuals and maximum generations number set to 5,000. Best results are in bold

Dimension	20	50	100	200
$\Delta = 10$	5.98E−01	9.96E−01	9.96E−01	1.49E−03
	(3.40E+00)	(6.97E+00)	(9.91E+00)	(1.27E−02)
$\Delta = 20$	**3.99E−01**	3.63E−07	**2.83E−07**	**1.33E−03**
	(2.79E+00)	(3.61E−06)	**(1.64E−06))**	**(3.78E−03)**
$\Delta = 50$	5.98E−01	**4.37E−09**	1.01E+00	1.04E−01
	(3.40E+00)	**(4.00E−08)**	(9.90E+00)	(8.62E−01)
$\Delta = 100$	8.99E−01	4.89E−01	2.34E+00	3.78E+00
	(3.98E+00)	(4.95E+00)	(1.40E+01)	(2.22E+01)
*$\Delta = 200$	1.00E+00	2.51E+00	3.36E+00	1.01E+00
	(4.34E+00)	(1.08E+01)	(1.58E+01)	(5.33E+00)
$\Delta = 500$	8.90E+00	8.90E+00	3.31E+01	9.99E+01
	(1.15E+01)	(2.61E+01)	(1.22E+02)	(3.41E+02)

Table 19.2 Final results (average final best fitness and standard deviation on 100 runs in brackets) depending on Δ and dimension values for Rastrigin function with a population of 100 individuals and maximum generations number set to 5,000. Best results are in bold

Dimension	20	50	100	200
$\Delta = 10$	3.98E−01	**0.00E+00**	9.96E−01	**1.35E−05**
	(2.79E+00)	**(0.00E+00)**	(9.91E+00)	**(8.40E−05)**
$\Delta = 20$	**0.00E+00**	**0.00E+00**	**0.00E+00**	1.28E−03
	(0.00E+00)	**(0.00E+00)**	**(0.00E+00)**	(1.23E−02)
$\Delta = 50$	3.98E−01	9.96E−01	1.01E+00	3.00E−04
	(2.79E+00)	(6.97E+01)	(9.91E+00)	(2.90E−03)
$\Delta = 100$	1.99E−01	5.89E−06	4.85E−04	3.17E+00
	(1.98E+00)	(3.37E−05)	(2.51E−03)	(2.25E+01)
$\Delta = 200$	**0.00E+00**	4.98E−01	7.06E−03	2.18E+00
	(2.06E−08)	(4.95E+00)	(2.46E−02)	(1.98E+01)
$\Delta = 500$	2.18E+00	4.35E+00	1.61E+01	4.83E+01
	(6.90E+00)	(1.59E+01)	(4.40E+01)	(1.50E+02)

is important and more the big values of Δ also gets the optimum. The exact same observations can be made regarding the Griewank problem from Tables 19.6, 19.7, 19.8, 19.9 and 19.10, in Fig. 19.6, excepting that $\Delta = 500$ generations reaches the optimum.

Figures 19.7, 19.8 and 19.9 show the distribution of Δ values getting best solutions for Rastrigin and Griewank apart. Regarding to the first graph, the most effective Δ value is 20 generations with a score up to 27% of the results. Then, in second position,

Table 19.3 Final results (average final best fitness and standard deviation on 100 executions in brackets) depending on Δ and dimension values for Rastrigin function with a population of 300 individuals and maximum generations number set to 5,000. Best results are in bold

Dimension	20	50	100	200
$\Delta = 10$	**0.00E+00**	**0.00E+00**	**0.00E+00**	**0.00E+00**
	(0.00E+00)	**(0.00E+00)**	**(0.00E+00)**	**(0.00E+00)**
$\Delta = 20$	**0.00E+00**	**0.00E+00**	**0.00E+00**	**0.00E+00**
	(0.00E+00)	**(0.00E+00)**	**(0.00E+00)**	**(1.92E−08)**
$\Delta = 50$	**0.00E+00**	**0.00E+00**	**0.00E+00**	$8.13E{-}06$
	(0.00E+00)	**(0.00E+00)**	**(0.00E+00)**	$(7.97E{-}05)$
$\Delta = 100$	**0.00E+00**	**0.00E+00**	**0.00E+00**	$1.70E{-}05$
	(0.00E+00)	**(0.00E+00)**	**(1.12E−08)**	$(1.10E{-}04)$
$\Delta = 200$	**0.00E+00**	**0.00E+00**	$3.72E{-}06$	$4.68E{-}04$
	(1.48E−08)	**(0.00E+00)**	$(3.70E{-}05)$	$(4.11E{-}03)$
$\Delta = 500$	$1.31E{-}01$	$3.44E{+}00$	$1.56E{+}01$	$8.41E{+}01$
	$(8.10E{-}01)$	$(1.12E{+}01)$	$(4.32E{+}01)$	$(1.92E{+}02)$

Table 19.4 Final results (average final best fitness and standard deviation on 100 executions in brackets) depending on Δ and dimension values for Rastrigin function with a population of 500 individuals and maximum generations number set to 5,000. Best results are in bold

Dimension	20	50	100	200
$\Delta = 10$	**0.00E+00**	**0.00E+00**	**0.00E+00**	**1.37E−13**
	(0.00E+00)	**(0.00E+00)**	**(0.00E+00)**	**(7.78E−13)**
$\Delta = 20$	**4.38E−13**	**0.00E+00**	**0.00E+00**	**1.05E−15**
	(4.36E−12)	**(0.00E+00)**	**(0.00E+00)**	**(7.39E−15)**
$\Delta = 50$	**2.26E−13**	**3.84E−12**	**7.64E−14**	**2.88E−09**
	(2.24E−12)	**(3.82E−11)**	**(7.60E−13)**	**(2.86E−08)**
$\Delta = 100$	**1.44E−09**	$1.90E{-}07$	**0.00E+00**	**4.18E−10**
	(1.43E−08)	$(1.37E{-}06)$	**(0.00E+00)**	**(4.03E−09)**
$\Delta = 200$	$9.03E{-}02$	$7.84E{-}03$	$6.59E{-}01$	$6.65E{+}00$
	$(8.98E{-}01)$	$(7.77E{-}02)$	$(6.51E{+}00)$	$(6.48E{+}01)$
$\Delta = 500$	**1.11E−09**	$3.66E{-}01$	$5.91E{-}01$	$2.18E{+}01$
	(1.10E−08)	$(2.56E{+}00)$	$(5.73E{+}00)$	$(8.08E{+}01)$

$\Delta = 10$ gets 25% and $\Delta = 50$ obtains 21% for the third place. Regarding to Fig. 19.8, we get the same trio but equally distributed, with 19% of obtained results. If we had to chose one Δ value it would be 20 since it gets the larger distribution for both problems (22%) in front of $\Delta = 10$ and $\Delta = 50$ with 21% and 19% respectively.

In conclusion, to solve high dimension optimization problems, it seems to be better to increase the population up to 500 individuals at least.

Table 19.5 Final results (average final best fitness and standard deviation on 100 executions in brackets) depending on Δ and dimension values for Rastrigin function with a population of 800 individuals and maximum generations number set to 5,000. Best results are in bold

Dimension	20	50	100	200
$\Delta = 10$	**0.00E+00**	**0.00E+00**	**0.00E+00**	**0.00E+00**
	(0.00E+00)	**(0.00E+00)**	**(0.00E+00)**	**(0.00E+00)**
$\Delta = 20$	**0.00E+00**	**0.00E+00**	**0.00E+00**	**0.00E+00**
	(0.00E+00)	**(0.00E+00)**	**(0.00E+00)**	**(0.00E+00)**
$\Delta = 50$	**0.00E+00**	**0.00E+00**	**0.00E+00**	**0.00E+00**
	(0.00E+00)	**(0.00E+00)**	**(0.00E+00)**	**(0.00E+00)**
$\Delta = 100$	**0.00E+00**	**0.00E+00**	**0.00E+00**	**0.00E+00**
	(0.00E+00)	**(0.00E+00)**	**(0.00E+00)**	**(0.00E+00)**
$\Delta = 200$	**0.00E+00**	**0.00E+00**	$4.49E-01$	$8.67E-05$
	(0.00E+00)	**(0.00E+00)**	$(4.47E+00)$	$(6.95E-04)$
$\Delta = 500$	$1.56E-06$	$2.77E-05$	$8.07E-01$	$1.11E+00$
	$(1.55E-05)$	$(2.73E-04)$	$(5.73E+00)$	$(1.06E+01)$

Table 19.6 Final results (average final best fitness and standard deviation on 100 executions in brackets) depending on Δ and dimension values for Griewank function with a population of 50 individuals and maximum generations number set to 5,000. Best results are in bold

Dimension	20	50	100	200
$\Delta = 10$	$1.15E-03$	**0.00E+00**	**0.00E+00**	$1.19E-03$
	$(1.15E-02)$	**(0.00E+00)**	**(0.00E+00)**	$(1.57E-03)$
$\Delta = 20$	**0.00E+00**	$4.27E-04$	**0.00E+00**	$3.00E-04$
	(0.00E+00)	$(6.65E-04)$	**(0.00E+00)**	$(6.31E-04)$
$\Delta = 50$	$2.65E-03$	**0.00E+00**	**0.00E+00**	$1.52E-03$
	$(2.64E-02)$	**(0.00E+00)**	**(0.00E+00)**	$(3.30E-03)$
$\Delta = 100$	$1.70E-03$	**0.00E+00**	**0.00E+00**	$4.92E-04$
	$(1.69E-02)$	**(0.00E+00)**	**(0.00E+00)**	$(1.71E-03)$
$\Delta = 200$	$8.90E-03$	$2.80E-06$	$2.70E-05$	$2.89E-02$
	$(4.00E-02)$	$(2.57E-05)$	$(1.97E-04)$	$(6.34E-02)$
$\Delta = 500$	$3.02E-03$	$7.13E-04$	$4.25E-02$	$3.32E-01$
	$(1.78E-02)$	$(2.84E-03)$	$(1.43E-01)$	$(3.06E-01)$

19.3.2 Sensitivity of Operators Rates

For this experiment, we performed the same tests, using the same parameter settings except for the mutation and crossover rates, they are set to 1. This means that MEA is performing a crossover and a mutation each time from a generation to another. The same graphic illustrations, from Figs. 19.9, 19.10, 19.11, 19.12, 19.13 and 19.14.

Table 19.7 Final results (average final best fitness and standard deviation on 100 executions in brackets) depending on Δ and dimension values for Griewank function with a population of 100 individuals and maximum generations number set to 5,000. Best results are in bold

Dimension	20	50	100	200
$\Delta = 10$	**0.00E+00**	**0.00E+00**	**0.00E+00**	$1.67E-05$
	(0.00E+00)	**(0.00E+00)**	**(0.00E+00)**	$(2.75E-05)$
$\Delta = 20$	**0.00E+00**	**0.00E+00**	**0.00E+00**	$7.91E-05$
	(0.00E+00)	**(0.00E+00)**	**(0.00E+00)**	$(9.14E-05)$
$\Delta = 50$	$9.10E-04$	**0.00E+00**	**0.00E+00**	$2.45E-04$
	$(9.06E-03)$	**(0.00E+00)**	**(0.00E+00)**	$(4.26E-04)$
$\Delta = 100$	**0.00E+00**	**0.00E+00**	**0.00E+00**	$1.11E-03$
	(0.00E+00)	**(0.00E+00)**	**(0.00E+00)**	$(2.04E-03)$
$\Delta = 200$	**0.00E+00**	**0.00E+00**	$1.97E-08$	$1.12E-03$
	(0.00E+00)	**(0.00E+00)**	$(6.05E-08)$	$(5.97E-03)$
$\Delta = 500$	$2.95E-03$	$2.73E-04$	$2.36E-03$	$1.18E-02$
	$(1.51E-02)$	$(1.94E-03)$	$(1.69E-02)$	$(1.05E-01)$

Table 19.8 Final results (average final best fitness and standard deviation on 100 executions in brackets) depending on Δ and dimension values for Griewank function with a population of 300 individuals and maximum generations number set to 5,000. Best results are in bold

Dimension	20	50	100	200
$\Delta = 10$	**0.00E+00**	**0.00E+00**	**0.00E+00**	**0.00E+00**
	(0.00E+00)	**(0.00E+00)**	**(0.00E+00)**	**(0.00E+00)**
$\Delta = 20$	**0.00E+00**	**0.00E+00**	**0.00E+00**	**0.00E+00**
	(0.00E+00)	**(0.00E+00)**	**(0.00E+00)**	**(0.00E+00)**
$\Delta = 50$	**0.00E+00**	**0.00E+00**	**0.00E+00**	$1.49E-06$
	(0.00E+00)	**(0.00E+00)**	**(0.00E+00)**	$(2.82E-06)$
$\Delta = 100$	**0.00E+00**	**0.00E+00**	**0.00E+00**	**0.00E+00**
	(0.00E+00)	**(0.00E+00)**	**(0.00E+00)**	**(2.07E-08)**
$\Delta = 200$	**0.00E+00**	**0.00E+00**	**0.00E+00**	$2.56E-04$
	(0.00E+00)	**(0.00E+00)**	**(0.00E+00)**	$(4.77E-04)$
$\Delta = 500$	**0.00E+00**	**0.00E+00**	**0.00E+00**	$1.62E-07$
	(0.00E+00)	**(0.00E+00)**	**(0.00E+00)**	$(6.47E-07)$

Globally, results are clearly modified, making MEA applies crossover and mutation always used, decrease its performance.

Figure 19.9 shows that MEA reaches optimum only for population of size 50 individuals both for Rastrigin and Griewank problems. Then, Figs. 19.10 and 19.11 confirm that only a population of 50 individuals is able to reach the optimum for $\Delta = 10$, 20 and 50, unlike other population sizes.

Table 19.9 Final results (average final best fitness and standard deviation on 100 executions in brackets) depending on Δ and dimension values for Griewank function with a population of 500 individuals and maximum generations number set to 5,000. Best results are in bold

Dimension	20	50	100	200
$\Delta = 10$	**0.00E+00**	**0.00E+00**	**0.00E+00**	**0.00E+00**
	(0.00E+00)	**(0.00E+00)**	**(0.00E+00)**	**(0.00E+00)**
$\Delta = 20$	**0.00E+00**	**0.00E+00**	**0.00E+00**	**0.00E+00**
	(0.00E+00)	**(0.00E+00)**	**(0.00E+00)**	**(0.00E+00)**
$\Delta = 50$	**0.00E+00**	**0.00E+00**	**0.00E+00**	**0.00E+00**
	(0.00E+00)	**(0.00E+00)**	**(0.00E+00)**	**(0.00E+00)**
$\Delta = 100$	**0.00E+00**	**0.00E+00**	**0.00E+00**	**0.00E+00**
	(0.00E+00)	**(0.00E+00)**	**(0.00E+00)**	**(0.00E+00)**
$\Delta = 200$	**0.00E+00**	**0.00E+00**	**0.00E+00**	**0.00E+00**
	(0.00E+00)	**(0.00E+00)**	**(0.00E+00)**	**(0.00E+00)**
$\Delta = 500$	**0.00E+00**	**0.00E+00**	**0.00E+00**	**0.00E+00**
	(0.00E+00)	**(0.00E+00)**	**(0.00E+00)**	**(0.00E+00)**

Table 19.10 Final results (average final best fitness and standard deviation in brackets on 100 executions) depending on Δ and dimension values for Griewank function with a population of 800 individuals and maximum generations number set to 5,000. Best results are in bold

Dimension	20	50	100	200
$\Delta = 10$	**0.00E+00**	**0.00E+00**	**0.00E+00**	**0.00E+00**
	(0.00E+00)	**(0.00E+00)**	**(0.00E+00)**	**(0.00E+00)**
$\Delta = 20$	**0.00E+00**	**0.00E+00**	**0.00E+00**	**0.00E+00**
	(0.00E+00)	**(0.00E+00)**	**(0.00E+00)**	**(0.00E+00)**
$\Delta = 50$	**0.00E+00**	**0.00E+00**	**0.00E+00**	**0.00E+00**
	(0.00E+00)	**(0.00E+00)**	**(0.00E+00)**	**(0.00E+00)**
$\Delta = 100$	**0.00E+00**	**0.00E+00**	**0.00E+00**	**0.00E+00**
	(0.00E+00)	**(0.00E+00)**	**(0.00E+00)**	**(0.00E+00)**
$\Delta = 200$	**0.00E+00**	**0.00E+00**	**0.00E+00**	**0.00E+00**
	(0.00E+00)	**(0.00E+00)**	**(0.00E+00)**	**(0.00E+00)**
$\Delta = 500$	**0.00E+00**	**0.00E+00**	**0.00E+00**	**0.00E+00**
	(0.00E+00)	**(0.00E+00)**	**(0.00E+00)**	**(0.00E+00)**

Comparing Δ distributions from Figs. 19.12, 19.13 and 19.14, the decrease in effectiveness the lack of results make the choice of a Δ well distributed less obvious. Figure 19.12 present $\Delta = 20$ the most efficient (50% of distribution) in front of $\Delta = 10$ and $\Delta = 50$ with 25%. Figure 19.13 shows an equality between $\Delta = 10, 20, 50, 100$ with 25%. Finally, Fig. 19.14 summarizes previous figures, and there is a Δ value which is better than the other, $\Delta = 20$ (33%), then $\Delta = 50$ (25%) and $\Delta = 10$ (25%). Yet, it can be noticed that the same Δ values reach the optimum

Fig. 19.4 Evolution of the number of times the optimum is reached according to population size. In grey, results get from Rastrigin problem, in black those from Griewank problem

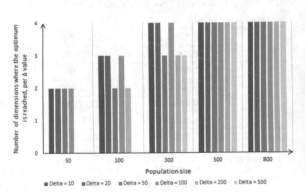

Fig. 19.5 Considering Rastrigin problem, evolution of the number of times each Δ reaches the optimum according to the population size. Four dimensions are tested: 20, 50, 100, 200

Fig. 19.6 Considering Griewank problem, evolution of the number of times each Δ reaches the optimum according to the population size. Four dimensions are tested: 20, 50, 100, 200

and get the best distributions despite results globally less effective with a crossover and a mutation rates set to 1.

Fig. 19.7 Distribution of best solutions per Δ, **a** Rastrigin problem, **b** Griewank problem

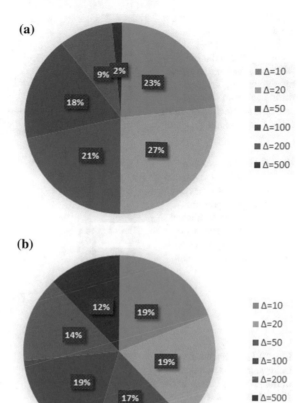

Fig. 19.8 distribution of the best solutions per Δ for Rastrigin and Griewank problems

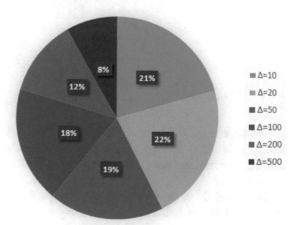

Fig. 19.9 Evolution of the number of times the optimum is reached according to population size with crossover and mutation rates equal to 1. In grey, results get from Rastrigin problem, in black those from Griewank problem

Fig. 19.10 Considering Rastrigin problem, evolution of the number of times each Δ reaches the optimum according to the population size. Four dimensions are tested: 20, 50, 100, 200 with crossover and mutation rates equal to 1

Fig. 19.11 Considering Griewank problem, evolution of the number of times each Δ reaches the optimum according to the population size. Four dimensions are tested: 20, 50, 100, 200 with crossover and mutation rates equal to 1

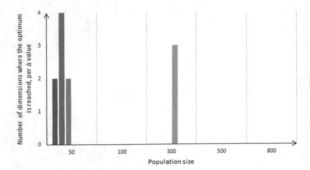

19.4 Conclusion

In this paper, the recall the principle of MEA and its operating steps were presented. This approach's key point is the dynamic adaptation of evolutionary operators applied. For this, the maximum a posteriori (MAP) based on population's diversity is used.

Then, a study of the size sensitivity was performed on well-known optimization problems. We concluded that it has an important impact on the performances since more the population size is high, more final results are improved. Beside, we have

Fig. 19.12 Considering Rastrigin problem, distribution of the best solutions per Δ with crossover and mutation rates equal to 1

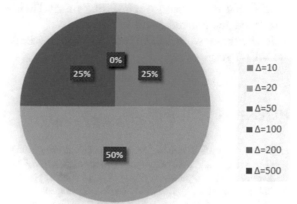

Fig. 19.13 Considering Griewank problem, distribution of the best solutions per Δ with crossover and mutation rates equal to 1

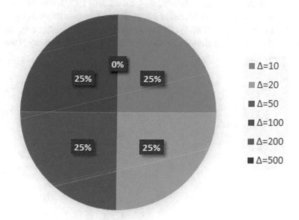

Fig. 19.14 Considering both problems, distribution of the best solutions per Δ with crossover and mutation rates equal to 1

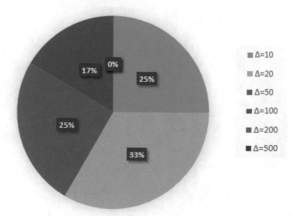

noticed that it affects also the setting of Δ. Then, a test on operators' rates was performed and pointed out its importance.

In our future work will be focus on studying impact of sequence of strategies on the MEA's results.

References

1. Oppacher, F., & Wineberg, M. (1999). The shifting balance genetic algorithm: Improving the GA in a dynamic environment. In *Proceedings of the First Annual Conference on Genetic and Evolutionary Computation, Orlando, USA* (Vol. 1, pp. 504–510).
2. Shimodaira, H. (1997). DCGA: A diversity control oriented genetic algorithm. In *Proceedings of Nineth IEEE International Conference on Tools with Artificial Intelligence, Newport Beach, USA* (pp. 367–374).
3. Tsutsui, S., Fujimoto, Y., & Ghosh, A. (1997). Forking genetic algorithms: GAs with search space division schemes. *Evolutionary Computation, 5*(1), 61–80.
4. Ursem, R. K. (2002). Diversity-guided evolutionary algorithms. In *Proceedings of the Seventh International Conference on Parallel Problem Solving from Nature, Granada, Spain* (pp. 462–471).

Printed in the United States
By Bookmasters